Fungal–Plant Interactions

Fungal–Plant Interactions

Susan Isaac

CHAPMAN & HALL
London · New York · Tokyo · Melbourne · Madras

Published by Chapman & Hall, 2–6 Boundary Row, London SE1 8HN

Chapman & Hall, 2–6 Boundary Row, London SE1 8HN, UK

Chapman & Hall, 29 West 35th Street, New York NY10001, USA

Chapman & Hall Japan, Thomson Publishing Japan,
Hirakawacho Nemoto Building, 7F,
1–7–11 Hirakawa–cho, Chiyoda–ku, Tokyo 102, Japan

Chapman & Hall Australia, Thomas Nelson Australia,
102 Dodds Street, South Melbourne, Victoria 3205, Australia

Chapman & Hall India, R. Seshadri,
32 Second Main Road, CIT East, Madras 600 035, India

First edition 1992

Typeset in 10/11½ Palatino by Falcon Typographic Art Ltd, Edinburgh
Printed in Great Britain at The University Printing House, Cambridge

ISBN 0 412 35390 3 (HB)   0 412 36470 0 (PB)

A catalogue record for this book is available from the British Library

Library of Congress Cataloging-in-Publication data
Isaac, Susan.
Fungal–plant interactions / Susan Isaac.
p. cm.
Includes bibliographical references and index.
ISBN (invalid) 0–412–35340–3. — ISBN 0–412–36470–0 (pbk.)
1. Plant–fungi relationships. I. Title.
QK604.I73 1991
589.2′04524–dc20 91–27839 CIP

For my family

Contents

Acknowledgements

I am indebted to many people for their help in the preparation of this book. Particularly to those who generously provided photographs, reprints and preprints, and who kindly gave me permission to use copyright material in the text, including Profs R.J. Howard, G. Lysek, R. Marchant, G. Pegg, D. Read and Drs D. Anglesea, P.G. Ayres, A.J. Beale, J.H. Carder, G. Carroll, K. Clay, M. Coffey, H.A. Collin, A. Crossley, R.I. Crute, A. Donovan, J. Farrar, J. Gay, V. Gianinazzi-Pearson, G. Hadley, K. Hardwick, S. Humphreys, K. Ibrahim, S.F. Lowry, J.A. Lucas, P.T. Lynch, R. Machado, P. Mason, R. B. Muse, R.J. O'Connell, O. Petrini, D.S. Shaw, P.T.N. Spencer-Phillips, J. Stone, J. Süske and Miss S. J. Edwards.

I gratefully acknowledge the help, encouragement, enthusiastic discussions and persistent questioning that I have enjoyed with many colleagues and students around the world, especially Profs L. Ferenczy, D.H. Jennings, G. Lysek, J.F. Peberdy, G. Pugh, S.T. Williams, A.J.S. Whalley, Drs P.G. Ayres, J. Bailey, J. Croft, J. Fisher, G.N. Greenhalgh, M.O. Moss, and others. They have, collectively, motivated me to explore further into the microbial world, constantly increasing my sense of wonder.

My thanks are also due to C.J. Veltkamp, B.T.J. Isaac and A.J. Tollitt for assistance with photography and to Joan Brown and Linda Marsh for making the preparation of the manuscript more enjoyable.

Finally, very special thanks must extend to my family for their tolerance, constant support and encouragement.

Preface

The primary aim of this book is to stimulate interest in the many complex and varied relationships which occur between fungi and plants, by an integrated presentation of fungal physiology and plant physiology which are often treated as separate and distinct disciplines. It has been aimed mainly at undergraduates already having some knowledge of biology, microbiology and physiological processes. In my teaching I find it increasingly necessary to draw information together, in a digestible form, for undergraduate students who may have a good background understanding of metabolism in microbial systems, or of the functioning and ecology of plants, but not both, and it is with these people in mind that I have prepared the manuscript. Some topics are included owing to my personal fascinations, and also to provide insight into the development of current theories and to explain the exciting potential of modern techniques which are now being applied to the experimental investigation of fungal and plant responses. Wherever appropriate, review articles and chapters that I have found stimulating have been cited, so that reference to these will give an easy route through to other more specialised orginal papers and act as sources for further reading.

The first chapter provides some basic information concerning physiological aspects of fungal growth and development, relating this to the means by which fungi invade and exploit their environment. Chapter 2 examines higher plants as a substrate for fungi, the consequences of fungal activities in agricultural situations and the ways in which detrimental effects can be minimised or controlled. The ways in which various fungi invade and colonise plant tissues and the strategies which are used to combat the defensive responses of the plant are explained in Chapter 3. Chapter 4 examines the physiological and biochemical influences which are exerted during invasion and disease establishment and the consequences to plant growth and development. Chapter 5 considers mutualistic symbioses, covering mycorrhizas, lichens and endophytic relationships. Chapter 6 presents some modern approaches to plant breeding for resistance and to increasing our understanding of the molecular aspects of fungal–plant interactions.

Many times during the preparation of this book I have deliberated, and marvelled, at the extreme elegance and efficiency by which these biological systems function and interact.

Susan Isaac

Fungal life-style 1

INTRODUCTION

The fungi are a very large and diverse group of organisms which have a unique life-style. They have world-wide distribution and successfully exploit many different habitats. They are extremely variable in form and versatile in the ways they solve the problems posed by the environments they inhabit.

The vegetative phase may be limited, occurring either as single cells (yeasts), or small, relatively undifferentiated thalli (aquatic Chytrids), or may be more extensive and obvious macroscopically. For most species vegetative growth is as filamentous, tubular hyphae which grow by extension at the tips and branch to form mycelial colonies, spreading over and through the substrate. The fungi have true nuclei (eukaryotic) and exhibit a variety of nuclear phases during their lifecycles. More differentiated structures are formed at reproduction. Dispersal occurs by means of spores which may be produced asexually or sexually. Many species produce more than one type of spore.

Perhaps the most important distinguishing feature of the fungi is their need to obtain preformed organic compounds as nutrients. Fungi are not able to photosynthesise but are heterotrophic for carbon. Enzymes are released from the hyphae on to the substrate for the external hydrolysis of potential nutrients. Solubilised materials are subsequently absorbed, by transport through the membrane, and used for the metabolic processes involved in growth and development. Fungi have a very important role in nutrient cycling in the environment particularly since many have a saprotrophic life-style, obtaining nutrients by degradation of organic matter. Such species break down and decompose very recalcitrant substrates, using the resources obtained and also increasing the availability of such materials to other organisms. Many fungi interact with plants, providing supplies of vital nutrients which increase the productivity of those plants and generally improve plant health, particularly in harsh environments. Others use the resources of living plants directly for their own needs causing damage, disease and even the death of those plants.

CLASSIFICATION

The fungi are placed into taxonomic groups by morphological characteristics, particularly those relating to spores, e.g. modes of spore formation and the structures which bear the spores. In the future, with the advent of new technology and systems for making very precise measurements, other non-morphological schemes may be adopted, or used to provide useful supplementary characters for clarification in those areas of greatest debate and controversy.

The application of names to the categories into which the fungi are divided, and the order of these, is governed by the principles and rules of the International Code of Botanical Nomenclature in order to prevent the use of confusing or misleading terminology. A number of different schemes have been proposed, indeed the whole area of fungal taxonomy has attracted a great deal of attention. The scheme published by Ainsworth (1973) and adopted by Webster (1980) is the most generally accepted, and has largely been used here. It is very important to realise that there is still a great deal of disagreement, especially in some particular cases.

The fungi are grouped into two divisions, based on the presence or absence of a cell wall (Table 1.1). The Myxomycota, or slime moulds, which contains over 600 species, are wall-less forms. The Eumycota, or true fungi, of which there are over 60 000 species (over 75 000 if those which form lichens are included) have cell walls composed of various polysaccharides. A detailed understanding of taxonomy is not necessary for many purposes but some comprehension of the ways in which species are grouped is important in terms of ordering such a large and varied body of information and also in appreciating the great diversity of life-styles which are adopted by these organisms.

Myxomycota (slime moulds)

This group of slime moulds may in fact be more closely related to the protozoa than to the true fungi, but these species are often studied by both mycologists and plant pathologists. The vegetative form is as a plasmodium, a multinucleate body which does not have a cell wall. This plasmodium displays amoeboid movement and migrates across the substratum engulfing dead organic matter (including bacterial and yeast

Table 1.1 The major taxonomic divisions of fungi

Taxonomic group	*Vegetative phase*	*Septa*	*Cell wall*	*Spores*	
				Asexual	*Sexual*
Mycota (slime moulds)					
Myxomycota	Plasmodium	None	None	Zoospores	Zoospore fusion
Eumycota (true fungi)					
Mastigomycotina	Filamentous, diploid	Aseptate	Cellulose, glucan	Zoospores	Oöspores
Zygomycotina	Filamentous, haploid	Aseptate	Chitin, chitosan	Sporangiospores	Zygospores
Ascomycotina	Filamentous, haploid	Septate	Chitin, glucan	Conidia	Ascospores
Basidiomycotina	Filamentous, diploid	Dolipore septum, clamp connections	Chitin, glucan		Basidiospores
Deuteromycotina	Filamentous, haploid	Septate	Chitin, glucan	Conidia	No sexual stage known

Table 1.2 The four classes within the Myxomycota (slime moulds)

Wall-less forms (probably more related to the protozoa than the fungi)	
Acrasiomycetes:	Cellular slime moulds. Vegetative amoebae aggregate prior to reproduction. e.g.*Dictyostelium discoideum*
Hydromyxomycetes:	Net slime moulds. Slimy filaments parasitic on marine vegetation. e.g. *Labarinthula* spp.
Myxomycetes:	True slime moulds. Multinucleate, naked plasmodium with amoeboid movement. Sexual reproduction by zoospore fusion. e.g. *Physarum polycephalum*
Plasmodiophoromycetes:	Endoparasitic slime moulds. Biotrophic parasites on plants. Zoospores motile. e.g. *Plamodiophora brassicae*

cells) as nutrient materials, and eventually gives rise to resting spores. In favourable conditions motile zoospores are released from these spores. There are four classes within this division (Table 1.2).

Most species in this division are saprotrophs but the Plasmodiophoromycetes includes species which induce economically important plant diseases. These pathogens, which draw nutrients directly from living host tissues, attack the roots and young shoots of higher plants growing in poorly drained soil. *Plasmodiophora brassicae* causes club root of brassicas (e.g. cabbage) and *Spongospora subterranea* causes powdery scab of potato. Owing to the highly resistant spores which are produced these pathogens are very difficult to eradicate from contaminated soils.

Eumycota (true fungi)

The Eumycota are further subdivided into five divisions (Table 1.1), the Mastigomycotina, Zygomycotina, Ascomycotina, Basidiomycotina and Deuteromycotina.

Mastigomycotina

There are three classes in this group (Table 1.3) within which there is a great range of diverse morphology. The vegetative phase, or thallus, is haploid and this may be very simple in form (Chitridiales) with limited differentiation, or may occur as more extensive mycelium. The diploid

Table 1.3 The subdivision Mastigomycotina

Aseptate mycelium. Motile zoospores. Very variable morphology, many parasitic forms.

Chytridiomycetes:	Wall chitin and glucan. Single, posterior, whiplash flagellum
Chytridiales:	Thallus may be single cell or may be differentiated into rhizoids with one or more sporangia, e.g. *Olpidium brassicae*
Blastocladiales:	Haploid and diploid thalli, alternate generations. Extensive thallus with rhizoids and reproductive structures, e.g. *Allomyces arbuscular*
Monoblepharidales:	Branched hyphae. Sexual reproduction oöspores, e.g. *Monoblepharis polymorpha*
Hyphochytridiomycetes:	Wall cellulose and chitin. Single anterior tinsel flagellum
Oömycetes:	Wall glucan and cellulose. Two flagellae (posterior whiplash, anterior tinsel). Diploid. Mainly aquatic forms
Saprolegniales:	Water moulds. Homothallic aseptate mycelium, e.g *Saprolegnia parasitica*
Leptomitales:	Oögonium with single oöspore
Lagenidales:	Aquatic, parasitic on algae
Peronosporales:	Often parasitic on land plants, e.g. *Pythium* spp. (damping-off disease), *Bremia lactucae* (downy mildew), *Phytophthora* (late blight), *Albugo* (white rust)

phase is restricted to a single cell except in the Oömycetes where the main vegetative phase is diploid. Asexual zoospores are produced, which are motile dispersal units. Zoospore structure is used as a major feature in the classification of the Mastigomycotina and species are placed in subdivisions according to zoospore structure. The cell walls of these fungi are unusual amongst the fungi in that they are composed from glucan and cellulose, or glucan and chitin.

Most of the Mastigomycotina are aquatic saprotrophs or inhabit waterlogged soils; however, two classes contain species which are pathogenic on higher plants. These proliferate most readily in damp situations and require a film of water for infection to occur. Most of these species attack roots and below-ground stems causing damping-off disease, blight and soft rot. Seedlings are especially vulnerable during the early stages of development, when they are least mature and most susceptible. Amongst the Chytridiales, *Olpidium brassicae* attacks cabbage roots and, additionally, is known to transmit viruses to the host plants during infection. The Oömycetes also include some very important plant pathogenic species, e.g. *Pythium*, which causes damping-off disease and root rots,

Table 1.4 The subdivision Zygomycotina

Aseptate, haploid mycelium. Asexual spores contained in sporangia (non-motile). Sexual spores, zygospores.

Zygomycetes:	Wall components chitin and chitosan
Mucorales:	Saprophytes in soil, e.g. *Mucor rouxii*, *Rhizopus* spp., *Absidia glauca*. Includes Endogonaceae (form V–A mycorrhizae, e.g. *Glomus mosseae*, *Endogone* spp.)
Entomophthorales:	Mainly insect parasites, e.g. *Basidiobolus ranarum*
Trichomycetes:	Parasitic in guts of arthropods

Phytophthora, the causal agent of potato blight and the downy mildew fungi (e.g. *Peronospora* and *Bremia*).

Zygomycotina

The vegetative phase of species within this group is more extensive, haploid mycelium, lacking cross walls (aseptate, coenocytic). Non-motile, asexual spores are formed on specialised hyphae arising from the vegetative mycelium. For sexual reproduction to occur the fusion of two hyphae, often from different but compatible strains (heterothallic), is required and gives rise to the fusion of haploid nuclei. As a result thick-walled resting spores, known as zygospores, are produced and following meiosis outgrowth of haploid, vegetative mycelium resumes. The major wall components of these fungi are chitin and chitosan.

There are relatively few species (about 750) within this group. Most of the Mucorales (Table 1.4) are saprotrophic, living in soils and on animal dung (coprophilous fungi), e.g. *Mucor mucedo*, *Pilobolus kleinii* and *Thamnidium elegans*. Some of these fungi are weak pathogens of plants and species of *Rhizopus* cause soft rots of fruits and vegetables, often creating problems during the transportation and storage of produce. The family Endogonaceae includes fungi which form vesicular-arbuscular mycorrhizal associations with higher plant roots, e.g. *Glomus mosseae*, *Gigaspora* spp. and *Endogone lactiflua*. These species have not yet been grown in pure culture. The Entomophthorales includes parasites of animals and insects.

Ascomycotina

The Ascomycotina is the largest group within the Eumycota (about 28 500 species). The vegetative phase of these fungi usually occurs as septate,

Table 1.5 The subdivision Ascomycotina

Septate (perforated) hyphae. Vegetative phase haploid. Sexual spores (ascospores) in ascus. Cell wall components: chitin and glucan.

Hemiascomycetes:	Single-walled asci (unitunicate), naked
Endomycetales:	Budding or fission yeasts, e.g. *Saccharomyces cerevisiae*, *Schizosaccharomyces pombe*
Taphrinales:	Mycelial forms parasitic on plants, e.g. *Taphrina deformans*
Plectomycetes:	Unitunicate asci, enclosed in cleistothecium
Eurotiales:	Asci small, non-explosive, e.g. *Aspergillus* and *Penicillium*
Erysiphles:	Biotrophic plant pathogens, forming haustoria only in epidermal host cells, e.g. powdery mildew, *Erysiphe graminis*
Pyrenomycetes:	Unitunicate asci, explosive spore dispersal through apical pore
Sphaeriales:	Asci within perithecia, e.g. *Nectria galligena*, *Claviceps purpurea* and *Ceratocytis ulmi*
Discomycetes:	Unitunicate asci, hymenium disc-shaped (ascocarp) and freely exposed, explosive dispersal
Pezizales:	Asci open by operculum, e.g. *Peziza aurantia* and *Helvella crispa*
Helotiales:	Asci open by slit (inoperculate), e.g. *Sclerotina fructigena*
Tuberales:	Hymenium enclosed below ground (truffles), e.g. *Tuber aestivum*
Lecanorales:	Lichen-forming species, living in symbiosis with algae, e.g. *Cladonia pyxidata* and *Peltigera canina*
Loculoascomycetes:	Bitunicate asci, outer wall thin, inner wall extending rapidly prior to spore discharge
Pleosporales:	Asci in pseudothecia, with pseudoparaphyses, e.g. *Pleospora herbarum*
Dothideales:	Asci in pseudothecia, without pseudoparaphyses

haploid mycelium which may become quite extensive, although species of single-celled yeasts are also included in this group (Table 1.5). The septa which are formed often remain open and allow functional exchange of materials between cells towards apical regions of hyphae, only becoming closed as the cells age. The cell walls are composed primarily from microfibrils of chitin and glucan. Extremely large numbers of asexual spores, known as conidia, are formed, usually on specialised hyphae arising from this mycelium. Ascospores (sexual spores) are formed following the fusion between two cells containing one or more nuclei. Since nuclear fusion does not take place immediately, a dikaryon is formed and both nuclei divide together, separating and passing to daughter

Table 1.6 The subdivision Deuteromycotina

Fungi forming only conidia; no sexual stages formed (Fungi Imperfecti)	
Blastomycetes:	Yeast forms, *Sporobolomyces*, *Blastomyces dermatitidis*
Hyphomycetes:	Mycelium with conidia on hyphae or conidiophores, some aquatic forms. *Alternaria triticina*, *Arthrobotrys oligospora* (nematophagous)
Coelomycetes:	Mycelium with conidia in pycnidia, *Colletotrichum lindemuthianum*, *Septoria nodorum*

cells. The dikaryophase is formed only at reproduction. Nuclear fusion occurs, followed by meiosis, in a cell which will form an ascus and contain the eight haploid ascospores. These give rise to haploid vegetative mycelium. Many asci may be formed and these are grouped together in an ascocarp of haploid hyphae. The taxonomy of this group is based upon the arrangement and morphology of the asci which contain the sexual spores (Table 1.5). A hypha which contains genetically different nuclei is called a heterokaryon and it is interesting to note that heterokaryosis has been demonstrated in the higher fungi (Ascomycotina, Deuteromycotina and Basidiomycotina). This provides these organisms with a high degree of genetic variability.

Many species of the Ascomycotina are saprotrophic and are widespread. Some are important plant pathogens, e.g. *Erysiphe graminis* (powdery mildew), *Ceratocystis ulmi* (Dutch Elm disease) and *Claviceps purpurea* (ergot). Members of the Lecanorales (Table 1.5) are lichen-forming fungi which associate symbiotically with algae, e.g. *Xanthoria* spp. and *Peltigera* spp.

Deuteromycotina

The Deuteromycotina (often referred to as the Fungi Imperfecti) are known to occur only in a vegetative mycelial form which gives rise only to asexual spores, known as conidia (anamorph). No sexual reproductive phase (teleomorph) has been identified. These fungi are grouped together (Table 1.6) as a matter of convenience (Hawksworth *et al.*, 1983). Many species have characteristics similar to those of the Ascomycotina although asexual forms are missing, and some share common features with members of the Basidiomycotina. In the absence of sexual reproduction parasexual recombination provides a means for the introduction of genetic variation.

Many of the Deuteromycotina are aquatic fungi and many are saprotrophs inhabiting soils with high levels of plant debris. Some species are predacious on nematodes, entrapping, obtaining nutrients from and

Table 1.7 The subdivision Basidiomycotina

Vegetative mycelium dikaryotic and septate (dolipore septa). Basidiospores formed on basidia. Cell wall components: chitin and glucan.

Hymenomycetes:	Basidia formed on hymenium. Mushrooms and toadstools
Holobasidiomycetidae:	Basidium undivided
Agaricales:	Fleshy basidiocarps with gills, pores or spines, e.g. *Agaricus bisporus* and *Boletus edulis*
Aphyllophorales:	Woody basidiocarps, including bracket fungi, e.g.*Piptoporus betulinus* and *Serpula lacrymans*
Dacrymycetales:	Basidium forked, e.g. *Dacrymyces stillatus*
Phragmobasidiomycetidae:	Basidium divided by septum. Jelly fungi
Tremellales:	Often wood-rotting species, e.g. *Tremella frondosa*
Auriculariales:	Gelatinous fruit bodies, some pathogenic species, e.g.*Auricularia aurantia*
Gasteromycetes:	Basidia enclosed at maturity. Puff balls, e.g. *Lycoperdon perlatum* and earth stars, e.g. *Geastrum triplex*
Teliomycetes:	Basidiocarps lacking. Simple septa
Ustilaginales:	Plant parasites. Smut fungi, *Ustilago nuda*
Uredinales:	Biotrophs forming haustoria on plants. Rust fungi, e.g. *Puccinia graminis*

sporulating on these small worms. Others are plant pathogens which are seed borne and cause some serious foliar diseases, e.g. *Septoria nodorum* which infects wheat plants, *Septoria apiicola* which causes leaf spot of celery and *Alternaria solani* and *A. dauci* which cause blight of tomato and carrot, respectively.

Basidiomycotina

In the Basidiomycotina the vegetative mycelial phase is dikaryotic and gives rise to basidiocarps (fruiting bodies), which carry the haploid basidiospores. These spores are formed on the basidiocarp, sometimes held on a specialised layer known as a hymenium. Nuclear fusion occurs in some cells (basidium initials) and is followed immediately by meiosis. The resultant nuclei pass into the basidiospores. After dispersal the spores germinate and give rise to homokaryotic hyphae (primary phase mycelium). Contact between two compatible homokaryons results in hyphal fusion and nuclear migration to form a dikaryon (secondary phase mycelium). The vegetative dikaryon can carry recessive genes and complementation can occur. Additionally, nuclei can be exchanged by hyphal fusion (anastomosis) between adjacent heterokaryons which provides for rapid genetic alterations. Basidiomycetes have complex mating systems. The secondary, vegetative mycelium is septate, each hyphal compartment containing two nuclei. The cross walls have an unusual and characteristic pore (dolipore septum), through which cytoplasmic continuity is maintained but which probably allows only limited exchange of materials between hyphal compartments. The cell walls are composed from chitin, present as a structural polymer, and amorphous glucan.

The basidiocarps may be quite large (mushrooms, toadstools, puff balls, jelly fungi) and are easily recognisable macroscopically. The taxonomic classification of this group is based on basidiocarp morphology (Table 1.7). A great many species are saprotrophic, and are often considered as agents of decay, e.g. the wood-rotting fungi, *Serpula lacrymans* and *Phanerochaete velutina*. Some species form mycorrhizal associations with tree roots, e.g. *Boletus chrysenteron* and *Suillus bovinus*. Additionally the group includes a number of very destructive plant pathogens, e.g. the rusts (*Puccinia graminis*), smuts (*Ustilago maydis, Tilletia caries*) and tree rots (*Armillaria mellea*).

FUNGAL MORPHOGENESIS AND PHYSIOLOGY

VEGETATIVE GROWTH AND DEVELOPMENT

The vegetative forms of fungi are extremely diverse, ranging from the simple, single-celled yeasts to those with more complex, and often more differentiated or extensive filamentous structures. Most fungi have a vegetative phase which is composed of cylindrical filaments, known as hyphae. The growth of hyphae occurs only by extension at the extreme apex although a single hypha may proliferate over considerable distances. Each hypha exhibits polarity of growth, may be more or less branched and may have cross walls or septa which delimit individual compartments. Many hyphae together, usually known as mycelium, make up the thallus or colony of a fungus. If hyphal fragments become detached from the parent mycelium they are capable of independent existence, and are able to form new colonies. Fungal hyphae have a very characteristic structure and organisation which enables this group of organisms to exploit an extremely wide range of habitats very successfully.

Hyphal ultrastructure

The majority of fungal hyphae are tubular structures about 1–15 μm wide which show polarised growth, extending at the apex, proliferating across and through the substrate, often exhibiting invasive, penetrating growth by virtue of this morphology. It has been shown, by light and electron microscopy, that the distribution of cellular organelles is different in apical regions from the rest of the hypha and that there is longitudinal differentiation within each filament.

The fungal protoplast is bounded by a plasma membrane which lines the cell wall, pressed closely to it by the turgor pressure of the cytoplasm. This membrane is composed from lipids and protein, together with some small amounts of carbohydrate. Phospholipids and sphingolipids are the most important components in fungal membranes. In the lower fungi the fatty acid components are saturated, whereas in higher fungi these are polyunsaturated. Membrane proteins are important in nutrient transport and also for their role as membrane-associated enzymes. The carbohydrate components, and particularly the glycoproteins, are located on the outer membrane surface and probably have a role in cellular recognition.

All growing hyphae have aggregations of vesicles at the apices and at growing regions such as branch points, developing clamp connections, sites of germ tube emergence, etc. It is now generally accepted that these vesicles have a role in cell wall growth and hyphal extension. Apical vesicles have been identified in thin sections viewed under the electron microscope and have been shown to be of two types after conventional fixation. Some vesicles are within the size range 100–300 nm diameter and others, usually termed microvesicles, are smaller and in the range 30–100 nm diameter. In relatively recent studies using the technique of freeze substitution to prepare material for electron microscopy, some vesicles have been shown to stain more densely than others.

Most vesicles are seen in the extreme apical regions of hyphae (the apical 30–40 μm), and also at sites where branch points occur. A few have been identified in subapical cytoplasm sometimes associated with membranous structures which appear to correspond to Golgi bodies. It has been suggested that vesicles arise from membranous structures in subapical regions and are transported through the cytoplasm to the hyphal apex.

The arrangement of vesicles within the tip region has also been shown to be variable within the different taxonomic groups of fungi (Grove, 1978). In the Oömycetes both the larger vesicles and microvesicles are randomly distributed throughout the apical zone. In the Zygomycetes the larger vesicles are aligned in a layer close to the plasma membrane and the microvesicles are more randomly distributed subapically, behind this layer. In the Ascomycete fungi the larger vesicles are aligned near to the extreme apical wall with the microvesicles in a cluster behind. A similar arrangement is seen in the Basidiomycotina with a dense, spherical mass behind the outer layer of vesicles which line the apical dome. This dense body was first observed in hyphae viewed using the light microscope and was given the name Spitzenkörper. Such structures

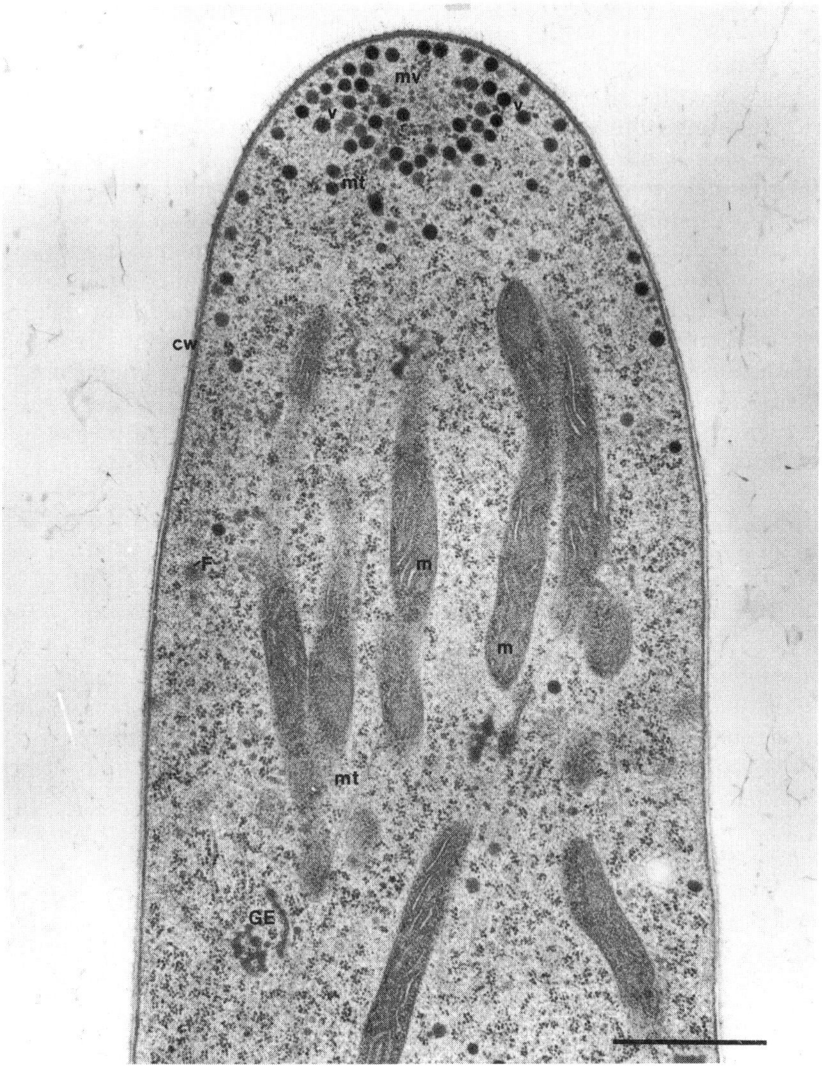

Fig. 1.1 Transmission electron micrograph to show a median section through a hyphal tip of *Fusarium acuminatum,* freeze-substituted and treated with D_2O + DMSO (45 min). The plane of sectioning is parallel to the longitudinal axis of the cell. Mitochondria (m) lie behind the apex, in the central region of the cell. The Spitzenkörper is composed of a mass of apical vesicles (v) with microvesicles (mv) clustered centrally, with numerous microtubules (mt). Cell wall (cw); filasome (F); mitochondria (m); Golgi-like endomembrane cisternae (GE). Bar represents 1 μm. Reproduced from Howard and Aist (1980), courtesy of Prof. R. J. Howard and the Rockefeller University Press.

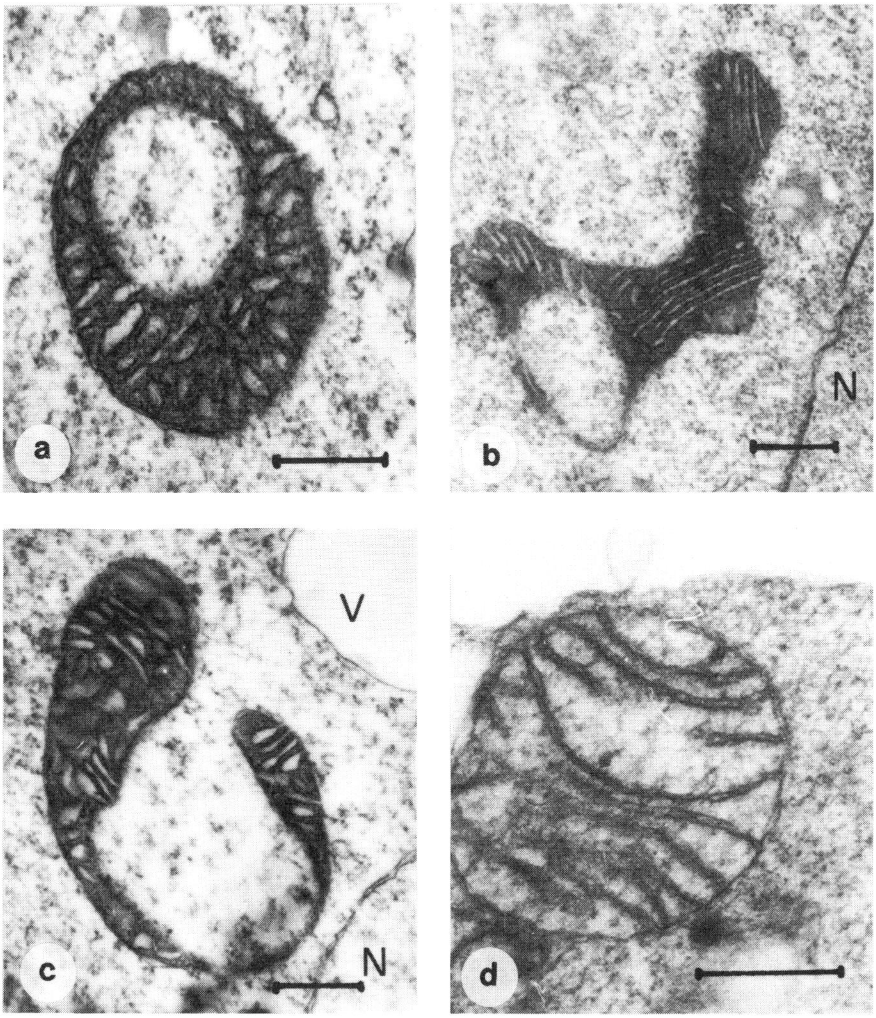

Fig. 1.2 Transmission electron micrographs to show the variability of mitochondial form in *Aspergillus nidulans*. (a, b, c) Mature mitochondria from subapical regions. (d) Mitochondrion from apical region of hypha. Nucleus (N); vacuole (V). Bars represent 0.25 μm.

have not been observed in non-growing hyphae and are always seen to be orientated to the direction of hyphal growth. In freeze-substituted material, viewed using high voltage electron microscopy, it has been shown (Fig. 1.1) that this region contains a core zone of microvesicles, probably corresponding to the Spitzenkörper, surrounded by larger vesicles (Hoch and Howard, 1980).

These apical vesicles are always associated with the presence of ribosomes. Other organelles do not occur in this region. Using the techniques of freeze substitution, microtubules, previously most heavily implicated as functioning during nuclear division, have been detected in cytoplasm at the extreme apex and in the Spitzenkörper region (Howard and Aist, 1979; Hoch and Howard, 1980). From studies with agents which interfere with tubulin it seems likely that these microtubules have some role in the transport of cytoplasmic vesicles to the apical dome (Gooday, 1983). Microfilaments (4–8 nm diameter) have also been identified in freeze-substituted hyphae (Howard, 1981), associated with vesicles and at other regions of cytological activity. The composition of these is not clear, although the main component is thought to be the protein, actin. The exact role of microfilaments in cellular organisation has not been identified although they may have a role in cytoplasmic streaming and may be components of the fungal cytoskeleton.

Although mitochondria are not observed at the extreme apex many are present in abundance just subapically to the vesicular zone. These are often rather irregular in form (Fig. 1.2) and may be elongated and aligned longitudinally in the hypha. Mitochondria contain DNA and ribosomes which are different from those in the main body of the fungal cell. The numbers and structure of mitochondria and the cristae which are contained may change throughout the life of a fungal cell.

Nuclei are present subapically to the region where mitochondria first occur. The nuclei are bound by a double membrane, or nuclear envelope, which is perforated by pores. A nucleolus is often visible in thin sections, as a densely staining region. Fungal nuclei are generally small and not always easy to detect in intact hyphae. Nuclei maintain a central position within hyphae but are also seen to move, migrating through the hyphal network, probably guided by actin filaments.

Much membranous material, or endoplasmic reticulum is also present within actively growing hyphae. Endoplasmic reticulum is continuous with the nuclear envelope and may allow the exchange of materials within the cytoplasm. The occurrence of Golgi apparatus is still a subject for much discussion, although vesicles which are eventually transported to the apex do appear to arise from such endomembranes. In fungi where Golgi bodies have been described it is seen that the appearance and activities of this membrane system change throughout the life of the fungus.

Few vacuoles are seen in actively growing regions but as the hypha matures and ages, more vacuoles occur and coalesce in subapical regions. It has been suggested that these act as storage areas for nutrient materials (e.g. amino acids and polyphosphates) although more recently (Klionsky *et al.*, 1990) vacuoles have been recognised as complex organelles with a range of functions in cellular metabolism. Fungal vacuoles have been

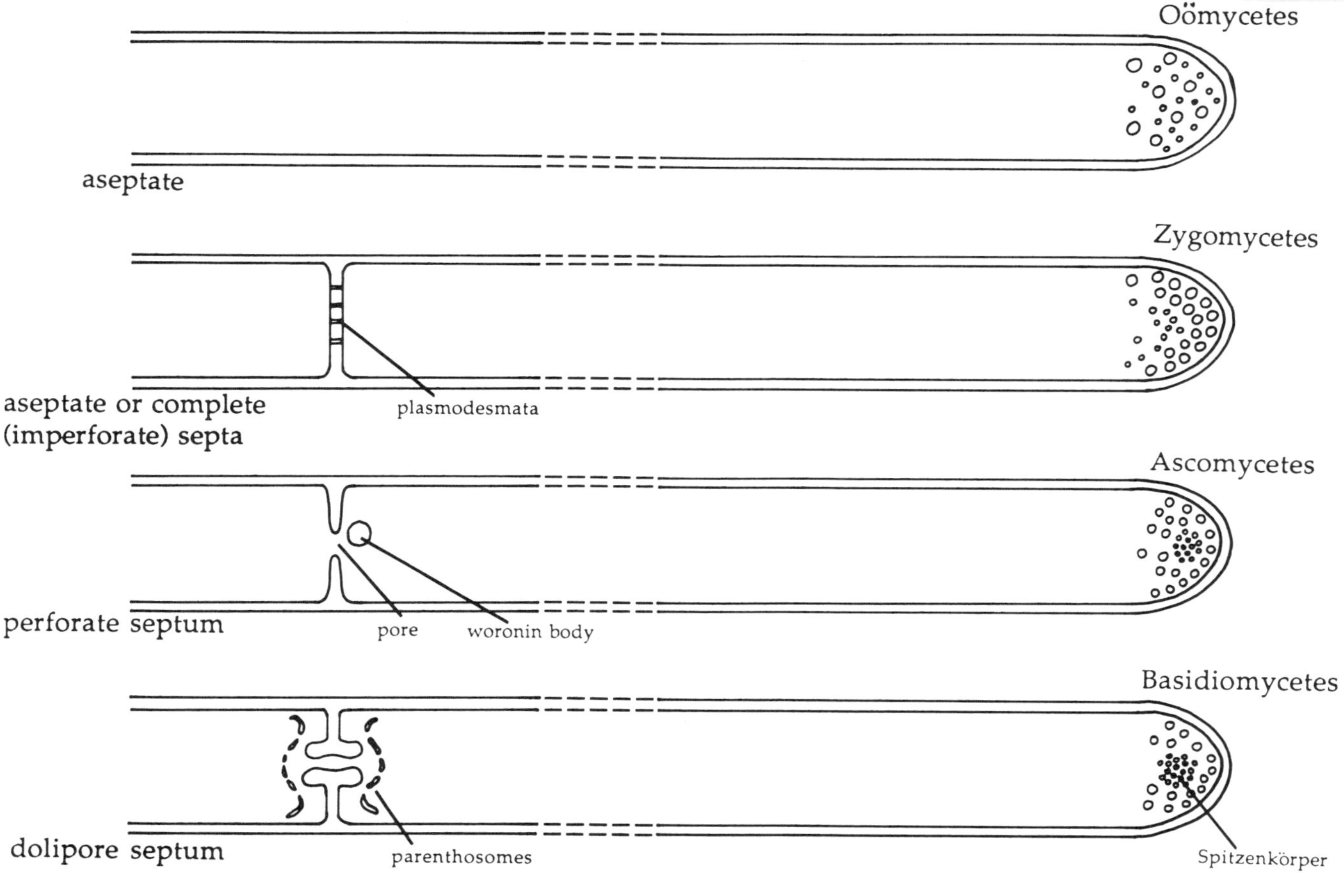

Fig. 1.13 Simplified diagram to show the major pathways of glycolysis and fermentation. The EMP system is shown in the centre (heavy arrows); the PP pathway to the right (light arrows) and the ED system to the left (dotted arrows). Many of these reactions are reversible.

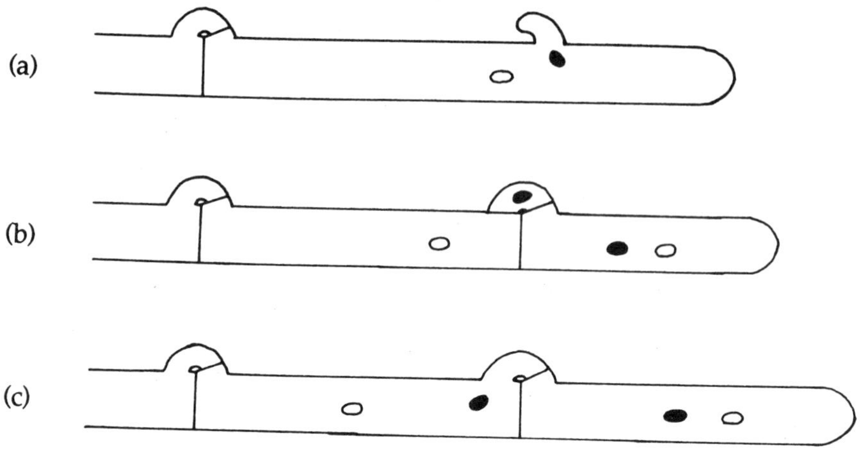

Fig. 1.4 Clamp connections typical of Basidiomycete fungi. (a) Diagrammatic representation of clamp formation in dikaryotic hypha; lateral branch production (clamp) directed away from apex. Nuclear division is simultaneous and one daughter nucleus passes into the branch. (b) Cross walls delimit apical compartment, which contains two compatible nuclei. (c) Clamp fuses with subapical compartment and nucleus passes in. (d) Scanning electron micrograph of clamp connection (arrowed). Bars represent 1.25 μm.

shown to be involved in degradative processes and to contain hydrolases (e.g. proteinase, carboxypeptidases and aminopeptidases) in quantities which vary throughout the development of the fungus. Fungi are able to adapt readily to different nutrient conditions and it has been shown that vacuoles contain many enzymes likely to be active in the internal recycling of peptides. These organelles probably have a role in adaptation to changes in nutrient status, together with important functions in the internal (homeostatic) regulation of cytosolic ion concentration, maintenance of pH and osmoregulation. In relatively aged mycelium very large volumes of hyphae may be occupied by vacuoles. Nuclei and other organelles often become sequestered into pockets of cytoplasm in these regions, or may be transported to more metabolically active areas.

In general, as hyphae age and develop, septa are laid down just subapically, dividing the hypha into compartments and probably increasing the mechanical strength of the hyphal construction (Fig. 1.3). In the lower fungi (Mastigomycotina and Zygomycotina) hyphae are non-septate or coenocytic and do not usually have cross walls. In these species, only extremely aged regions or damaged portions of a hypha are isolated by the formation of cross walls. In the higher fungi (Ascomycotina and Basidiomycotina) septa are laid down more regularly. In the Ascomycotina cross walls grow inwards from the cell wall, but these hyphae remain essentially coenocytic. The central region, or pore (0.1–0.5 μm diameter),

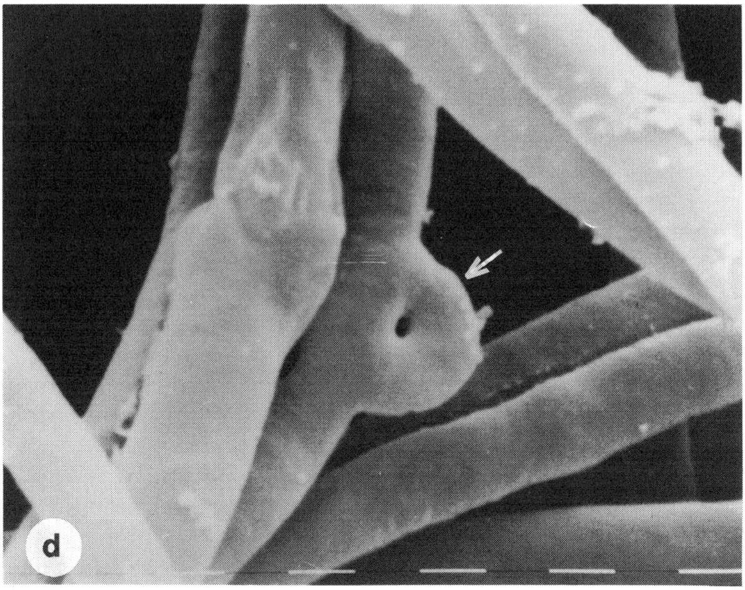

Fig. 1.4d

in the septum is not closed which allows cytoplasmic continuity between adjacent compartments. Organelles, including relatively large structures, such as mitochondria and nuclei, are able to migrate and pass through these pores, in addition to the normal cytoplasmic and nutrient flows along the hypha. Woronin bodies are small, dense, spherical organelles which appear to be membrane bound and are associated with such septal pores. Woronin bodies are normally located near to pores and are believed to have a role in regulating the cytoplasmic flow between hyphal compartments. In senescent areas these may eventually act as seals, effectively cutting off areas which are relatively devoid of cytoplasm.

Septa formed in dikaryotic Basidiomycotina are also perforate but with a rather more complex arrangement at the central pore, known as the dolipore septum. The central region of the septum is flanged and elongated, giving rise to a narrow channel (up to 0.2 μm diameter) by which the cells remain interconnected. Additionally there is a hemispherical perforated pore cap, composed of membrane and arranged in the cytoplasm, on either side of the septal flanges. Also in the Basidiomycotina clamp connections may form (Fig. 1.4), apparently as a bypass around septa. A side branch forms from the compartment on the apical side of the septum which grows towards that compartment which is immediately subapical to it. The branch tip then becomes anastomosed with that compartment, just behind the dividing septum. Nuclei pass from one compartment to another via such clamp connections but eventually these passages are also blocked by a septum.

Fungal cell wall extension

The fungal cell wall is a rigid and complex structure which endows protection on the enclosed protoplast and also determines the morphology of the fungus concerned. The wall is dynamic and a multilayered structure which is laid down at the apex. Although the main features of wall structure are determined at the apex, thickening and development continues as growth proceeds. As the wall matures, so materials are added to the structure, rigidifying it and adding to the mechanical strength. The wall also appears to form a barrier to the passage of molecules into and out of fungal cells (Peberdy, 1989a). However, as well as this largely protective role, the activities of some important enzymes have been shown to be wall bound. In some cases, e.g. invertase, acid phosphatase, the largest proportion of the activity of these enzymes in fungal cells can be detected in the wall fraction and/or the periplasmic space. Additionally, the wall may have a different structure and overall composition at different stages of the life cycle of a species. Spore walls are often greatly thickened and more resistant than growing hyphal walls. Later during development, more changes will occur again as the mycelium ages and becomes senescent.

Cell wall architecture and organisation

Conventional fixation techniques for electron microscopy demonstrated (Grove, 1978) that the plasma membrane is completely surrounded by electron-dense and more electron-lucent materials in layers of variable thickness and recently, freeze substitution techniques have shown that there may be more layers than were first realised. It is clear that the wall is much thinner and more plastic (Burnett, 1979) at the extreme apex (primary wall) where it is laid down, than in more distal regions (secondary wall). However, the multilayered characteristic can be seen at the apical dome. Distally the layers become thickened and individual fibrils also become larger. Electron microscope studies in conjunction with enzyme dissection and chemical techniques (Hunsley and Burnett, 1970) have provided much of the available information concerning the physical and biochemical nature of the wall components.

It is clear that most materials incorporated into the lateral wall are also present at the hyphal tip. In general the fibrillar, skeletal components are laid down adjacent to the membrane and amorphous materials external to these. It is important to remember that each wall layer is interlinked to the adjacent layers (Fig. 1.5). In *Neurospora crassa* secondary wall

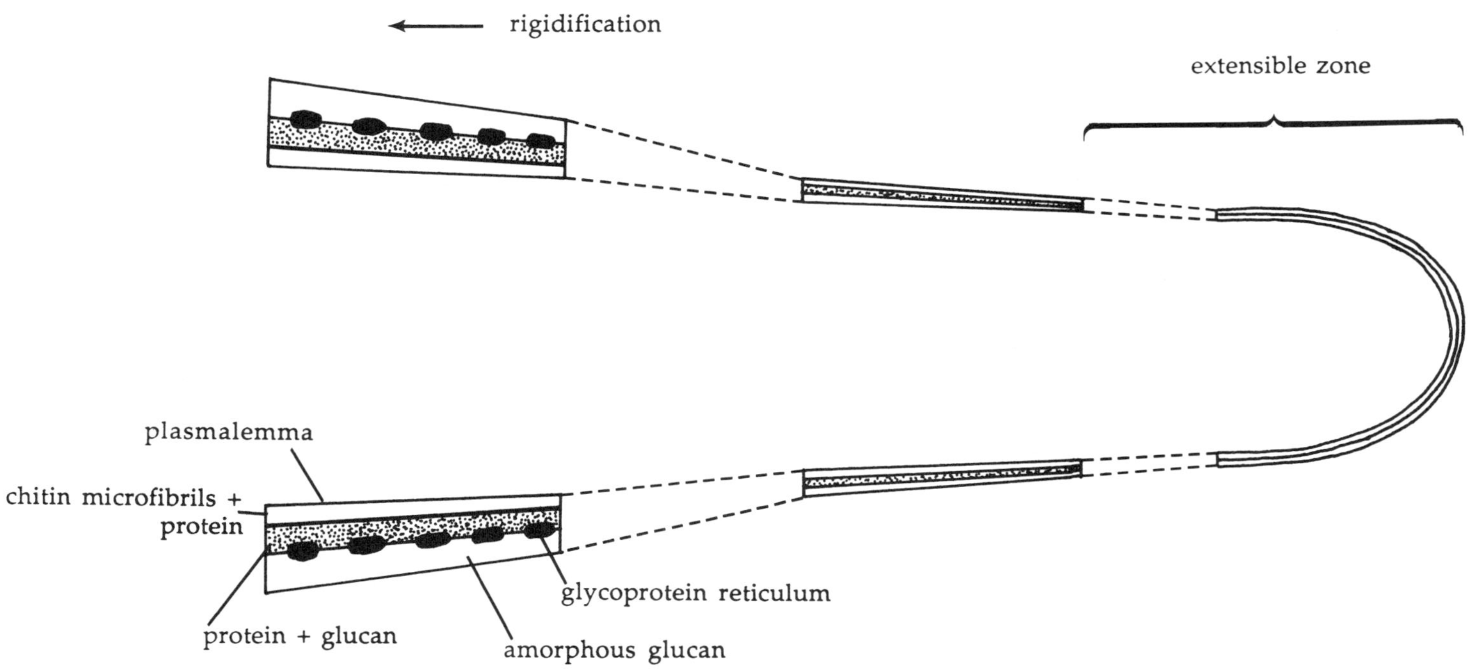

Fig. 1.5 Arrangement of the principle wall layers in *Neurospora crassa*, and the influence on wall rigidity and extensibility (after Hunsley and Kay, 1976).

thickening has been shown to be significant (Trinci and Collinge, 1975). The primary wall was recorded as 50 nm thick over the apical 50 μm but 3 mm from the tip where secondary growth had occurred the wall was 275 nm thick. A reticulum of glycoprotein has also been identified in this species, which is laid down subapically, embedded in the wall layers, and this probably functions as a strengthening component. It has also been shown, by the use of fluorescent brighteners which bind to specific wall components, that some materials which are accessible at the tip, are covered in subapical regions. Additionally autoradiographic studies have shown that labelled wall precursor compounds (^{3}H-glucose, ^{3}H-*N*-acetylglucosamine) are predominantly incorporated at the apex.

Cytoplasmic material can be isolated from actively growing hyphae and released into osmotically stabilised solution as protoplasts, by the use of lytic enzyme mixtures which degrade wall components. Providing that such protoplasts contain a nucleus each has the potential to regenerate a new cell wall, so that it is evident that the biochemical machinery for wall synthesis remains intact in these isolated units of cytoplasm. In the presence of a carbon source, a proportion of the protoplasts from filamentous fungi regenerate and grow out over the surface of solid medium in a few hours (Peberdy, 1979; Isaac, 1985). It can be seen that the first wall materials to be laid down at regeneration are the fibrillar materials. Each regenerating protoplast forms a network of fibrils around the membrane and at that stage the protoplast remains fragile and osmotically sensitive. Subsequently amorphous material is added and normal, polarised, hyphal growth is resumed.

Cell wall components

(a) Polysaccharides

The main components of fungal walls are polysaccharides which constitute 80–90% of the dry mass of the wall. These occur as either fibrillar materials which have a structural role, or as more amorphous components which act as packing materials (matrix), surrounding and embedding the microfibrils. It is interesting that a correlation has been found (Bartnicki-Garcia, 1968) between the chemical composition of the cell wall polysaccharides and the taxonomic classification of fungi, which has been based mainly on morphological features.

Much of the carbohydrate appears as glucans, which are a group of D-glucose polymers which are linked by different glycosidic bonds. β-(1, 3)-linked glucan occurs, with side branches which often have β-(1, 6)-linkages. It seems likely that this component is responsible for much of the cross-linking within the wall structure. This component has a wide distribution amongst the fungi and has been found in almost all groups

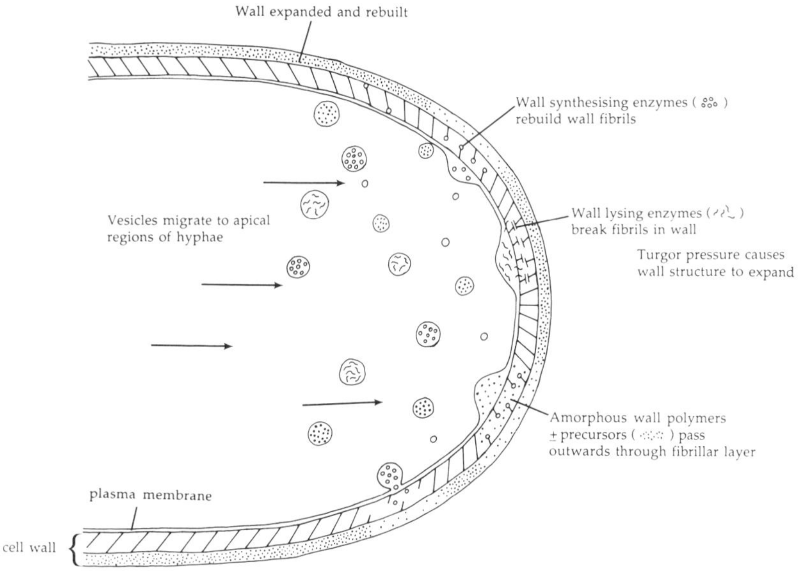

Fig. 1.6 Hypothetical representation of unit wall synthesis in hyphal tip growth. Vesicles transport wall polymers, precursor compounds and lytic enzymes to the apical dome where wall synthesis occurs (adapted from Bartnicki-Garcia, 1973).

except the Zygomycotina. Some α-(1, 3)- and α-(1, 4)-linked glucans also occur, particularly in the Ascomycotina and Basidiomycotina. All these non-cellulosic glucans occur as both microfibrils and matrix materials in fungal walls and are often complexed together. It is not always clear exactly which linkages are present or in what proportions and the extent of molecular branching is difficult to determine owing to chemical changes which occur on isolation or extraction of these components.

In some fungal species glucans constitute a very large proportion of the wall dry mass. In these fungi, the Oömycetes, the major component which forms up to 45% of the wall polysaccharide, is cellulose, a polymer of β-(1, 4)-linked glucan. In these species the cellulose component often occurs in conjunction with other linkages and may have branches containing β-(1, 3)- and β-(1, 6)-links. These fungi also contain the unusual amino acid hydroxyproline, in association with the cell walls.

For the majority of filamentous fungi, however, an important and major cell wall component is chitin, which may constitute up to 60% of the dry material of the wall. Chitin is an unbranched polymer of β-(1, 4)-linked N-acetylglucosamine units. It occurs as microfibrils, usually 10–30 nm in diameter but which may be relatively long. Chitin is usually laid down in the innermost wall region as a layer adjacent to the plasma membrane. In the Zygomycotina, chitin occurs in association with chitosan, a deacetylated form of chitin which lacks the acetyl groups. A large proportion of chitosan is found in the Mucorales.

Other polymers of galactosamine, polyuronides, mannans, galactose, fucose and xylose have also been found in varying amounts and often in complexes with proteins (Rosenberger, 1976).

(b) Proteins and lipids

Polysaccharides occur in fungal cell walls together with lesser amounts of lipid (up to 10%) and protein (1–12%), with some of these components being complexed together. Cell wall proteins are present as structural materials, often linking polysaccharides (glycoproteins). Some are also present as enzymes associated with the cell wall, such as phosphatases and proteases. These may function as hydrolytic enzymes modifying nutrient materials prior to uptake into the cells, or may be involved in cell wall synthesis.

Lipids occur in the cell wall matrix and are usually saturated fatty acids and phospholipids. Some inorganic ions, calcium and magnesium are also associated with wall components. In aged mycelium pigments, such as melanin, may account for a large proportion of the wall material (up to 20% dry weight). These probably function as protective materials, increasing resistance to attack by hydrolytic enzymes.

Cell wall synthesis

It has been suggested that the vesicles which aggregate at growing regions provide a means by which cell wall and membrane precursors, lytic enzymes (e.g. chitinase and β-glucanases) and enzymes involved in the synthesis of cell wall polymers, are delivered to the sites of wall assembly. These vesicles probably arise from subapical membranous organelles, pass through the cytoplasm towards hyphal tips, probably directed by microfilaments, and then fuse with the plasma membrane in the apical dome. Individual vesicles probably have different contents. Some are more electron dense than others, in some cases fibrillar materials are contained and some appear to carry more amorphous components. Lytic enzymes which degrade wall polymers have been detected in association with cell wall fractions of hyphae, although there is no direct evidence for their involvement in hyphal growth. Tip growth is considered to involve a highly dynamic balance between the lysis and synthesis of wall polymers in conjunction with turgor pressure. A model (Fig. 1.6) for the events which constitute a unit of cell wall growth has been proposed (Bartnicki-Garcia, 1973). Lytic enzymes are released through the plasma membrane from vesicles and these weaken the fibrillar components within the structure of the wall. The high internal turgor pressure then expands that region of the wall. Subsequently precursors of wall polymers and synthetic enzymes, which are delivered to the plasma membrane in the same way, are released into the wall to re-pack and rigidify that expanded zone.

Increased understanding of the patterns of synthesis for individual cell wall components has provided some evidence of molecular mechanisms by which wall synthesis may be localised and therefore for the control of morphogenesis in fungi.

(a) Chitin synthesis

In terms of the synthesis of individual wall components most is known concerning the biosynthesis of chitin. To some extent our understanding of the control mechanisms involved in chitin synthesis has led to theories which have implications concerning the morphogenesis of fungi. Information has been pieced together from work with yeasts, and species which are predominantly filamentous (Farkas, 1979). In fungal walls chitin occurs as fibrils embedded in matrix material and is cross-linked to proteins and polysaccharides (e.g. chitosan, mannan and glucans). The biochemical pathway by which chitin is synthesised is shown in Fig. 1.7. The final reaction in this pathway is catalysed by the enzyme chitin synthetase:

$$(\beta\text{-(1,4)-GlcNAc})_n + \text{UDP-GlcNAc} \xrightarrow[\text{GlcNAc}]{\text{Mg}^{2+}} (\beta\text{-(1,4)-GlcNAc})_{n+1} + \text{UDP}$$

requiring Mg^{2+} and GlcNAc as cofactors and GlcNAc probably acts as a primer. The reaction is inhibited by polyoxin D, a pyrimidine antibiotic, which is an analogue of UDP-GlcNAc and competes with the substrate for the active site on the enzyme.

The enzyme chitin synthetase has been detected in homogenates of fungal hyphae, and high specific activities have been found in association with membrane fractions. Indeed, fungal protoplasts have been used successfully for the preparation of chitin synthetase and additionally the use of Concanavalin A, to preserve membrane integrity, has shown that the plasmalemma is the site of chitin synthetase activity in yeasts (Farkas, 1979). It has been suggested (Bartnicki-Garcia, 1980) that microvesicles, termed chitosomes, are the means by which the enzyme is transported to the sites of wall synthesis. Such microvesicles were first detected in preparations of chitin synthetase but chitosomes also occur in the cytoplasm as membrane-bound microvesicles (40–70 nm diameter). Coiled microfibrils have been detected within chitosomes which suggests that these may be the site of chitin synthesis (Bartnicki-Garcia and Bracker, 1984). Each chitosome is thought to be responsible for the synthesis of one chitin microfibril.

Much of the chitin synthetase activity in growing hyphae is present in an

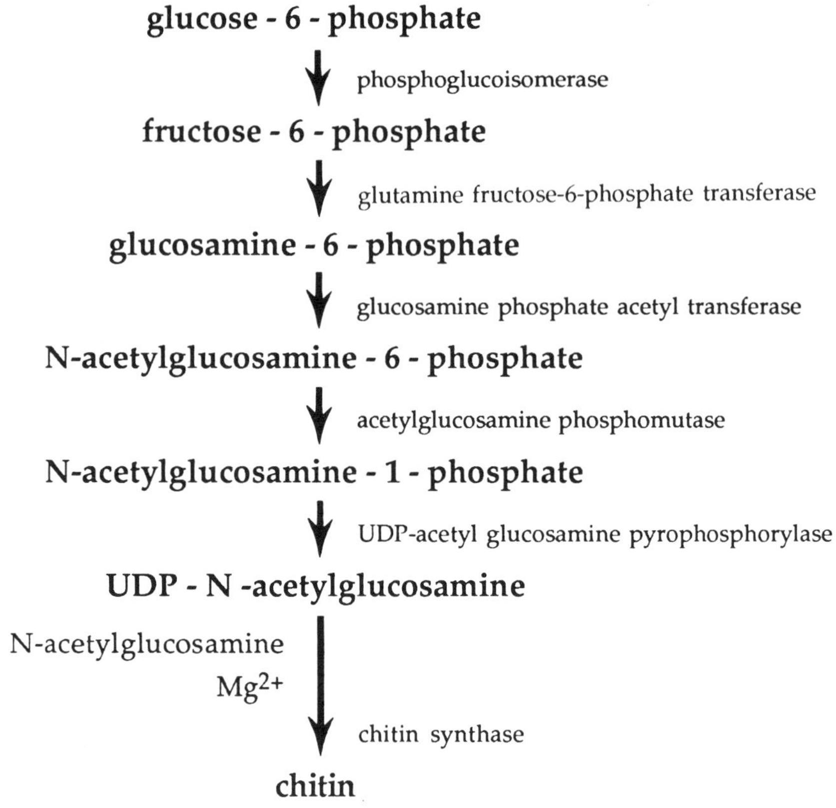

Fig. 1.7 Biochemical pathway of chitin synthesis.

inactive or zymogen form which can be activated by protease treatment. Such proteolytic activation of chitin synthetase has been demonstrated for preparations from a range of fungal species (Bartnicki-Garcia *et al.*, 1978; Cabib *et al.*, 1979). Chitin synthetase is probably delivered to the plasma membrane at the apex in the inactive form, in chitosomes. It has recently been found (Florez-Martinez *et al.*, 1990) that 90–96% of the chitin synthetase in purified chitosomes from *Mucor rouxii* was in the zymogenic form. Activation of the zymogen occurs subsequently and chitin microfibrils are probably formed at the plasma membrane or just external to it. Some zymogen remains associated with the plasma membrane, becoming progressively subapical as more new materials are added at the extreme apex. This zymogen may be activated later, at branch formation or as microfibrils are thickened within the wall structure. This suggests that both spatial and temporal regulation of chitin synthesis operates in the intact fungus.

The preparation of protoplasts from a filamentous fungal hypha represents longitudinal fractionation of that hypha. Protoplasts liberated from actively growing hyphae by means of lytic enzymes are released first from apical regions and subsequently from progressively subapical zones (Isaac, 1985). Protoplasts released from tip regions of *Aspergillus nidulans* hyphae were shown to contain predominantly active chitin synthetase and very little of the zymogen form. Those released from more subapical regions, however, contained a larger proportion of zymogen-form enzyme which was activatable by proteolytic treatment (Isaac *et al.*, 1978). This information supports a model in which active chitin synthetase is located at hyphal tips and the zymogen form is present subapically.

(b) Glucan synthesis

Less information is available concerning the biosynthesis of glucans although a system similar to that for chitin might be expected to operate for the insertion of microfibrils into the wall structure. Although most wall glucans are β-linked microfibrillar components very little is known concerning the molecular mechanism by which these are synthesised. Indeed a range of the types and proportions of glycosidic bonds which are present in cell walls have been found. It is likely that the synthesising enzymes are largely inactivated by those treatments which have been used to isolate them. Glucan-synthesising enzymes have been found in cell wall and membrane fractions (e.g. β-(1,3)-glucan synthase) but cellulose microfibril synthesis has not yet been demonstrated. It is likely that apical vesicles carry glucan matrix materials to tip regions.

(c) Localized cell wall lysis

The accepted model for cell wall growth relies heavily on the balance between the lysis and synthesis of polymers integral to that wall structure. Lytic enzymes probably have a very important role in localised degradation and wall softening, associated with the precise positioning of newly synthesised components. Endogenous lytic enzyme activities have been detected and shown to be bound to cell wall and protein fractions of hyphae (Rosenberger, 1979).

Membrane-bound cellulase activity has also been detected in preparations from *Achlya ambisexualis* (Hill and Mullins, 1980). More recently, chitinase activities have been detected in homogenates of *Mucor mucedo* mycelium (Humphreys and Gooday, 1984). Some chitinase activity was detected in supernatants from mycelial homogenates and attributed to that enzyme normally present in vacuoles. Chitinase activity was also associated with the membrane fraction and occurred in an apparently zymogenic form which was activated by protease treatment. This suggests that chitinase activity may be controlled by a similar mechanism to chitin synthetase in growing hyphae.

Dimorphism

Some fungi have the ability to change their growth habit from single, almost spherical, yeast-form cells, to exhibit linear growth, predominantly forming hyphal filaments. A number of intermediate forms are distinguishable (Stewart and Rogers, 1978), although it is the transition from unicells to multicellular or coenocytic filamentous forms which is most usually considered and has been most researched. The switch between morphological forms can be effected in culture by the manipulation of environmental conditions. Dimorphic species (e.g. *Mucor racemosus, M. rouxii, Candida albicans* and *Blastomyces dermatitidis*) are often pathogenic on animals and man in the yeast form, causing important mycoses, and have therefore attracted much attention. However, these species have also been studied with a view to increasing the understanding of differentiation in fungi and the mechanism by which conversion between forms is effected (Stewart and Rogers, 1978, 1983).

In dimorphic fungi a number of different factors influence the conversion between morphological forms. For many years these species were divided by their responses to environmental conditions, which trigger form conversions, and it is now clear that a range and combination of factors are responsible for the promotion of a particular growth form. In some cases a rise in the temperature at which the cells are grown will cause the switch in growth form, e.g. from 20 or 25°C to 37°C causes yeast-like growth in the animal pathogens *Paracoccoides brasiliensis* and *Blastomyces dermatitidis*. The main effect of this is an alteration in the composition of the cell walls; the major wall polysaccharides are different in the two forms. In *P. brasiliensis* the yeast-form cell walls contain more α-(1,3)-linked glucan, whereas mycelial-form walls are composed primarily of β-(1,3)-linked glucan together with smaller amounts of chitin and protein. The arrangement of the wall polymers, together with the localised action of those lytic enzymes involved in continued wall synthesis, is thought to control the overall plasticity of the wall and to determine the resultant morphology.

Dimorphism in *Candida albicans* is influenced by factors other than temperature changes. In this case the morphological switch has been attributed to changes in the chemical bonds within the cell wall structure, which are brought about by differences in carbon metabolism and effected in culture by different nutrient substrates. Growth on a glucose medium results in the formation of yeast-phase cells but mycelial growth occurs on less easily metabolised substrates such as starch or glycogen. Large amounts of freely available glucose provide high levels of reducing power in the cells, in the form of $NADH_2$ and $NADPH_2$. This affects the binding of the protein component of the cell wall. The disulphide bonds between the glucoprotein–mannan complexes are reduced to sulphydryl groups. This results in greater plasticity within the structure of the wall and therefore the yeast form is maintained. Additionally the yeast-phase cells have been shown to possess higher levels of protein disulphide reductase, the enzyme which reduces disulphide bonds. However, studies using mutant strains of this organism have shown that this is not the only control mechanism for dimorphism in *C. albicans.*

There are a number of differences between yeast-phase cells and hyphae (Cutler and Hazen, 1983) and much research interest has centred around the mechanism and regulation of chitin synthesis in dimorphic species. *M. rouxii* and *C. albicans* have been studied extensively and in both species more chitin synthetase activity was detected in hyphal forms than in yeast-form cells. In yeast-phase cells of *M. rouxii* mainly the zymogen form of the enzyme was found. In *C. albicans* more active enzyme was detected in the hyphal form than in the yeast form. However, in that species the proteolytic zymogen activation system may be different. Comparisons between the chitin contents of other dimorphic species show that differences which can be detected are not correlated with the morphological form of the fungus (Dow and Rubery, 1977).

In other species, such as *Mucor*, it is the degree of aeration and the supply of CO_2 which influence the conversion between yeast and mycelial phases. Changes in the gaseous environment, the energy-providing substrate (carbon metabolism) and the nitrogen source available for utilisation, have been investigated in order to explain dimorphism in biochemical terms. It has been suggested that mycelial-phase growth occurs in conditions where oxidative metabolism takes place but yeast-phase cells form under fermentation conditions. However, the situation is a complex one and form conversion must be the result of the changing balance and interactions between a number of controlling factors. More recently it has been shown that a relationship exists between the concentration of cAMP in *M. rouxii* and cell morphology. Yeast cells grow in the presence of CO_2, but the addition of dibutyryl cAMP induces aerobically cultured mycelium to grow as yeast-phase cells. Endogenous cAMP levels are low in mycelium but higher in yeast cells. It has been suggested that cAMP may affect the overall pattern of wall synthesis by interaction with other molecules. This effect may be brought about by alterations in the transcription of genes, or by the activation of protein kinases specifically involved in wall synthesis. cAMP may also interact with microtubules, which regulate precursor supply to sites of wall synthesis and changes in polarity, and other activities at the plasma membrane, which may result in morphological alterations.

Establishment and maintenance of polarity

Polarity is established prior to fungal spore germination and leads to the outgrowth of the germ tube from a particular site. Vesicles aggregate precisely at that region and are presumed to participate in the formation and emergence of the germ tube, providing precursors and wall components under the driving force of the internal turgor pressure. Once it is established the polarity is maintained and active extension is restricted to the extreme hyphal apex. It is not yet clear exactly how this polarity is set up or what factors control and drive the transport and aggregation of vesicles. The mechanism of hyphal branching and the means by which branch points are initiated are also not known.

In the filamentous fungus *Neurospora crassa*, measurements of ionic currents, made using intracellular microelectrodes, showed a membrane potential of −200 mV (Slayman and Slayman, 1962), which was generated by an electrogenic proton-translocating ATPase located in the plasma membrane (Gradmann *et al.*, 1978). These workers also showed that there was a gradient of membrane potential along the hyphae which was highest in the tip region, and strong enough to move charged molecules. X-ray microanalysis of hyphae of the marine fungus *Dendryphiella salina* showed that apical regions contained less potassium than subapical regions and the converse for sodium (Galpin *et al.*, 1978). This work led to the proposal that ATPase, located in the membrane subapically, drives an internal potassium current towards the apex (Jennings, 1979).

Recent research has increased awareness of the role of electrical currents in cellular control mechanisms. Jaffe and Nuccitelli (1974) developed the vibrating probe as an experimental system to investigate transcellular electrical currents and showed that these currents were associated with the polarity of cell growth. Transcellular electrical currents have been identified in the tip growth of algae, pollen grains, plant roots and root hairs, as well as in fungal hyphae. Representatives from all the major taxonomic groups of fungi have now been shown to generate electrical currents around their hyphal tips (Gow, 1984). The positive charges always enter at the apex and leave subapically, at a distance further back from the tip (Fig. 1.8). Sites of future outgrowths, such as points of germ tube emergence or the formation of branches can be predicted from the establishment of an influx of current at that position. Significantly, non-growing hyphae and non-polarised cells do not generate transcellular currents.

It is thought that polarity may be established by the generation of electrophoretic or electro-osmotic fields within hyphae (Jaffe, 1977; Nuccitelli,

1983). In this way charged molecules and particles may be mobilised and organised in the hyphal filament. However, in the water mould *Achlya bisexualis*, Kropf *et al.* (1983) have shown that branching reduces, or even reverses, the inward current at the hyphal apex and yet extension continues at the same rate. It has been postulated that in this case (Kropf *et al.*, 1984; Gow *et al.*, 1984) the endogenous electrical current is carried by protons and that apical growth correlates with the transcellular proton flux. The inward passage of protons at the tip is coupled with the transport of amino acids, sugars and ions and is dependent on the presence of amino acids. Protons may be carried into hyphae by the electrogenic co-transport (symport) with amino acids. The outward current is thought to be generated by an electrogenic ATPase in the membrane which extrudes protons. The possible spatial separation of symporters, occurring predominantly at the tip, and extrusion pumps, with more subapical location, would result in the formation of a proton concentration gradient by which polarity would be maintained. It is highly likely, however, that there is more than one mechanism for the generation and maintenance of electrical current flow through hyphae.

The arrangement of the cytoskeleton may be influenced by such electrical currents which may in turn affect the organisation of vesicular aggregates (Luther *et al.*, 1983) and so influence branching patterns or hyphal orientation. Actin, which is present in microfilaments, has been implicated in the polarisation of growth (Adams and Pringle, 1984) and possibly in the redistribution of vesicles prior to branching. Treatment of *Achlya bisexualis* with cytochalasins caused increased branching probably as the result of microfilament disruption (Harold and Harold, 1986). It is interesting to note that the application of endogenous electrical fields to various fungal species, in culture, caused dramatic changes in the polarity of growth (McGillivray and Gow, 1986). However, no universal pattern of response was detected, with some species exhibiting increased branching and growth towards the anode and others towards the cathode. The reason for this is not clear. Long hyphae became orientated perpendicularly to the electrical field which was thought to minimise the changes in membrane potential caused by the application of the exogenous currents.

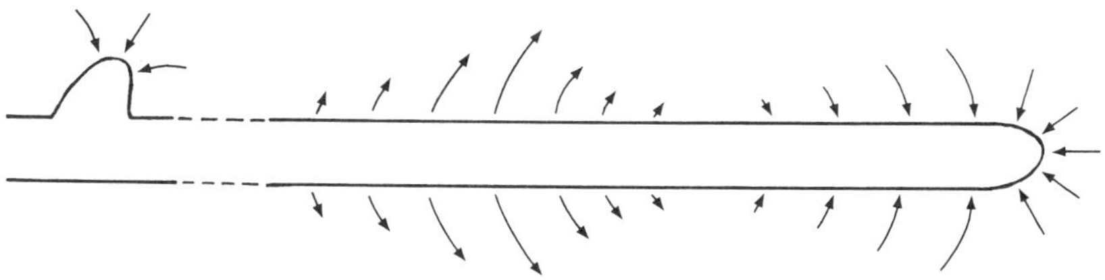

Fig. 1.8 Diagrammatic representation of the pattern of transcellular electrical currents around hyphae (redrawn after Gow, 1984).

Relatively small but localised changes in pH may also influence differentiation and the translocation of vesicles (Harold, 1982), perhaps under the influence of proton ionophores. Growing hyphae of *Achlya bisexualis* have been shown to create a longitudinal pH gradient in the external medium (Kropf *et al.*, 1984). The medium in the vicinity of the extreme apex was more alkaline than the surrounding medium and that close to the subapical regions was more acidic. Cytoplasmic pH was alkaline (pH 8.0) during filamentous growth in strains of *Candida albicans* which were able to exhibit dimorphism (Stewart *et al.*, 1988). Additionally, the application of a plasma membrane ATPase inhibitor prevented both cytoplasmic alkalinisation and dimorphism. It was therefore suggested that the increase in cytoplasmic pH, which always accompanied the switch to filamentous growth, may have been attributable to the activation of proton pumping by ATPase in this fungus. There is also some evidence which implies that calcium ions may have a role in the regulation of branching and structural dimorphism in fungi.

Hyphal and colony extension

Fungal hyphae extend by proliferation at the apex and subsequently by the formation and extension of branches. This growth habit has some advantages, for example cells do not pile up on each other but hyphae grow out to invade unexploited substrates in three dimensions, allowing the utilisation of nutrients which would otherwise only become available gradually, by diffusion. Additionally, hyphal forms are variable throughout the fungi as a group, especially in terms of diameter, branching patterns, the types and frequency of septation and therefore the lengths of compartment which are formed. Fungi are very adaptable and growth is constantly influenced by the environment. As a consequence, the mechanisms involved in branch formation, initiation, branching frequency and the way in which hyphal growth patterns are integrated and controlled in mycelial colonies are not yet thoroughly understood.

Growth is usually defined as the increase in mass of an organism under a given set of conditions. Fungal species have the capacity to grow easily in either solid or liquid medium and growth can be considered at the colony or hyphal level. Growth rate is modified by the environment and the assessment of the kinetics of growth are a useful means by which induced changes can be compared. For unicellular species (e.g. yeasts) the measurement of cell number is useful and can be made relatively easily for populations growing in liquid medium. With filamentous forms, where cell separation does not occur the determination of dry weight at intervals, by destructive harvest, allows the change in mass to be monitored. A typical sigmoidal curve is usually obtained (Fig. 1.9). Immediately following

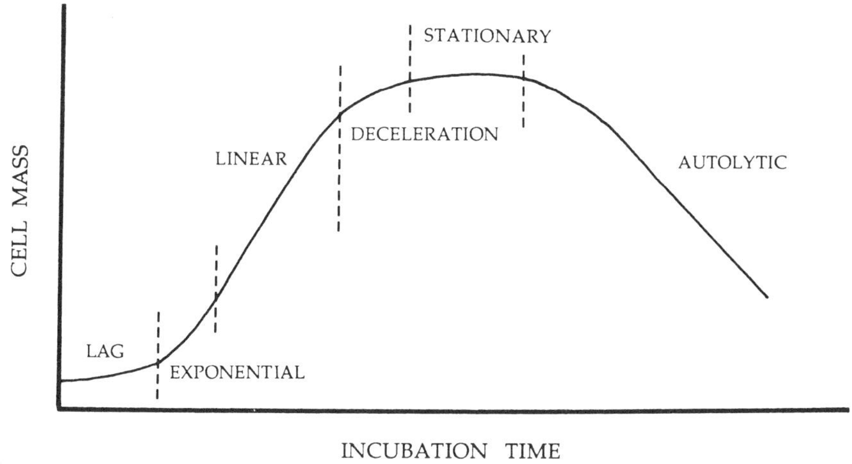

Fig. 1.9 Generalised growth curve.

inoculation there is an initial lag phase during which the cells become adapted to the environment and so proliferation and the increase in mass is very limited. Growth rate gradually increases and when the maximum growth rate for the conditions is achieved the population is said to be in exponential growth. At this stage all nutrients are present in excess, the rate of growth is constant and is termed the specific growth rate (usually designated μ). The specific growth rate is defined as the number (N) or mass (M) of cells produced by a known amount of organism in a given time (t):

$$\frac{dM}{dt} = \mu M$$

Alternatively when integrated this expression gives the exponential function:

$$Mt_2 = Mt_1 \, e^{kt}$$

where Mt_1 is the mass at time t_1, Mt_2 is the mass at time t_2, the start of exponential growth, k is the specific growth rate constant and e is the base of natural logarithms (2.72). This is also often referred to as the logarithmic phase of growth because a plot of log M, over this time period gives a straight line with a slope of μ. Different environmental conditions will change the specific growth rate which can be achieved in culture. This growth rate cannot be supported indefinitely in a batch culture and as nutrients become limiting the growth rate declines. The further depletion of nutrients combined with the gradual accumulation of toxic metabolites and inhibitory compounds in the culture leads to the marked deceleration of growth. After the maximum yield has been achieved, for the given conditions, the culture passes into stationary phase. The cells are still viable at this stage and often use reserve substances for maintenance. Subsequently the death, or autolytic, phase occurs during which the cells gradually die and lyse and therefore both mass and cell number decline.

In a continuous culture fresh medium is supplied constantly at a known flow rate, so that the nutrient concentration can be maintained, and spent medium and cells are collected continuously so that toxic metabolites do not accumulate. In such a system exponential growth may theoretically continue indefinitely, provided that the balance between medium inflow and collection is appropriate. This type of culture system can be used very successfully, to obtain large quantities of cells or products for the unicellular yeasts, but for some filamentous fungi continuous cultures are difficult to maintain for practical reasons. The heterogeneous nature of the growth forms can cause cultural problems, especially when hyphal filaments become entangled around stirring apparatus and monitors in the culture vessel or if extracellular polysaccharides are produced which tend to be sticky. Batch cultures are more usually used for large-scale industrial fermentations.

Hyphal growth

The extension of germ tubes from fungal spores has been shown to occur in phases (Trinci, 1971a). Initially there is a lag phase during which germ tube outgrowth and synthetic processes become established. Subsequently exponential growth is achieved and may be at a very high rate because this growth is supported by the endogenous nutrient reserves from the fungal spore and is influenced less by the supply of exogenous nutrients. In most species which have been investigated (Trinci, 1971a) the exponential growth of germ tubes was shown to be greater than the equivalent specific growth rate measured in submerged liquid culture. As the nutrient reserves in the spore are depleted the growth rate reduces and growth becomes linear, a phase which may be very long lasting.

On a solid substrate hyphae grow, proliferating and branching away from the original inoculation point, in all directions. As well as surface growth some hyphae grow into the medium (submerged growth) and some grow away from the substrate to become established as aerial hyphae which often become differentiated into spore-bearing structures. In stirred liquid medium growth can easily occur in all directions and filamentous forms grow as spherical colonies under such conditions. Growth within the central region of such pellets can quickly become limited. Owing to the density of hyphal filaments in the outer zones the inward diffusion of nutrients and oxygen is slow so that levels rapidly decline and toxic products increase in concentration in the central regions. In a static liquid culture growth may be submerged or, more usually, hyphae grow on the surface of the liquid in a manner analogous to that on a solid medium.

Colony growth

A fungal colony growing on solid medium will proliferate radially from the inoculum. Exponential growth cannot be supported for very long (e.g. $\pm 100\mu m$ for *Aspergillus nidulans*) and leading hyphae then extend into uncolonised regions at a linear rate. The hyphal apices alone contribute to extension growth. The region of a colony which contributes to radial extension, in this way, is known as the peripheral growth zone (Trinci, 1971b) and is defined as the shortest length of hypha which permits radial extension at a rate equal to the specific growth rate in liquid culture. Hyphae in the central regions of a colony contribute to an increase in biomass and may grow aerially or become differentiated but are not involved in colony extension. However, the radial extension of a colony (K_r) is a function of the width of the peripheral growth zone (ω) and the specific growth rate of that culture (μ):

$$K_r = \omega\mu$$

The peripheral growth zones of leading hyphae are a constant length under a given set of conditions. However, any factor which influences the frequency of branching will influence the width of the peripheral growth zone and will therefore influence the relationship between K_r and μ. It can be seen therefore, that direct comparisons between radial extension rates of colonies growing under different nutritional conditions, where branching patterns are altered, or between species which have different growth habits, are not meaningful. The rate of spread may be fast with few branches on nutrient-poor media and some species may produce more aerial mycelium.

The average length of hypha which is required to support tip growth can be determined from the hyphal growth unit, which has been defined as the ratio between the total length of hypha formed by a colony and the number of tips (Caldwell and Trinci, 1973).

$$\text{hyphal growth unit } (G) = \frac{\text{total length of mycelium}}{\text{number of hyphal tips}}$$

Initially, after spore germination, this ratio increases exponentially but later, once a colony has become established and when branch formation is more or less continuous the hyphal growth unit tends towards a constant value (Trinci, 1974). If more cytoplasm is formed than is needed to support growth of the existing tips then branching occurs and new tips are formed. The actual length of the hyphal growth unit is specific to the particular species or strain of the fungus in question. It has also been shown that the mean rate of hyphal extension (E) is a function of the hyphal growth unit and the specific growth rate for that fungus (Steele and Trinci, 1975)

$$E = G\mu$$

Maximum extension occurs at the extreme apices and it is quite probable that the rate of extension actually declines across the peripheral growth zone with increasing distance from the hyphal tips and the colony margin (Bull and Trinci, 1977). Experimental measurements have shown that aseptate fungi (e.g. *Rhizopus stolonifer*) usually have very long peripheral growth zones whereas for those fungi which form complete septa (e.g. *Basidiobolus ranarum* and *Geotrichum candidum*) this zone must be limited to the length of the apical compartment. In species which form incomplete septa and in which some communication between compartments is maintained (*Aspergillus nidulans*), linear extension continues for a short time after septation (Fig. 1.10), which suggests that the subapical compartments do contribute to radial extension for a while.

The duplication cycle

During the growth of budding yeasts there is a very definite and reproducible cycle, marked by the separation of the cells, i.e. buds from mother cells. This cell cycle takes into account the replication of DNA which must accompany the formation of daughter cells. After cell separation a new individual cell will pass through a period of consolidation or gap (G_1) during which cells may increase in size, which is followed by a period of DNA synthesis (S) and bud initiation, a second gap (G_2), and then mitosis (M) prior to cell division and separation. The timing of these phases is dependent on the environmental conditions and the state of the cells involved. G_1 is the most variable phase. Only cytokinesis (division of cytoplasm prior to cell separation) need be complete for the cycle to begin again.

The metabolic events, such as nuclear division and septation, which occur within a growing filamentous hypha must be co-ordinated and integrated, probably by mechanisms similar to those occurring in unicellular organisms. Even though cell separation does not occur a regularity in those events has been observed within the apical compartments of a range of different fungal species. This has been termed the duplication cycle (Trinci, 1978).

In the leading hyphae of uninucleate or binucleate species (e.g. *Schizophyllum commune*) the apical compartment increases in length and the nuclei migrate towards the hyphal tip, maintaining a central position in the compartment. The nuclei then divide synchronously and subsequently the compartment is divided in half by the formation of a single septum. Clamp connections form and each of the new cells contains a pair of nuclei. This is a regular process (Fig. 1.11) which suggests that there is a relationship

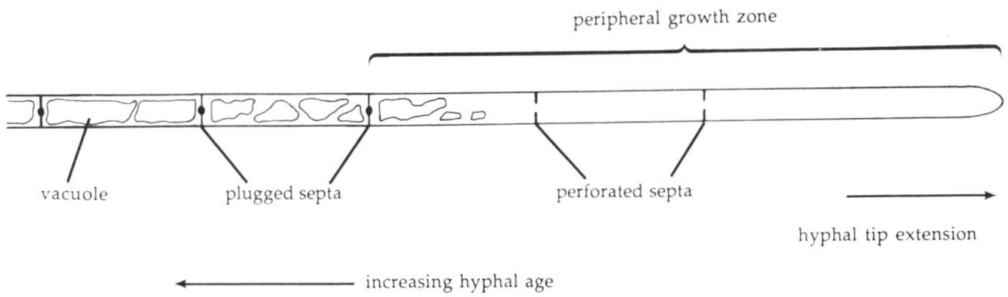

Fig. 1.10 Diagrammatic representation of the peripheral growth zone and increasing vacuolation of ageing hyphae. Hyphal compartments contribute to apical extension until septa are closed. After plugging materials cannot pass to the apical zone.

between the volume of cytoplasm and the DNA content within the apical compartment.

A similar regularity is seen in coenocytic species (e.g. *Aspergillus nidulans*), but where large numbers of nuclei are involved there is a wave of nuclear divisions starting with the nuclei near the apex and passing distally along the hypha. Septum formation divides the apical compartment in half and the more distal compartment is then divided further by the formation of several more septae. As the cytoplasm doubles in volume nuclei divide to restore the ratio with DNA content. A precise control mechanism appears to operate (Trinci, 1978) and since there is a wave of cell division through a hypha some form of cytoplasmic factor may be involved which relies on cytoplasmic mixing. Additionally it is interesting to note that while there is good synchrony of nuclear division in an individual hyphal tip there is no synchrony between different apical compartments.

Branching

Fungal colonies display apical dominance, but the mechanism by which this is controlled is not yet understood. If hyphae are subjected to osmotic shock or a change in the osmotic potential of the environment, branching patterns may be disrupted and altered. Increased branching frequency may result. This indicates that the potential for new wall synthesis exists along the entire length of a hypha but, under normal conditions must be under very precise control. Branches usually arise first at some distance subapically and are often formed immediately behind septa. Prior to branch formation vesicles aggregate at the point of outgrowth and it has been suggested that a septum may encourage the collection of such vesicles. The mechanism by which branch formation is initiated must hold the key to the developmental pattern within a colony and together with the regulation of polarity must control the spatial distribution of hyphae with respect to each other and the substrate.

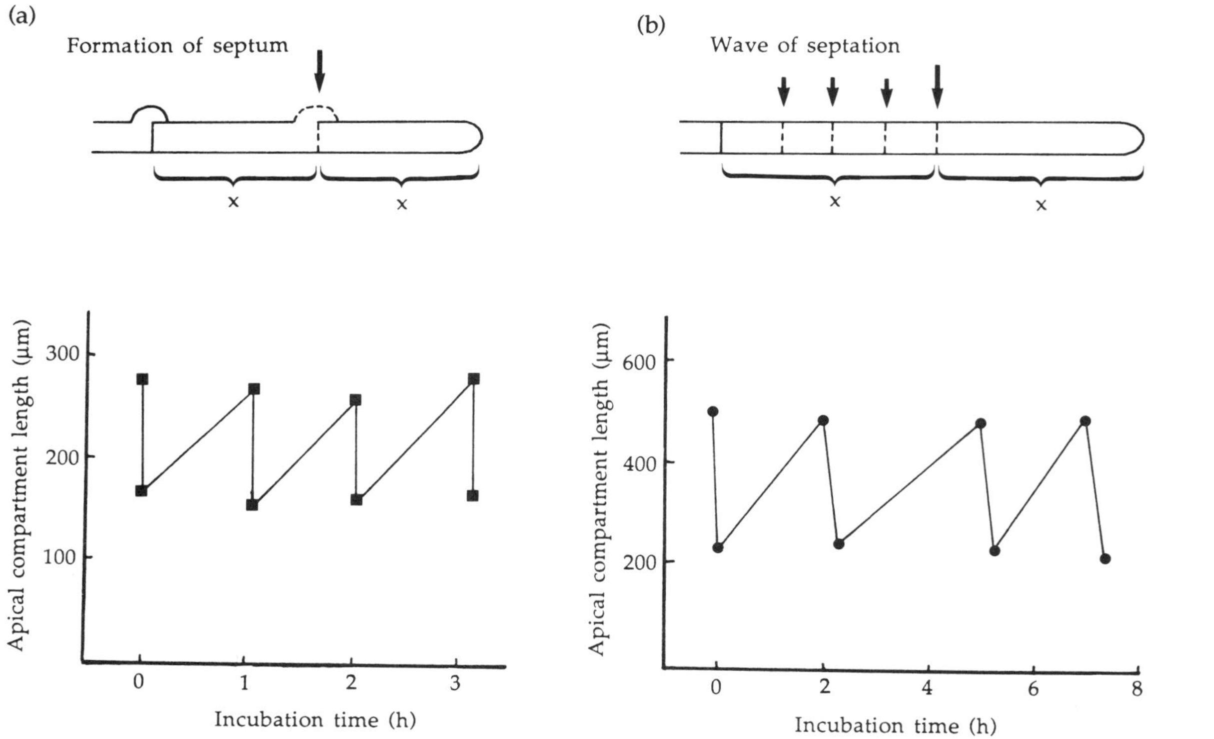

Fig. 1.11 Fungal duplication cycle (after Trinci, 1978). (a) Leading hypha of *Schizophyllum commune* (dikaryotic); nuclear division in the apical compartment precedes the laying down of a septum which divides that cell in half. (b) Leading hypha of *Aspergillus nidulans* (coenocytic); a wave of nuclear division in the apical compartment precedes the laying down of a number of septa, the most apical of which divides that cell in half.

During early colony growth branches are produced at 90° to the parent hypha, and growth is directed radially. However, the angle subsequently reduces and all branches diverge from one another. Later formed branches are much finer than the first branches and their growth is less directed. It has been suggested (Prosser, 1983) that these changes may occur owing to the response of hyphae to a concentration gradient for nutrients so that leading hyphae therefore grow out into fresh medium. Changes in branch angle may also result from chemotropic response to metabolites accumulating from other hyphae or to reduced oxygen concentrations in close proximity to other growing hyphae (Robinson, 1973). This may be viewed as some form of avoidance mechanism by newly formed branches, which orients and prevents them growing in very close proximity to those branches already present.

It has recently been suggested that for *Achlya bisexualis* branching and chemotropism may be related responses and may be controlled by the presence of specific amino acids (Schreurs *et al.*, 1989) in the external environment. In this hypothesis receptors, located on the hyphal surface, are activated by specific amino acids, particularly methionine. These receptors would then localise the aggregation of vesicles, containing cell wall precursor compounds, and give rise to branch formation or to a change in the direction of tip growth. These receptors appear to respond very sensitively to low concentration gradients in the growth medium, with branch points always occurring towards the highest concentration.

NUTRITION AND PHYSIOLOGY

Fungi are heterotrophs and must obtain organic compounds from the surrounding substrate to support their growth and development. The various means by which this is achieved, and by which fungi exploit their environment, are diverse and make these organisms both interesting and unusual. A very wide range of complex substrates can be used as nutrients by the fungi. However, before these can be taken up materials must first be degraded outside the cells, by the release of extracellular hydrolytic enzymes through cell walls. Simpler, smaller molecules are then transported into the cell and either converted into energy or used for synthetic processes.

Fungi produce a very wide range of enzymes for the dissolution of substrates, e.g. proteases, depolymerases, amylases, cellulases and pectic enzymes, which are released into the environment. Some are normally produced by the cells (constitutive enzymes) and are always present in growing colonies, others are formed in response to the presence of particular substrates (inducible enzymes). The breakdown of materials by the action of such enzymes is extremely efficient and fungi are a very successful group of decomposer organisms. The use of large-scale cultures has also led to the widespread exploitation of fungal enzymes on a commercial basis, for industrial processes.

The means by which extracellular enzymes are released from hyphae are not yet clear. However, it is likely that synthesis takes place within the main body of the cytoplasm and the enzymes are transported to the plasma membrane in vesicles, for release through the structure of the wall. Much enzyme activity is associated with growing hyphal apices, but it is also likely that some enzymes are predominantly released subapically and much enzyme activity is associated with the periplasmic space.

Membrane transport processes

The passage of materials into, and out of, fungal hyphae is mediated by the selective permeability of the plasma membrane which lies on the inner surface of the cell wall and is a highly effective barrier to many molecules. Fungi are able to select nutrients from a mixture taking up some molecules preferentially. Several processes are normally involved.

Non-mediated diffusion

Some molecules pass through the plasma membrane by passive diffusion, entering a cell by movement down an electrochemical gradient, which uses no metabolic energy. The rate of diffusion is related to the difference in concentration across the membrane and slows down as equilibrium is reached. This applies primarily to molecules that are lipid-soluble, probably owing to the high lipid content of the membrane structure itself.

Facilitated diffusion

Other molecules, such as sugars, amino acids, ions and vitamins also pass through the membrane down an electrochemical gradient, but are aided by carrier proteins. These carriers, also known as permeases, act as regulators for both the uptake and release of molecules and are located on the plasma membrane. These are generally highly specific and there are different carriers for different groups of molecules. In addition some will bind preferentially to a particular substrate. If there is a large concentration difference across the membrane for a particular molecule this facilitated diffusion will be relatively fast at first, but as the concentration difference becomes less so the rate of uptake will slow down (Michaelis–Menton kinetics).

Active transport

Fungi are also able to transport nutrients against electrochemical gradients, although this does require the use of metabolic energy. ATP is used to pump molecules across the plasma membrane. It is this ability that allows fungi to take up particular nutrients even when the concentration within a hypha is already greater, sometimes much greater, than that in the surrounding environment. This is also a carrier-mediated process. The hydrolysis of ATP, by an ATPase located within the plasma membrane, provides energy which drives the passage of molecules through the membrane via a carrier protein. Similarly it is thought that some energy may be used to transport hydrogen ions out of the cell subapically, thus creating a concentration gradient for H^+. Outside, these H^+ ions bind with carrier molecules, increasing the affinity for substrate molecules. These can then be co-transported, with the H^+, across the membrane into the cell.

Exploitation of nutrient sources

Fungi normally flourish in moist habitats where nutrient diffusion occurs freely and where membrane permeability can be maintained. Nutrient materials are absorbed by growing hyphae and are therefore gradually depleted in the immediate vicinity. However, the fungi are generally good scavengers, and as main nutrient sources are used up other lesser, or more complex compounds, may be effectively absorbed. Indeed the production of inducible enzymes may be repressed until supplies of more easily assimilated compounds have been depleted. Additionally, hyphal tips grow out into regions containing fresh nutrients, previously unexploited by that mycelium. It is likely, however, that in the natural environment fungi often face very patchy distribution of nutrients in their immediate surroundings. Colonies are able to alter branching frequency, hyphal diameter, formation of new septa and so on, to give rapidly spreading growth patterns and to reach new pockets of substrate as efficiently and rapidly as possible. Some species can transport nutrients from one region of a colony to another, often over considerable distances. Alternatively, fungi also have a range of adaptive mechanisms for overcoming nutrient stress by escape, as for example, by the formation of resting states or the efficient dispersal of propagules.

As a group, fungi have a very wide habitat range and are able to use a very wide range of substrates as nutrient sources. As individuals, however, although some species are highly versatile and adaptive others have very specific requirements for long-term growth and development. Although a great deal of information is available concerning the utilisation of individual nutrient components and different chemical forms of components the fungi normally grow on complex, mixed substrates in natural situations.

Carbon

Carbon-containing compounds are required to provide both sources of energy and also the basic molecules for biosynthesis. In general the fungi are able to use an extremely wide range of carbon compounds, e.g. hexoses, pentose sugars, organic acids, sugar alcohols, hydrocarbons, amino acids, proteins and polysaccharides, many being the only compounds which are needed. Only a small number of species are able to fix CO_2 and a very few can use it as the sole source of carbon. However, not all these compounds can be utilised effectively by all species, or at all stages of the life cycle. Some soil fungi have been shown to be oligocarbotrophic (Wainwright, 1988), growing and predominating in environments where carbon levels are extremely low.

Many of the more complex carbohydrates are very important nutrient sources but must be hydrolysed to smaller, simpler molecules and subunits prior to uptake into hyphae. Higher plants produce some extremely complex carbohydrates which form a very large pool of nutrient materials in the environment but which are persistent, being recalcitrant to degradation by virtue of their complexity. Saprophytic fungi, which rely on dead organic matter, are particularly efficient and active degraders of these compounds. The production of enzymes, sometimes a series of enzymes, for the degradation of such materials must confer a definite ecological advantage and open an enormous reservoir of nutrients to these species.

Cellulose, for example, is a major component of plant tissues, entering the ecosystem as leaf litter and forming a huge source of carbohydrate in the natural environment. Cellulose microfibrils are polysaccharides composed of chains of glucose molecules complexed together and are found in all plant cell walls. The degradation of cellulose is normally achieved by a number of different hydrolytic enzymes and fungi have a very important role in the degradation of cellulose. Additionally some fungi which attack and invade living plants, as pathogens, produce cellulose degrading enzymes (cellulases) in order to effect entry to the plant tissues.

Lignin is also a major and important component of plant tissues which occurs in thickened cells and as a consequence this is a very abundant source of carbon. It has a complex chemical structure and is composed of polymers of substituted alcohols, including *p*-coumaryl alcohol, sinapyl alcohol and coniferyl alcohol. The actual structure and composition varies considerably in different plants. Lignin is extremely resistant to degradation although many fungi, mainly Basidiomycetes, can use it as a carbon source. These species include some important plant pathogens, such as *Armillaria mellea* and also *Serpula lacrymans*, the dry-rot fungus, which can cause widespread problems in building timbers. The extent of lignin degradation is dependent on the carbon to nitrogen ratio, since adequate supplies of nitrogen are essential for continued growth, and also the moisture content within the material, both of which affect the degradative ability of these fungi.

Nitrogen

Most fungal species can use inorganic nitrogen sources as well as organic nitrogen-containing compounds (proteins and amino acids) for growth. Nitrogen is essential for the biosynthesis of complex molecules in cells, such as amino acids, proteins, nucleic acids and some vitamins. Although many nitrogen sources can easily be taken up into hyphae, often by diffusion, proteins must be broken down into smaller units for uptake, a process carried out by protease enzymes. Additionally some very specific carrier systems have been described, which allow amino acids to be effectively scavenged from a nutrient-poor environment. A wide range of fungi require particular, preformed, amino acids for growth and are able to utilise these most effectively in the presence of carbon-containing compounds. Many fungi are able to use nitrate as a nitrogen source, except some aquatic species (Saprolegniaceae) and some Basidiomycetes which appear to lack the enzyme nitrate reductase, and although nitrite can be used by some species it is toxic to others. Ammonium, however, can be used by most fungi. It is often taken up preferentially to nitrate and in some species the absence of ammonium ions prevents spore formation.

Macronutrients

Although inorganic nutrients are not required in such large quantities as carbon and nitrogen they form an extremely important part of fungal nutrition and as a consequence, are often termed macronutrients.

Potassium has an important role in osmotic regulation, the maintenance of turgor and in the transport processes of hyphae. The marine fungi in particular, take up K^+ and extrude Na^+ and H^+ which prevents the accumulation of Na^+, which might otherwise reach toxic levels in cells. Indeed the marine yeast, *Debaryomyces hansenii*, maintains a higher intracellular ratio of K^+/Na^+ than species which do not show salt tolerance. Potassium is also important in the activation of enzymes.

Fungi produce a range of phosphatase enzymes by means of which phosphates (PO_4^{3-}) are taken up into cells for incorporation into nucleic acids and phospholipids or storage as polyphosphates in vacuoles. Phosphates also have a primary role in the storage and transfer of energy. Although phosphorus is not always readily available in large amounts in some environments (particularly soil), it is unlikely to be limiting to fungal growth. The presence of phosphorous is important for the uptake of magnesium by fungal cells. Magnesium has a stabilising influence on membranes and is involved in the synthesis of membrane glycoproteins. It is required as an enzyme activator and cofactor and has a key role in the metabolism and stabilisation of ATP.

Sulphur is an important macronutrient as a component of proteins, amino acids, sulphydryl compounds and vitamins. The sulphur component of amino acids, particularly cysteine and methionine, affect the function of proteins and may have an important influence on the morphology and development of cells. The presence of calcium enhances fungal growth, functioning as an enzyme activator. It is required, to a greater or lesser extent, for reproduction, and is particularly important for some of the lower fungi, e.g. *Achlya* and *Phytophthora* spp.

Micronutrients

Trace elements are required for the normal functioning of cells but in much lower concentrations than the macronutrients; indeed higher levels are often toxic to fungal cells with drastic consequences. Additionally the presence of these components may act to stimulate activity within some biochemical pathways. Iron is required for inclusion in the cytochromes and in a range of pigments. However, ferric and ferrous iron are relatively insoluble and are therefore not easily assimilated from the environment. The fungi produce organic acids which chelate iron, and a range of siderophores which solubilise it and effect transport into the cells. Within the cells iron is bound to siderophores or polyphosphates for storage. Copper is required in micronutrient levels by fungi and functions as an activator of oxidative enzymes and as a component of pigments which are produced. It is often stored in cells by copper-binding proteins, some of which are enzymes that require Cu^{2+}, e.g. cytochrome oxidase. High levels of copper can be very toxic to fungi although some species can develop copper tolerance. Manganese affects activity of fungal enzymes and may have a role in the maintenance of membrane permeability. Sodium also affects membrane permeability and is important for transport processes. Fungi tend to accumulate zinc, which is required as an active component of a range of enzymes, e.g. alcohol dehydrogenase. Molybdenum acts as an enzyme activator and, together with iron, as a cofactor for nitrate reductase, a key enzyme in nitrogen metabolism.

Vitamins are frequently required by fungi but are needed in only very small quantities for growth. Most act as important coenzymes, e.g. thiamine, for which many fungi are deficient. Biotin has a vital role in lipid synthesis, and some species and strains also require riboflavin, nicotinic acid and folic acid.

Role of environmental factors

Water is very important for fungal growth; indeed there are a range of fungi that are truely aquatic and many species of lower fungi have motile spores which require water films for movement. Although most species can tolerate much lower moisture levels, water is required for the diffusion and uptake of nutrient materials and fungi generally thrive in situations where it is freely available. Drying out cannot be tolerated, and results in membrane damage. The lichenised fungi are the most tolerant to low moisture levels and are also the most adapted to the stresses of repeated wetting and drying cycles.

Most fungi are regarded as mesophilic, with temperature optima between 15 and 40°C. However, many species are tolerant of lower temperatures, some growing successfully at 3–4°C. Other species are more thermotolerant and can grow well at up to 50°C in composts.

Fungi tend to acidify the local environment as growth proceeds and are more tolerant of lower pH values than many other microbes. This can provide an ecological advantage in some circumstances. Additionally, the uptake of nutrient materials into hyphae is markedly affected by the pH of the surrounding environment and may therefore influence nutrient availability.

Most fungi do not have any absolute requirement for light and are able to grow in very poorly illuminated sites. In some species, however, exposure to some light, even at low intensity and for a short period, may stimulate sporulation. Additionally some species produce sporangiophores which respond to, and become orientated towards, a light source (e.g. *Phycomyces* and *Pilobolus*). *Pilobolus*, which grows on animal dung (coprophilous fungus), produces tiny sporangia, borne on vesicles which are eventually involved in spore dispersal. As the vesicles develop incident light is focused on to a band of carotene pigment at the base which results in the orientation of the sporangia towards the light.

Translocation

In addition to short distance movement of molecules in hyphae, many fungal species are able to transport nutrients over very considerable

Fig. 1.12 Mycelial strands of *Serpula lacrymans* extending over a nutrient-poor plaster surface in the corner of a damp cellar. Bar represents 30 cm.

distances, either through mycelium or by means of specialised strands or rhizomorphs. In some species this ability is essential for survival.

Translocation takes place in specialised hyphae and mycelial strands, often formed by wood-decomposing species, particularly under conditions of nutrient limitation. Strands, or cords, are common in Ascomycete and Basidiomycete species, e.g. *Serpula lacrymans*, the so-called dry-rot fungus (Fig. 1.12). The strands are relatively simple structures composed of hyphae which aggregate together and grow in parallel to each other, branching, intertwining and anastomosing but remaining orientated together rather than spreading as in normal hyphal growth. Rhizomorphs are also hyphal aggregations but these show a much greater degree of organisation and structure. They are often larger aggregates which may be pigmented and grow in a manner akin to plant roots, with a definite, differentiated form. There is a meristematic zone just behind the apex (Granlund *et al.*, 1984). Hyphae formed in that region contribute to the growth of the rhizomorph, giving rise to a cortical layer, an inner medullary region with a central cavity, or lumen, which is often associated with oxygen movements. The highly destructive plant pathogen *Armillaria mellea* gains its common name, the boot-lace fungus, from the rhizomorphs it forms underground and beneath the bark of invaded trees and shrubs.

Both these types of specialised structure may extend considerable distances, often several metres, into the surrounding environment, often passing through non-nutrient substrates, supported by the main body of mycelium from which they arose. Such structures provide a great ecological advantage over other fungi, aiding successful invasion and colonisation of new substrates, by virtue of inoculum potential.

In Basidiomycetes translocation of solutes most probably occurs by turgor-driven mass flow (Jennings, 1987). Such movements have been demonstrated in mycelium of *Serpula lacrymans* growing from wood blocks, over a non-nutrient Perspex layer (Brownlee and Jennings, 1982a,b). In a series of experiments radiolabelled isotopes were added to the medium surrounding the infected wood blocks. Rates of translocation through the mycelium were determined by the movements of radiolabel between two detectors, one located near the nutrient source and one nearer to the growing mycelial front. The addition of a metabolic inhibitor (sodium azide) reduced the rate of translocation, probably by preventing solute loading.

In this species cellulose breakdown provides a source of glucose which is taken up by the fungus and converted to trehalose (Jennings, 1984). As a result water passes into the hyphae and generates turgor pressure which drives solutes through the mycelium by mass flow. Trehalose is the main carbohydrate translocated. This is eventually converted to arabitol in colony margins which may be regarded as the main nutrient sink. Movements in the reverse direction (cyclosis) to osmotically driven mass flow may also occur in hyphae. It has been suggested that an electro-osmotic flow may function. This may be important in the maintenance of pH and ionic balance within hyphae.

Primary metabolism

Once nutrient materials have been absorbed from the environment molecules enter the biochemical pathways of the fungus and are degraded for energy production (catabolism) or to provide the basic materials for the biosynthesis of cellular metabolites and the structural components required for growth and maintenance (anabolism). Fungi are able to carry out these processes simultaneously. Many of the enzyme systems which are involved in such metabolic pathways are under very precise regulatory control and may be compartmentalised within hyphae.

The major metabolic pathways which operate in fungi are largely similar to those which have been described and characterised for mammalian, higher plant and bacterial systems. It is intriguing that although some early information was derived from yeast cultures the filamentous fungi have, until recently, yielded relatively little conclusive information concerning metabolic pathways. This may be due, in part, to the heterogeneous nature of the growth form, since crude mycelial homogenates are usually used

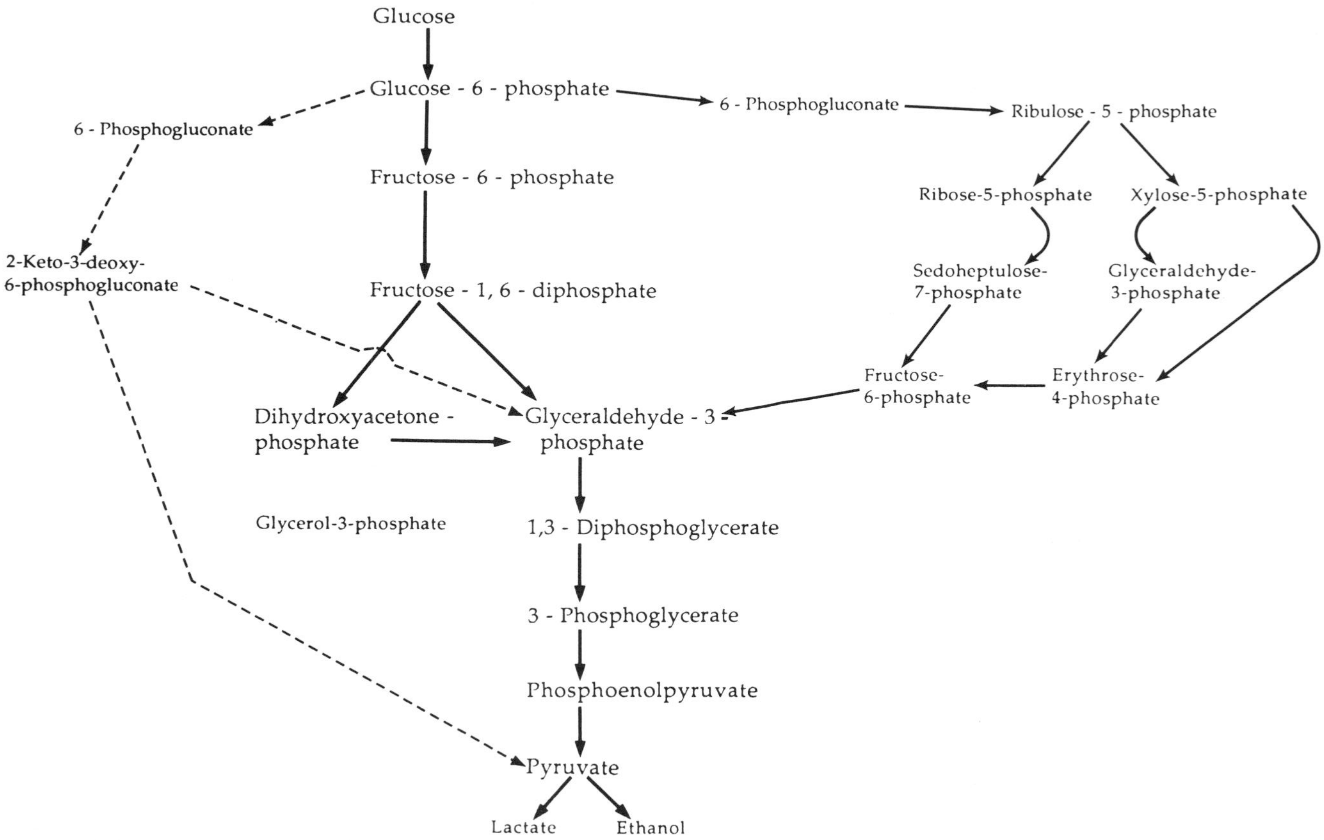

Fig. 1.13 Simplified diagram to show the major pathways of glycolysis and fermentation. The EMP system is shown in the centre (heavy arrows); the PP pathway to the right (light arrows) and the ED system to the left (dotted arrows). Many of these reactions are reversible.

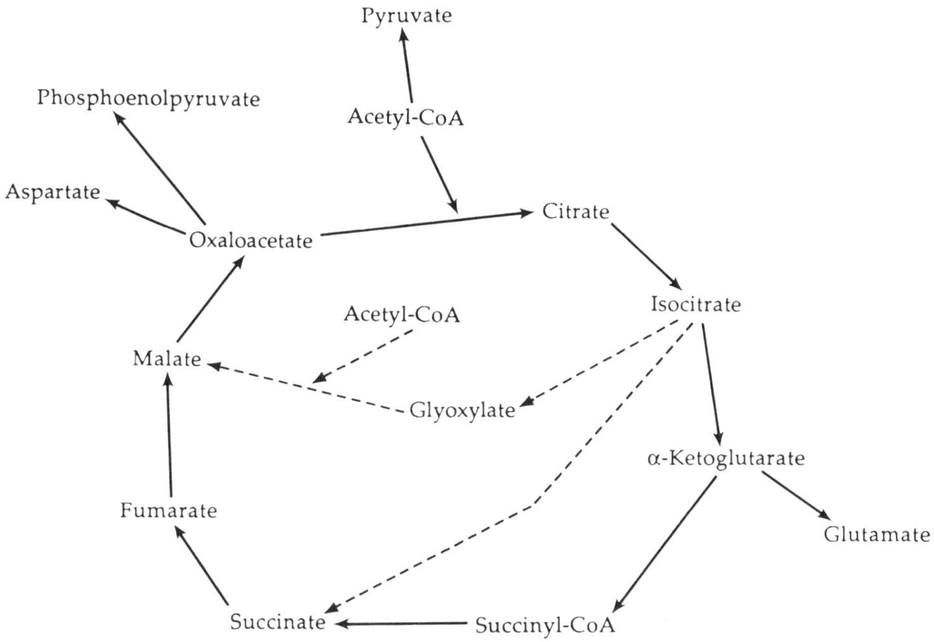

Fig. 1.14 Simplified diagram to show the tricarboxylic acid (TCA) cycle (heavy arrows) and the alternative glyoxylate pathway (dotted arrows).

to prepare material for analysis, and to the difficulties encountered in isolating intact, functional organelles.

Respiration

The energy for metabolism is obtained by the oxidation of organic materials. The general term glycolysis refers to the first stage in this process, the conversion of glucose to pyruvate with the concomitant production of energy, usually in the form of adenosine triphosphate (ATP). Three pathways of glycolysis (Fig. 1.13) have been described in detail and these provide precursors and intermediates for biosynthetic pathways, together with reduced forms of the pyrimidine nucleotides, nicotinamide adenine dinucleotide (NADH) and nicotinamide adenine dinucleotide phosphate (NADPH). However, in the fungi two main pathways have been shown to operate. Activities of those enzymes which catalyse the reactions of these pathways have been identified in fungal tissues and some evidence has also been gained from ^{14}C labelling studies (Cochrane, 1976).

The Embden–Meyerhof (EM) pathway is the most common and is usually regarded as universal among the fungi. Most carbon is metabolised via this pathway which provides ATP and pyruvate. The pentose phosphate

(PP) pathway has been identified in many fungi but is a less important pathway. This is essentially a branch of the EM route, which yields reducing power (NADPH) for synthetic processes. A pool of phosphates are formed from this pathway which are used for the synthesis of lipids and cell wall polysaccharides. Ribose is also produced by this route for use in the synthesis of nucleic acids. The Entner–Doudoroff (ED) pathway has been very tentatively identified in only a few fungal species and is of limited importance.

Pyruvate is completely oxidised to CO_2 and water in the presence of oxygen via the tricarboxylic acid (TCA) cycle (Fig. 1.14). Intermediates of this cycle provide carbon skeletons for a range of synthetic processes, in particular the precursors of amino acid synthesis. Key enzymes of the TCA cycle are associated with fungal mitochondria and have been detected in all the fungi examined (Kubicek, 1988). NADH and reduced flavin adenine dinucleotide (FADH) (from glycolysis) enter the electron transport chain. The transfer of electrons down the energy gradient releases energy to form ATP from ADP, the process known as oxidative phosphorylation. All these processes are very efficiently regulated by the ratios between the key enzymes and pyrimidine nucleotides (NADH/NAD). These pathways are similar to the processes which occur in higher plants and will be discussed again in that context in Chapter 4.

Fermentation

Anaerobic growth of fungi is unusual and proliferation in the absence of oxygen may be quite limited. It is clear that the nutrient requirements of fungi for growth under oxygen limitation may be very different from requirements under aerobic conditions, e.g. some species are unable to grow anaerobically unless vitamins or trace elements are supplied.

Fermentation is a very energy-inefficient process yielding 2 moles of ATP per mole of glucose as opposed to 36 moles under aerobic conditions. The EM pathway functions in both the aerobic and anaerobic catabolism of sugars in fungi (Cochrane, 1976). In the absence of oxygen, however, pyruvate may be converted to ethanol and CO_2. This is carried out mainly by species of yeasts but also by some other fungal species. Additionally, some of the lower fungi are capable of fermenting pyruvate to lactic acid. It has also been shown that some species are able to operate both fermentation pathways (e.g. *Rhizopus* spp.).

Gluconeogenesis

In addition to the degradation of carbon-containing compounds to obtain energy and carbon skeletons fungi are also able to synthesise carbohydrates. Essentially the processes of glycolysis are irreversible and therefore other pathways are used to allow the synthesis of polysaccharides. The term gluconeogenesis is used to describe the conversion of pyruvate to glucose. Some of the metabolic steps are linked to, and reactions shared with, the EM pathway with alternative routes to bypass irreversible steps. Gluconeogenesis is likely to occur in fungi because many species can use small molecules, e.g. amino acids, as carbon sources.

Anaplerotic pathways and the glyoxylate cycle

These are reactions which compensate for the removal of intermediates from the central metabolic pathways, which happens because many intermediates are also precursors in other biochemical routes, for example CO_2 fixation acts as an anaplerotic pathway to the TCA cycle (Fig. 1.14). CO_2 fixation or the glyoxylate cycle are pathways which compensate for the removal of TCA cycle intermediates during amino acid synthesis. The growth of fungal mycelium requires fixation of CO_2 to provide for the synthesis of C_4 intermediates and direct carboxylation reactions. Acetyl CoA is metabolised to isocitrate and then to succinate via the glyoxylate cycle, a bypass to the normal TCA cycle. Therefore lipids can be used for the synthesis of carbohydrate without the loss of carbon as CO_2.

Secondary Metabolism

During the active, exponential growth of a fungal culture nutrient uptake occurs at a maximum rate for that set of conditions and the culture is said to be in the balanced phase of growth. This cannot continue indefinitely and as nutrients become limiting the rate of proliferation slows down. The biomass of the culture may continue to rise however, as storage compounds (lipids, sugars) are laid down (storage phase). As soon as the growth rate declines from exponential the further depletion of nutrients prevents any increase in biomass (maintenance phase) and secondary metabolism begins. Subsequently, however, the biomass of the culture becomes reduced as the cells lyse and are digested (autolytic phase).

As specific nutrients become limiting (particularly nitrogen and phosphorus) intermediates of primary metabolism accumulate in cells, growth is restricted and activity is induced within secondary pathways (anabolic). Once the regulation of the primary metabolic pathways has broken down the secondary pathways come into operation, preventing any further buildup of primary intermediates, and as a result secondary metabolites are produced. These are released from hyphae, often very readily, although the mechanism and regulatory processes for this are not yet understood. Secondary compounds do not appear to have any direct role in the metabolism of the producer (Bell, 1981). They are not storage materials, nor are they likely to confer competitive advantage in the natural environment. Additionally, much of the metabolism is very wasteful of nutrients and energy. It seems likely, however, that the production of secondary compounds keeps the biochemical machinery of the fungus intact and functioning, to allow the resumption of primary metabolism immediately upon restoration of a positive nutrient status in the external medium. Secondary metabolism can also be considered as a form of biochemical differentiation in aged cultures which may accompany morphological differentiation (Moss, 1984).

Fungi produce a very wide range of secondary compounds which are biochemically diverse. These include antibiotics (e.g. penicillin and griseofulvin), plant hormones (e.g. gibberellin, cytokinin and ethylene), toxins, ergot alkaloids, lysergic acid, aflatoxins and pigments. Many such compounds are important to man, either because of the beneficial effects which can be derived from them, or owing to the detrimental effects they can produce. For example, antibiotics have a vital role in modern medicine, in combating bacterial infections. The production of mycotoxins, aflatoxins, carcinogens and mutagens in stored foodstuffs, can cause many dangerous, or often fatal, toxicoses in man and domestic livestock. Over 1000 secondary compounds from fungi have been characterised, although this is by no means an exhaustive list. In general, these are unusual compounds, both novel and specific to the producer species.

Intermediates of primary metabolism feed into the secondary pathways once metabolism has changed, indeed parts of these pathways and their branches are interrelated. However, relatively little is known concerning the regulation of secondary metabolism. Acetyl CoA is an important precursor for the synthesis of terpenoid and sterol compounds, via the mevalonic acid pathway and isopentenyl-pyrophosphate. Gibberellic acid (a diterpene), a higher plant hormone, is formed by this route, and is itself a precursor for the carotenoid pigments. Sterols (tetracyclic triterpenoids), e.g. lanosterol, ergosterol and cholesterol are also formed via acetyl CoA. The shikimic acid pathway gives rise to aromatic amino acids (phenylalanine, tyrosine and tryptophan) which are precursors for the formation of ergot alkaloids, lysergic acid, and the hallucinogens psilocin and psilocybin. The use of rye grain infected by *Claviceps purpurea* (ergot) has caused outbreaks of gangranous or convulsive ergotism in man over many centuries. However, the effects of alkaloids on the central nervous system and in the production of muscle spasms are now better understood and are used, in controlled doses, to great benefit in medicine. Also with medical applications, groups of compounds with antibiotic properties are produced by fungi from the Deuteromycotina. The penicillins (*Penicillium notatum, P. oxysporum*) and cephalosporins (*Cephalosporium acremonium*), are synthesised from non-aromatic amino acids. It is interesting that although the production of these compounds can be controlled in culture and forms an extremely important part of the pharmaceutical industry the details of the biochemical syntheses have only been elucidated recently.

Ageing

Ageing in fungi is a function of time but is also affected by environmental conditions. The position of a particular compartment in the colony and the nutrient status of the surrounding substrate determine the physiological age of that region.

The most recently formed parts, and therefore the youngest regions, of a hypha are at the apices. Morphological differentiation can be observed longitudinally along a hypha as ageing takes place and biochemical gradients must also occur. At the extreme apex there is a region containing only the vesicles which are associated with wall extension, no nuclei are included and very few mitochondria are found. In the immediate subapical region nuclei and mitochondria are present and active, with ribosomes, endoplasmic reticulum and some vesicles which are in transit to the tip. These regions of hyphae are very physiologically active, growing and extending into unexploited substrate. Materials which are formed in older regions are transported to the tip to support new growth. A colony may also form many branches and anastomoses such that the mycelium forms a network of interconnected hyphae through which nutrient materials, enzymes, precursor compounds and organelles may pass.

As colony proliferation proceeds, this part of a hypha becomes progressively further from the apex, owing to new growth at the tip, and may become cut off from extending, growing regions (Fig. 1.10). In older areas fewer vesicles are seen, but vacuoles form becoming more numerous distally and tending to increase in size in regions further away from the hyphal tip. The mitochondria which occur are smaller and more uniform in shape and nuclei may be sequestered into pockets of cytoplasm in very aged, vacuolated, regions. Large amounts of storage materials are located in these areas and storage granules are seen by electron microscopy.

The formation of septa is an irreversible process. Physiological ageing may be particularly dramatic if complete septa are laid down, or if perforate septa become blocked. Hyphal compartments are then separated from adjacent compartments. After septation has occurred no further nutrients can be lost to the tip and that compartment does not contribute further to growth unless branching occurs. Rejuvenation of a compartment can only take place by branch formation after septation. Additionally, once septa are closed a compartment may then be located in an area of substrate which has become nutrient depleted and may contain high levels of toxic or waste materials. Starvation conditions quickly lead to decreased physiological activity and to increased ageing.

Biochemical zonation must accompany the cytological changes which occur. It has been shown for *Neurospora crassa* that the rates of protein and nucleic acid synthesis do not change markedly throughout a hypha (Zalokar, 1959b). It is likely, however, that the types of protein synthesised will differ along the length of a hypha. Hyphal tips do not synthesise some enzymes, e.g. protease is not produced in *Aspergillus nidulans* hyphae until a length of 40μm has been attained. Cytochemical methods have confirmed that the apical region does differ biochemically from subapical areas. Hyphal tips were shown to contain more protein and nucleic acids as well as more protein-bound sulphydryl groups. This may represent transport of newly formed cytoplasm to hyphal tips, achieved by cytoplasmic streaming. Zalokar (1959a) also showed that enzyme activities in surface-grown hyphae and in deeply submerged hyphae differed, which indicates that the physiological processes carried out in these regions differs.

There must be a balance (Fencl, 1978) between the biochemical changes which are induced directly by ageing with time, and those changes induced by cultivation conditions and influences in the immediate environment. During the later stages of growth some regions of hyphae loose the ability for cellular maintenance and autolysis may begin.

Autolysis

Extremely aged hyphae may persist in a state of low metabolic activity for a considerable period of time, after which a fresh supply of nutrients might result in renewed growth. However, the progress of physiological ageing is greatly enhanced by nutrient exhaustion, especially depletion of those compounds which provide energy for synthetic processes, and also by the localised accumulation of toxic metabolites. In fact the physiological age of a hyphal compartment is determined by its position within a colony.

Eventually an imbalance in the normally highly regulated metabolic systems in an aged hypha, caused by internal or external factors, may result in autolysis or self-digestion. Initially this is manifest by the very extensive vacuolation of hyphae but subsequently organelles, particularly ribosomes and mitochondria become disrupted. To some extent materials degraded by such autolytic activity can be recycled and used for maintenance, so that starvation conditions can be tolerated, at least in the short term and some fungi are able to remain in stationary phase for a very considerable time. The autolytic breakdown of nitrogen-containing compounds within hyphae is particularly important for the maintenance and survival of fungi, particularly in habitats where nitrogen levels are restrictive to growth (e.g. wood decomposers). When the balance between wall synthesis and lysis is upset the chemical structure of the wall becomes weakened and gradually degraded. Eventually materials from these hyphae are released to the surrounding medium and are then available for uptake by other hyphal

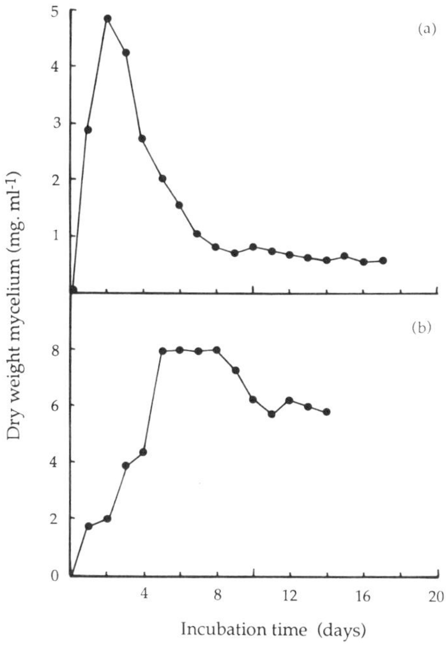

Fig. 1.15 Autolytic reduction in biomass in stirred liquid culture. (a) *Aspergillus nidulans*: over 85% reduction in dry weight occurred after 5 days autolysis. The residual culture fluid contained high levels of chitinase and β-glucanase activity (redrawn from Isaac and Gokhale, 1982). (b) *Fusarium tricinctum*: 25% autolysis occurred 10 days after the maximum dry weight for that culture was reached. High levels of β-glucanase were detected in residual culture fluid. No chitinase activity was detected (redrawn from Lynch *et al.*, 1985).

compartments in the vicinity. In a fungal colony growing on solid medium it is likely that hyphae in the central regions are undergoing autolysis as the new tips proliferate across new substrate (Trinci and Righelato, 1970). However, in such cultures autolysis does not occur synchronously but begins in isolated compartments at first.

In a batch liquid culture the onset of autolysis can be identified by the reduction in biomass which occurs. Autolysis is usually defined as the percentage decrease in dry weight from the maximum achieved in that culture. In some fungi very high levels of autolysis can be reached (Lahoz *et al.*, 1976; Isaac and Gokhale, 1982). For example, species such as *Penicillum oxalicum*, *Aspergillus nidulans* and *Neurospora crassa* may exhibit over 90% dry weight reduction from the maximum (autolysis) after a few days in culture (Fig. 1.15). Many other species, with representatives from all major taxonomic groups, also demonstrate autolysis although the time course of dry weight reduction is not usually so dramatic.

As the rate of dry weight loss reaches a maximum value enzymes which are usually associated with cell wall lysis can be detected in the culture medium. In fact autolytic enzymes, liberated into culture fluid in this way have been used to provide enzyme mixtures for the successful removal of the cell wall and production of protoplasts from fungi (Isaac, 1985). Other enzymes are also liberated from hyphae, e.g. proteases, invertase, esterase, acid and alkaline phosphatase. The patterns of release of these materials from hyphae during autolysis are characteristic but the extent to which this occurs as a consequence of hyphal lysis is not clear.

REPRODUCTION

Fungi reproduce and are dispersed by means of spores. In some fungi these are produced asexually, although in most species reproduction is a sexual process. However, many fungi are able to produce several sorts of spore, both sexual and asexual, sometimes in distinct phases of the lifecycle. The fusion of compatible nuclei, followed by meiosis, provides a mechanism for genetic variation among populations and is termed sexual reproduction. This gives rise to the production of perfect spores on the sexual form or teleomorph of the fungus. Where no nuclear fusion or meiosis occurs imperfect spores are formed, on the asexual form or anamorph of that species. These different states of a fungus may be given different names.

Some fungi which lack a true sexual cycle are very successful species and do show genetic variability. Genetic recombination does occur in these fungi, although in nature the process is very rare. In such cases fusion (anastomosis) between haploid hyphae which are genetically different, gives rise to the formation of a heterokaryon. A rare nuclear fusion event results in a diploid nucleus which divides and eventually forms diploid mycelium. Mitotic crossing over during division and subsequent haploidisation gives similar results to those of sexual recombination. This is termed parasexuality and is particularly important in members of the Deuteromycotina.

Amongst the fungi as a group, many different types of spores are formed and the structures are extremely varied. Some spores are simple and unicellular, others are multicellular and may be branched or highly ornamented. Some are motile by means of flagellae. Others are formed and held within equally diverse structures until they are released into the environment. Particular spore shapes, and their developmental pattern, are highly species-specific. As a result, the fungi are usually classified by the developmental pattern and morphological differences between the spores produced, or the development of those specialised structures which bear them.

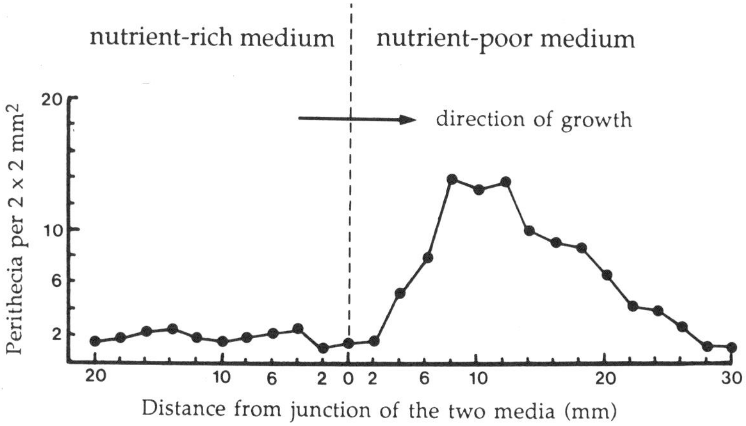

Fig. 1.16 The formation of perithecia in a *Podospora anserina* colony growing from a nutrient-rich medium (corn-meal agar) to a nutrient-poor medium (minimal agar) (redrawn from Lysek, 1976).

Spores also have an extremely important role in the life of fungi. They form the basis for dispersal, both in time and space. Some are very resistant resting structures by which the fungus may survive unfavourable environmental conditions, and may have very specific environmental and physiological requirements which must be met before germination and outgrowth can occur. Some spores are produced in enormous numbers and are small, specially designed structures, for efficient dispersal by wind, water, animals or insects. Fungi have very well-developed mechanisms for overcoming the inertia which holds them to the parent mycelium. Many species have very specialised explosive or expulsive mechanisms by which spores are liberated for dispersal. Other mechanisms, both mechanical and physiological ensure that the timing of spore release (time of day, time of year) is appropriate for future development. In general, fungal reproductive systems are extremely efficient.

Asexual reproduction

The formation of asexual spores most probably occurs as a response to nutrient depletion, especially to reductions in levels of available carbon, nitrogen or the carbon to nitrogen ratio, in the substrate. Nutrient depletion, or other conditions which bring about a cessation in the extension of hyphae (staling) over the substrate may result in sporulation. Following prior growth on a nutrient-rich medium (Fig. 1.16) just a few millimetres extension into nutrient-depleted conditions stimulated sporulation in *Podospora anserina* mycelium (Lysek, 1976). However, changes

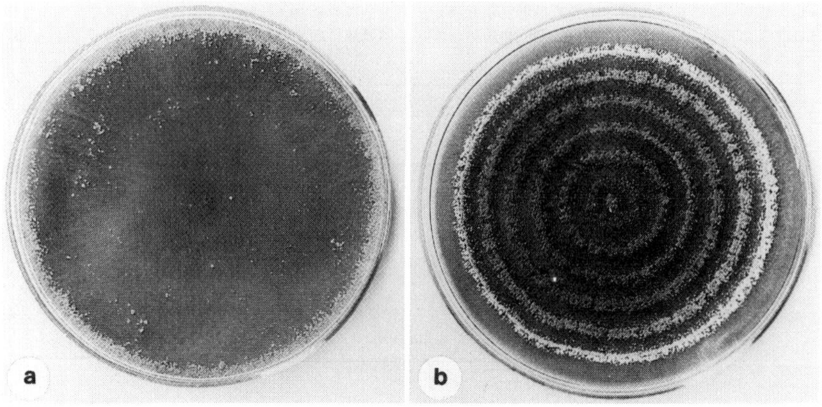

Fig. 1.17 *Trichoderma* cultures grown in a 10 h light/14 h dark regime. (a) Uniformly growing strain. (b) Rhythmically conidiating strain. Reproduced from Schrüfer and Lysek (1990), courtesy of Prof. Lysek and Cambridge University Press.

in other environmental conditions may also trigger spore production. In growing colonies of *Trichoderma* spp. exposure of hyphae to light prevents further longitudinal extension, and as a result those hyphae at the colony surface which are affected, sporulate. It has been shown that further colony growth continues in submerged hyphae and when these reach the surface extension ceases, stimulating the hyphae to sporulate. In a survey of 69 strains some isolates of *Trichoderma* (Schrüfer and Lysek, 1990) were shown to grow uniformly and sporulate at nutrient depletion but others were shown to sporulate rhythmically in response to alternating light/dark conditions (Fig. 1.17). Other rhythmic growth patterns have also been observed in fungi (Lysek, 1984). In other fungal species sporulation may be stimulated by the level or quality of the incident light received. Sporulation does not usually occur until after balanced growth has finished, but even then, does not follow automatically. In very general terms, the requirements for sporulation are more specific than those for vegetative growth.

Sporulation also represents a switch in energy metabolism with metabolic shifts occurring as a result of nutrient exhaustion in sufficiently aged mycelia. In vegetative hyphae the EM pathway of glycolysis predominates, converting glucose to pyruvate. In sporulating regions, however, the PP pathway is more active and presumably supplies more of the precursors which are required for spore formation. Enzymes of the TCA cycle have been shown to be present in high levels in sporulating cultures, which suggests that oxidative metabolism is necessary for spore formation. Storage compounds, such as lipids and polyols are usually laid down in spores, to support the early phase of germination and the resumption of growth.

In ultrastructural terms spores differ from vegetative cells in only a relatively few ways (Van Etten *et al.*, 1983). Most organelles are included although many are in a dormant, or semi-dormant state, particularly mitochondria, and amounts of endoplasmic reticulum may also be limited. Spores may be uninucleate or multinucleate but sporulation requires DNA synthesis and nuclear division together with RNA and protein synthesis. Spores do not usually contain vacuoles but the cytoplasm often contains large numbers of lipid bodies and glycogen granules. The water content of mature spores may be as low as 5% and is much lower than that of vegetative mycelium. It has been suggested that the internal concentrations of glycogen and trehalose may contribute to cytoplasmic dehydration. The spore wall usually differs, structurally and in the ratios of the chemicals contained, from the mycelial wall. It is thicker, often much thicker, may be pigmented and is more resistant to attack from enzymes or to desiccation.

Fungi produce two main types of asexual spores, conidia and sporangiospores, which have important roles in survival and dispersal. Chlamydospores, formed by aged mycelium, and sclerotia, which are resting structures, will also be considered in this section although these are not strictly asexual spores in the conventional sense.

Sporangiospores

(a) Zoospores

Zoospores develop within a sporangium often held terminally on a hypha and at maturity they are released from a pore in the sporangium wall (Fig. 1.18). These zoospores lack a true cell wall but have one or two flagellae, and are motile. They respire very rapidly, using up endogenous reserves, and actively swim through water films, attracted to particular materials by chemotaxis. Once a suitable substrate has been located the zoospores settle and encyst. The flagellae are withdrawn and a wall is formed around the spore very rapidly. A great deal of internal reorganisation then takes place. Organelles and chemical components which were compartmentalised, and maintained in a dormant state, are released. Massive biochemical changes take place and normal growth patterns are resumed. Germination and outgrowth follows very rapidly.

These spores are characteristic of members of the Mastigomycotina. Many of these are aquatic species, or inhabitants of soil, often living on plant and animal remains but also capable of causing very serious diseases of plants and fish, e.g. *Phytophthora infestans*, which causes late blight of potatoes, *Pythium* spp., which may cause damping-off disease of seedlings in poorly drained soils, *Saprolegnia*, *Blastocladiella* and *Achlya*.

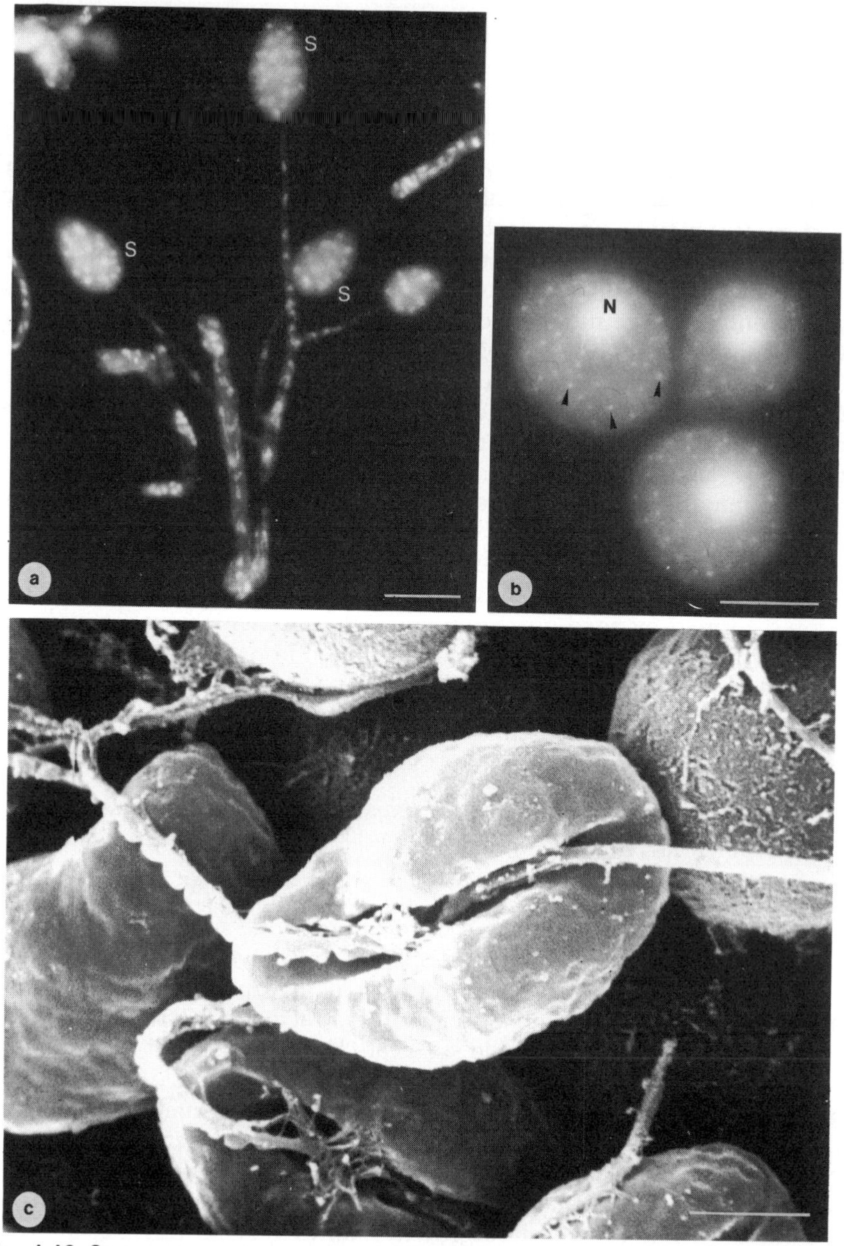

Fig. 1.18 Sporangia and zoospores of *Phytophthora*. (a) Sporangiophores and sporangia (S) of *Phytophthora infestans*. Fluorescence light micrograph of material stained with the fluorochrome DAPI to show DNA. (b) Isolated zoospores of *Phytophthora drechsleri*. Micrograph as above; large nuclei (N) and DNA in mitochondria (arrowed) visible. (c) Scanning electron micrograph to show zoospores of *Phytophthora drechsleri* immediately prior to encystment. Bars represent (a) 15μm; (b) 2.5μm; (c) 1 μm. Reproduced courtesy of Dr D.S. Shaw.

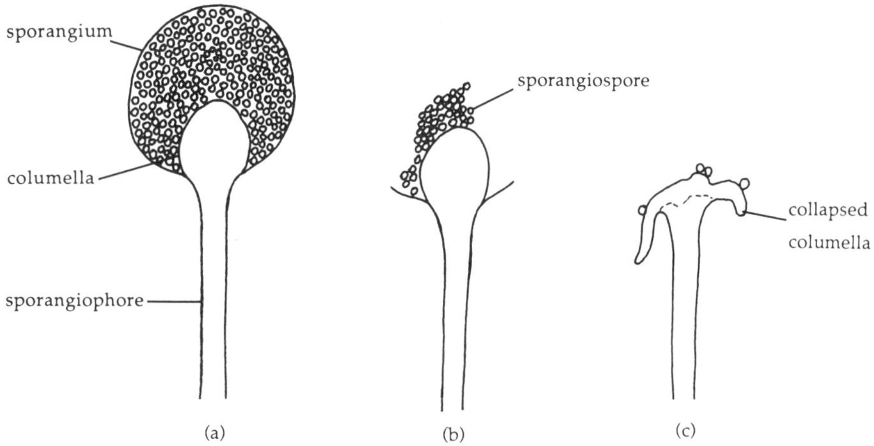

Fig. 1.19 Sporangiospore release in *Rhizopus*. (a) Intact sporangium containing sporangiospores. (b) Sporangial wall ruptured to release sporangiospores. (c) The collapse of the columella has liberated the remaining sporangiospores.

(b) Aplanospores

Sporangiospores which are not motile are produced within sporangia held terminally on specialised hyphae, called sporangiophores, which are held aerially, often high above the colony surface. Inside the sporangia the cytoplasm cleaves into spores and this region is cut off from the rest of the hypha by a septum, or columella. At maturity the sporangiospores are released by a variety of mechanisms which include the breakage of the sporangial wall (Fig. 1.19). In some cases spores are actively expelled by a rapid change in shape of the columella, whereas in others the spores are dispersed by wind or water droplets. These spores are characteristic of the Mucorales (Zygomycotina), e.g. *Rhizopus* and *Mucor*.

Conidia

Conidia are not enclosed within any structure as they are formed, and they are often produced in great profusion. There are an enormous number of different conidial types (Fig. 1.20) which are produced in a variety of different ways; they may be dry or slimy, multicellular, unicellular, highly pigmented and patterned. In some species conidia may be formed in a more enclosed situation, within fruiting structures such as acervuli or flask-shaped pycnidia (Fig. 1.20d). Fungi of the Ascomycotina, Deuteromycotina and some from the Basidiomycotina produce conidia. The development of the spores is very important in fungal classification, providing taxonomic criteria for the delimitation of species (Cole and Samson, 1979).

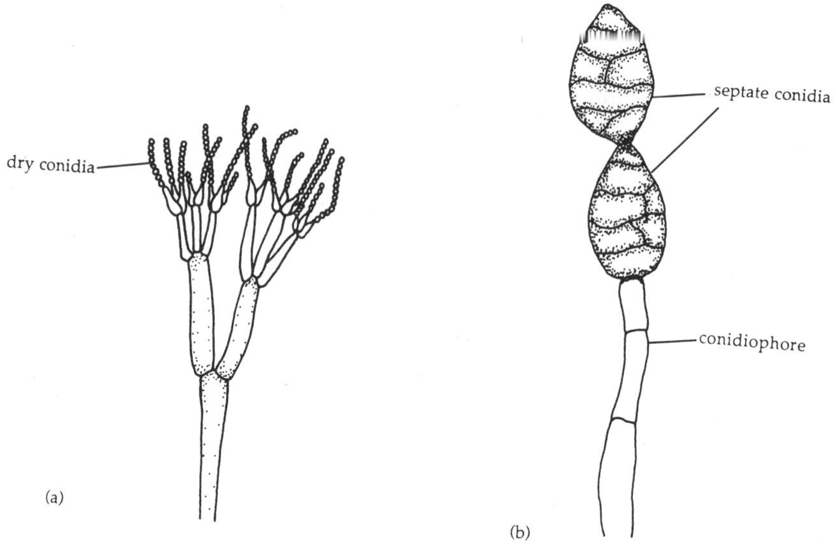

Fig. 1.20 Conidia. (a) *Penicillium*, chains of dry spores held aerially on conidiophores. (b) *Alternaria*, septate conidia. (c) *Fusarium culmorum*, septate conidia. Position of septum arrowed, bar represents 10 μm. Scanning electron micrograph reproduced courtesy of Dr S. F. Lowry and Prof. R. Marchant. (d) *Septoria apiicola*, long, thin, septate conidia contained within pycnidium formed in celery leaf tissue. Pycnidium (P); conidia (S); ostiole (O). Scanning electron micrograph reproduced courtesy of Dr A. Donovan. Bars represent 10 μm. [(c) and (d) opposite.]

(a) Blastic development

Blastoconidia are formed by a mechanism akin to hyphal tip growth and to budding which occurs in yeast cells (*Saccharomyces cerevisiae*). An area of the cell wall becomes plastic, probably as the result of lytic enzyme activity, and balloons out as the result of internal turgor pressure. This may occur from somatic cells or from specialised hyphae, called conidiophores and conidia may be produced singly or in chains. Fungi such as *Cladosporium* and *Aureobasidium* produce spores in this way.

Phialoconidia are formed by a similar developmental pattern, arising from a specialised cell known as a phialide. A number of conidia, each delimited by a septum, are produced from the tip of each phialide and may remain in chains until dispersal, with the most recently produced individuals nearest to the phialide. These spores are formed by many species, e.g. *Aspergillus, Penicillium, Trichoderma* and *Fusarium.*

(b) Thallic development

Thallic conidia are formed by fragmentation of hyphae, without any swelling, at the positions of septa. Once apical growth ceases, the wall

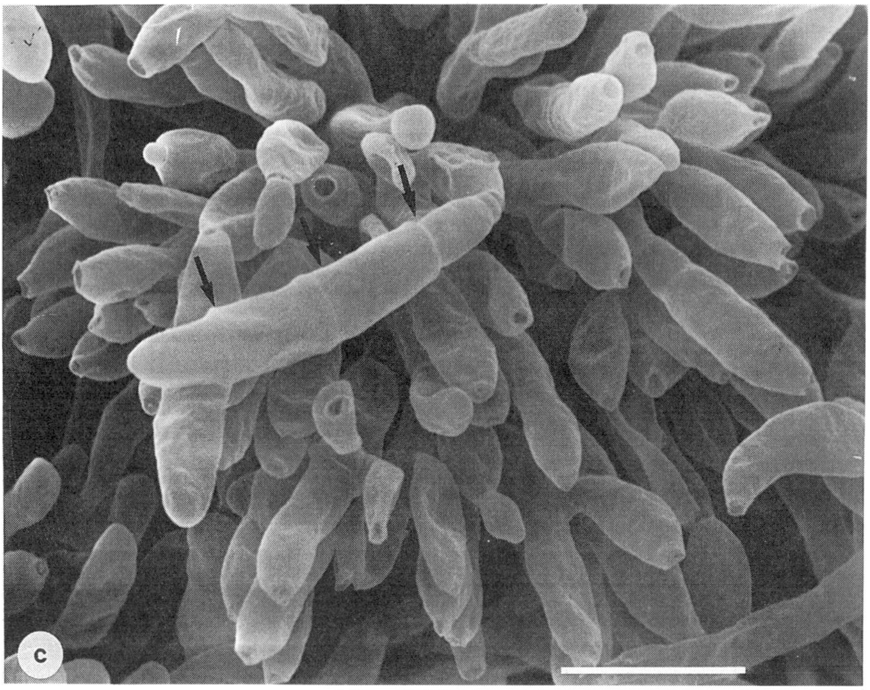

Fig. 1.20(c)

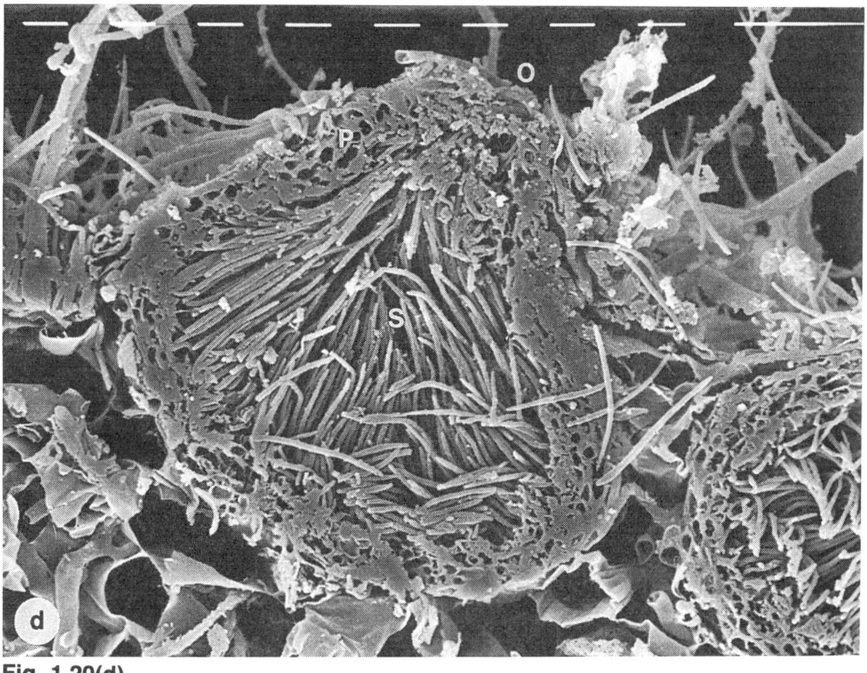

Fig. 1.20(d)

of the hypha thickens and septa are laid down. Chains of spores which remain together may be formed (hollothallic), or chains which eventually become disarticulated. If all wall layers are involved in spore production these are known as holoarthric spores but where the outer wall layer breaks to leave the inner layers around the spore these are enteroarthric spores.

Chlamydospores and sclerotia

As fungi become aged areas of hyphae gradually become vacuolated as materials are transported to the main, actively growing body of the mycelium. Pockets of cytoplasm, containing organelles, become sequestered behind septa, in otherwise empty hyphal walls. Eventually septa become closed and these regions are irreversibly cut off. In zones where sufficient cytoplasm remains, stores of glycogen and lipid are laid down. The walls in that area may become thickened and pigmented, giving rise to highly resistant chlamydospores. These may be released from the mycelium passively, by fragmentation, as autolysis occurs in other areas, and remain as persistent bodies in the environment for considerable periods of time. Chlamydospores are most usually formed by soil fungi, species for which the environment may often become nutrient deficient or hostile.

Sclerotia may also be formed. These are not strictly spores but are important to the fungus as survival structures. Indeed some sclerotia may persist for many years in soil. They are often produced by plant pathogenic fungi (*Claviceps, Sclerotinia* and *Sclerotium*), particularly species parasitic on plant roots and may be stimulated to germinate only in the presence of roots from the host species. Sclerotia are aggregations of hyphae (Fig. 1.21) which can be very variable in size (100μm–200mm). These aggregations have an organised structure with an outer layer, or rind, of thick-walled, pigmented hyphae, an outer cortex composed of cells which contain dense cytoplasm and an inner medulla containing very large hyphae which have a storage function. These cells contain many organelles and may be embedded within a matrix. Large quantities of fungal storage compounds and reserve materials are contained, such as trehalose, glycogen, mannitol and lipids. The structures are highly resistant to temperature extremes and to desiccation, probably by virtue of the muciage which is packed between the resting cells. This acts as a water store, aided by high levels of solutes which also help water retention. The matrix probably also absorbs water prior to germination and renewed outgrowth.

Sexual reproduction

Sexual reproduction provides a mechanism for the generation of genetic variability in a population. In general terms sexual reproduction occurs by

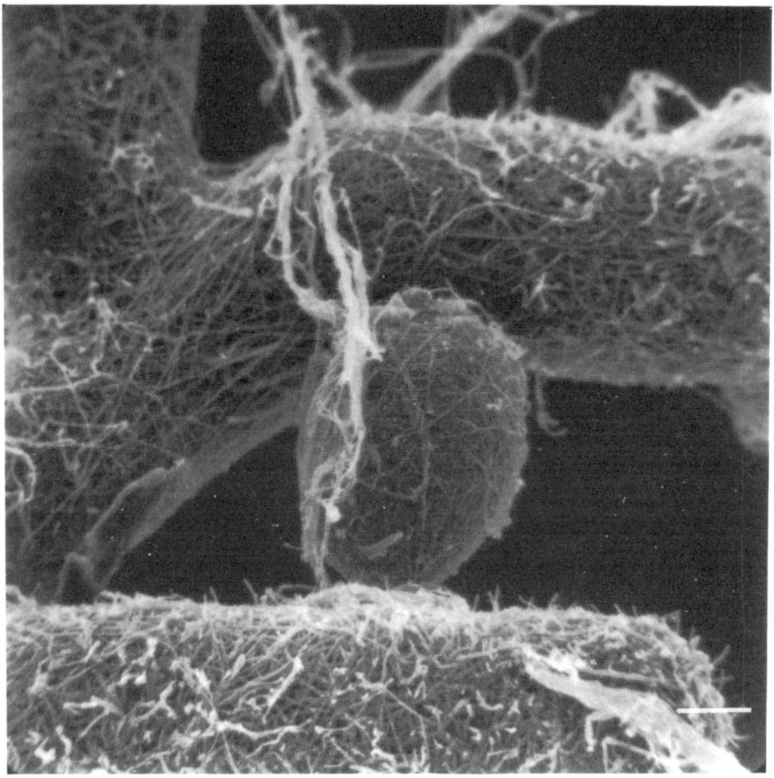

Fig. 1.21 Sclerotium between two branches of mycorrhizal roots of birch. Very few mycorrhizal fungi produce sclerotia. Bar represents 0.2 mm. Reproduced courtesy of Dr P. Mason and Dr A. Crossley.

the fusion between two sexual cells (plasmogamy), bringing together compatible nuclei. Sexual reproduction can occur within a single mycelium, without the need for interactions between different thalli (homothallism) or may only be possible if different thalli interact (heterothallism). Subsequent fusion between those compatible nuclei (karyogamy), may occur in specialised cells and is followed by meiosis, leading to the production of spores. In some cases different strains, or mating types, may be morphologically indistinguishable and are usually designated + and −. Where fusion between different thalli occurs this is a chance event. In other fungi distinct sexual structures are formed and male and female gametes are differentiated. These stages occur in many different ways in the fungi. In Basidiomycetes compatibility is under more complex genetic control which ensures outbreeding or inbreeding. Mating-type factors are controlled by one or two genes, with different alleles at the incompatibility loci. Mating may be prevented between strains which possess the same factors and so outbreeding occurs. Where common alleles occur at a locus the cross will be unsuccessful.

A great deal of information is available concerning the morphology of sexual spores and the developmental patterns which accompany their formation. However, the physiological mechanisms which trigger sexual reproduction are not well understood. It is clear that environmental influences are great and that requirements are fairly specific. It has also been shown that sexual reproduction is accompanied by a series of biochemical changes and represents a shift in metabolism. An increase in oxidative metabolism reflects the need for high levels of energy and reserve materials may be utilised. Increases in the synthesis of RNA and protein may also represent some internal recycling of nutrients. In some fungi it is clear that sexual reproduction is controlled, at least in part, by the production of hormones, by one mating type, which act as chemical attractants to gametes, from the other mating type, and also trigger sexual differentiation. The water mould *Allomyces* produces the hormone sirenin, which chemically attracts gametes and in *Achlya*, antheridiol and oogoniol cause sexual differentiation. A range of biochemical factors (e.g. glycoproteins), probably present at the cell surface, may determine the recognition of compatible mating types.

Sexual spores are very important as taxonomic characters and are used to distinguish the major fungal groups.

Oöspores

Oöspores are the sexual spores of Oömycete fungi (e.g. *Saprolegnia, Phytophthora* and *Pythium*), which arise from oöspheres formed within an oögonium. The cytoplasm within the young oögonium gives rise by cleavage to one or more oöspheres; nuclear division is meiotic. An antheridium, arising from an adjacent hypha or even from the same hypha, containing the haploid male nucleus, attaches to the oögonium. The male gametes enter the oöspheres via fertilisation tubes, and subsequently nuclear fusion occurs (Fig. 1.22). Food reserves, such as lipids, are laid down and stored in oil droplets and the oöspheres develop into oöspores, which are thick-walled, resistant, resting spores.

Zygospores

In the Zygomycetes (e.g. *Rhizopus, Phycomyces* and *Zygorhynchus*) sexually produced resting spores are known as zygospores. The gametangia grow towards each other and eventually fuse together to form the zygospore,

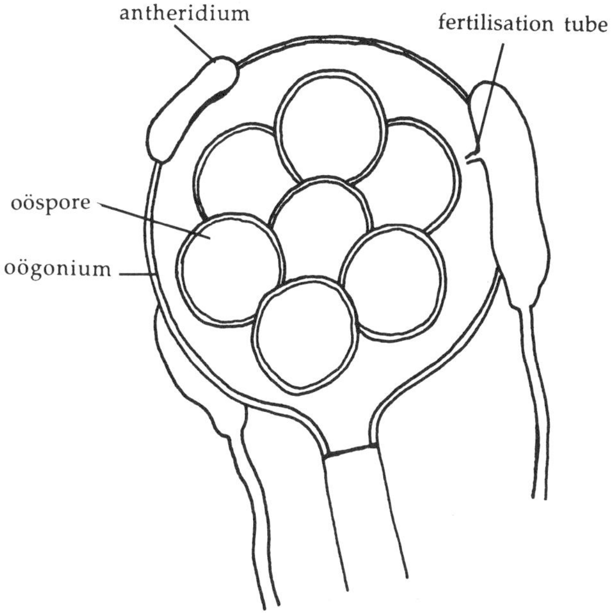

Fig. 1.22 Oöspore formation. In *Saprolegnia* antheridia attach to the oögonial wall. Fertilization tubes grow into the oögonium and haploid nuclei pass in to fertilise the eggs and give rise to oöspores.

inside which nuclear fusion and meiosis occur (Fig. 1.23). These spores have thick walls which are usually heavily pigmented and ornamented, and contain large amounts of reserve materials. Large zygospores are also characteristic of species that form vesicular–arbuscular (V–A) mycorrhizal associations with higher plant roots. Such spores can be isolated from soil and often occur in large numbers.

Ascospores

Fungi of the Ascomycotina form sexual spores called ascospores. Nuclear fusion occurs in the cell which will bear the spores and is followed by meiosis. The four resulting haploid nuclei divide by mitosis and give rise to eight haploid nuclei, each of which is eventually contained in an ascospore (Fig. 1.24). In most species these spores develop in an ascus, usually cylindrical in shape and having a thin wall (e.g. *Sordaria*). In many cases a buildup of turgor pressure within the mature ascus results in the forcible expulsion of these spores and to their dispersal. In some species of the Loculoascomycetes the ascus has a double wall (e.g. *Cochliobolus*,

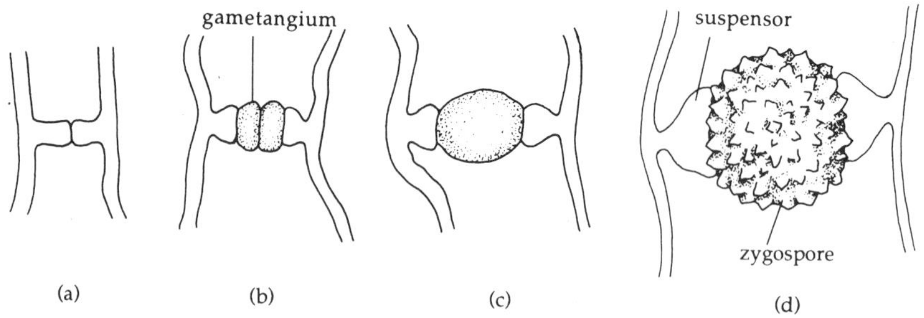

Fig. 1.23 Zygospore formation: (a) progametangia; (b) gametangia (containing haploid nuclei); (c) young zygospore (nuclear fusion); (d) mature zygospore (diploid).

Pyrenophora and *Ophiobolus*). At maturity the outer wall splits and the inner wall expands in size very rapidly and suddenly, releasing the spores to the environment. The ascospores themselves may be ornamented and pigmented, containing large reserves of sugars and lipids.

In most species the asci are enclosed by hyphae which may have a very definite and differentiated form, known as an ascocarp. In some species the asci are fully enclosed inside cleistothecia, some form flask-shaped perithecia which open at maturity, and some hold the asci in open cup-shaped arrangements known as apothecia. Throughout the Ascomycotina many different types of ascocarp are formed and the fruiting bodies are used in the classification of these fungi.

Basidiospores

In the Basidiomycete fungi sexually produced spores are termed basidiospores, which are formed on a basidium. Nuclear fusion occurs in the cell which will form the basidium, followed by meiotic division, and each of the resultant nuclei becomes incorporated into a basidiospore (Fig. 1.25). The basidiospores themselves may be pigmented and may have ornamented surfaces, but are less variable in form than ascospores.

Basidiospores are most usually carried on a structure called a basidiocarp, from which the spores are released by a number of mechanisms. Many, called ballistospores, are discharged by a flicking motion in the stalks, or sterigmata, by which the spores are attached to the basidium. During spore development a water droplet builds up at the point at which the spore is attached to the sterigma. Redistribution of this water droplet, to cover the whole spore surface, may provide sufficient energy for the spore to be liberated. However, many other mechanisms for spore discharge have also been developed by Basidiomycetes, including a range of explosive mechanisms. Many spores carry an electrostatic charge which may result in their being attracted to surfaces with an opposite charge, after they have been released. The basidiocarps which are formed by this group of fungi are often macroscopic and are the large fruiting structures, such as the agaric-type mushrooms and toadstools most usually associated with the fungi. An enormous range of basidiocarp structures are formed. The spores are often contained on a hymenial layer, held above the substrate, and it is from there that the spores are discharged and are carried away by air currents. The hymenium may be formed on gills (e.g. *Amanita* and *Agaricus*) or pores (e.g. *Boletus*) under the cap of the basidiocarp, may be completely enclosed (e.g. *Lycoperdon* spp., the puff-balls; *Tuber* spp., the truffles), or may have many other arrangements too numerous to mention here.

SPORE DORMANCY AND GERMINATION

Spores are the main mechanism by which fungi achieve dispersal and are also a means by which unfavourable environmental conditions may be overcome. The development of a spore represents a change from vegetative proliferation and the fungus enters a period of low physiological activity. Normal development is not resumed until polarity is re-established and outgrowth occurs at the initiation of a new developmental cycle.

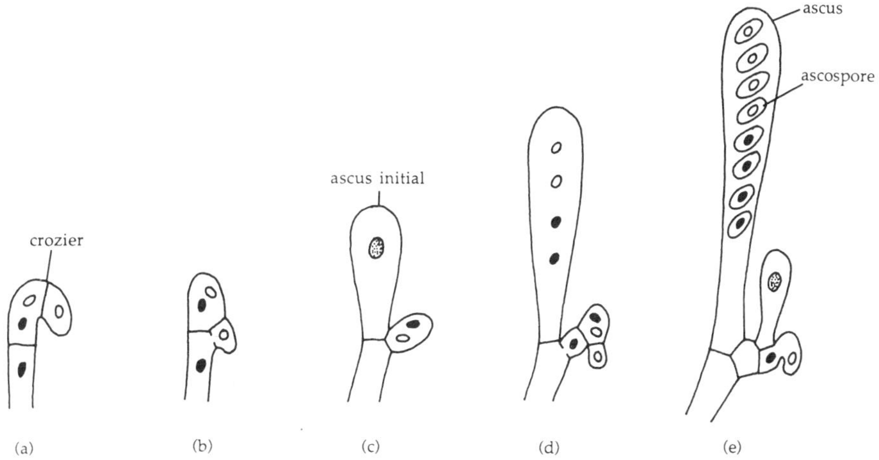

Fig. 1.24 Ascospore development: (a, b) apical cell of ascogenous hypha (dikaryotic) forms crozier (haploid nuclei); (c) nuclear fusion in ascus initial (diploid nucleus); (d) meiotic division in the ascus; (e) ascus containing eight ascospores (haploid).

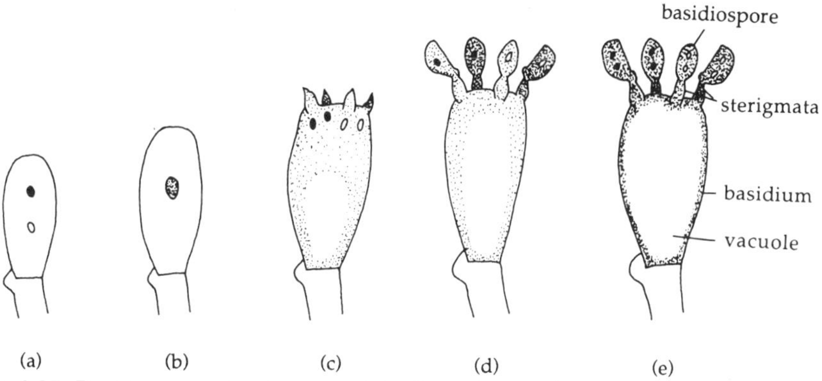

Fig. 1.25 Basidiospore development: (a) dikaryotic cell (haploid nuclei); (b) nuclear fusion (diploid nucleus); (c) meiosis; (d) mature basidium (uninucleate spores); (e) mitosis in basidiospores.

Dormancy

Dormancy is often considered to be a resting stage in the fungal lifecycle, and has been defined as a reversible interruption in the phenotypic development of the organism (Sussman, 1966). During dormancy metabolic activity of the spore is very low and although some respiratory activity may be detectable no synthetic processes occur at this stage. Some spores may germinate once favourable environmental conditions are restored (exogenous dormancy) but others do not germinate until dormancy has been broken by physiological ageing or other treatments (constitutive dormancy), ensuring dispersal and the presence of suitable conditions for vegetative growth.

Exogenous dormancy

For germination most fungi require a source of water. In general a higher level of moisture is needed for rehydration than is necessary for vegetative growth. Although storage compounds may support initial outgrowth of germ tubes nutrients are also needed, indeed some species have very specific requirements, and many spores are sensitive to pH and oxygen levels.

Compounds present in the external environment may also prevent germination. Microbes in the near vicinity may liberate inhibitory compounds which prevent nutrient uptake and the resumption of metabolism. Additionally, the presence of other active microbes may reduce the nutrient levels available and thereby delay germination. This is termed fungistasis and has been noted most within the soil environment.

Constitutive dormancy

Some spores may remain dormant for very considerable periods. This may be particularly important for some species, particularly pathogens. For example, spores of rust fungi (e.g. *Puccinia graminis*) require an overwintering period which probably enables spore survival. Outgrowth is timed to ensure the likely presence of host plants and promote infectivity.

If large numbers of spores occur together at a site, the percentage which achieve germination may be lower then if fewer spores were present. These spores contain self-inhibitors. Much research has been carried out into inhibitors from rust spores (uredospores) which act in very low concentrations (methyl-*cis*-3,4-dimethylcinnamate; methyl-*cis*-ferulate) to prevent germ tube emergence. Washing, or natural leaching of such spores releases the inhibition.

Constitutive dormancy may also be controlled and maintained by some form of metabolic block. Some spores may not be able to take up water or nutrients at this stage, although some materials can be taken up and a lack of permeability does not explain this dormancy (Van Etten *et al.*, 1983). Key enzymes may not be present in a functional form or may be compartmentalised in the spores. The separation of enzyme from substrate or from subunits may serve to maintain dormancy and has been demonstrated for a number of fungal spores.

Activation

Constitutive dormancy must be broken and the metabolism of the spores activated before germination is possible. Some environmental changes function as activation treatments. Some spores may germinate readily after periods of exposure to high (+40°C) or low (−5–0°C) temperatures. Such treatments may cause biochemical changes in key components of pathways or alter membrane permeability. Cycles of wetting and drying may also act as stimulants, especially in the presence of appropriate nutrient supplies. A combination of treatments is often most effective and probably reflects field conditions. However, some spores require a period of maturation prior to germination. Activation treatments have no effect until a period of physiological ageing has passed. It has been suggested that treatments which bring about spore activation may cause changes in the conformation of protein or lipid molecules in the plasma membrane. Such changes may therefore hold the key to the maintenance of dormancy in spores; however, it is also unlikely that any one universal mechanism operates.

Germination and outgrowth

Spore germination represents the resumption of metabolic activity and vegetative development. The first visible sign of germination is spore swelling. Initially water is taken up very rapidly, at first passively and then actively, until the spore becomes fully hydrated. Spores may swell to several times their dormant size. The outer wall layers may split and eventually, after polarity has been established, a germ tube grows out from the spore (Fig. 1.26). Once a germ tube has reached a length equal to the diameter of the swollen spore it is said to have germinated. Spores of some species may produce more than one germ tube.

New wall material must be synthesised by the activated spore before outgrowth occurs. The germ tube may arise by extension from the inner spore wall or alternatively, some spores lay down, and extend, a completely new wall layer. Vesicles aggregate at the site of germ tube emergence, once polarity has been established, providing wall precursor compounds and synthetic enzymes. It is likely that both enzymic degradation and physical force contribute to the rupture of the spore wall at germ tube emergence (Van Etten *et al.*, 1983).

Additionally, mitochondria increase in size and rates of respiration increase very dramatically. At this early stage spores use reserve materials (e.g. lipids and trehalose) to provide energy and structural molecules and the glyoxylate cycle (Fig. 1.14) is an important pathway in germinating spores. Key enzymes may be synthesised very rapidly, or compartmentalised enzymes released. RNA and protein synthesis also occur prior to germ tube emergence. All classes of RNA are synthesised during the initial stages of germination. DNA synthesis is important for continued outgrowth of the germ tube (*Ceratocystis ulmi*) but is not always a prerequisite for germination (*Neurospora crassa; Uromyces phaseolus*). As the physiological and biochemical processes become fully functional and integrated, germ tube emergence and subsequent outgrowth lead to the formation of vegetative colonies.

FUNGAL STRATEGIES

Much of the information which has been accrued concerning the growth and development of mycelium, physiological processes involved and the biochemical pathways which operate in fungi has been derived from the examination of growth in culture. Studies made under highly controlled, simplified and easily manipulated conditions have contributed a great deal to the understanding of fungal biology. It is clear that the fungi are highly versatile and adaptable, showing a great deal of plasticity in the developmental patterns which are expressed under different conditions. They are extremely successful organisms in the natural environment, competing and interacting with other organisms in a range of heterogeneous environments and often operating in circumstances which are restrictive and therefore suboptimal for growth and development.

Many fungi are well adapted to the niches they occupy; they are widely

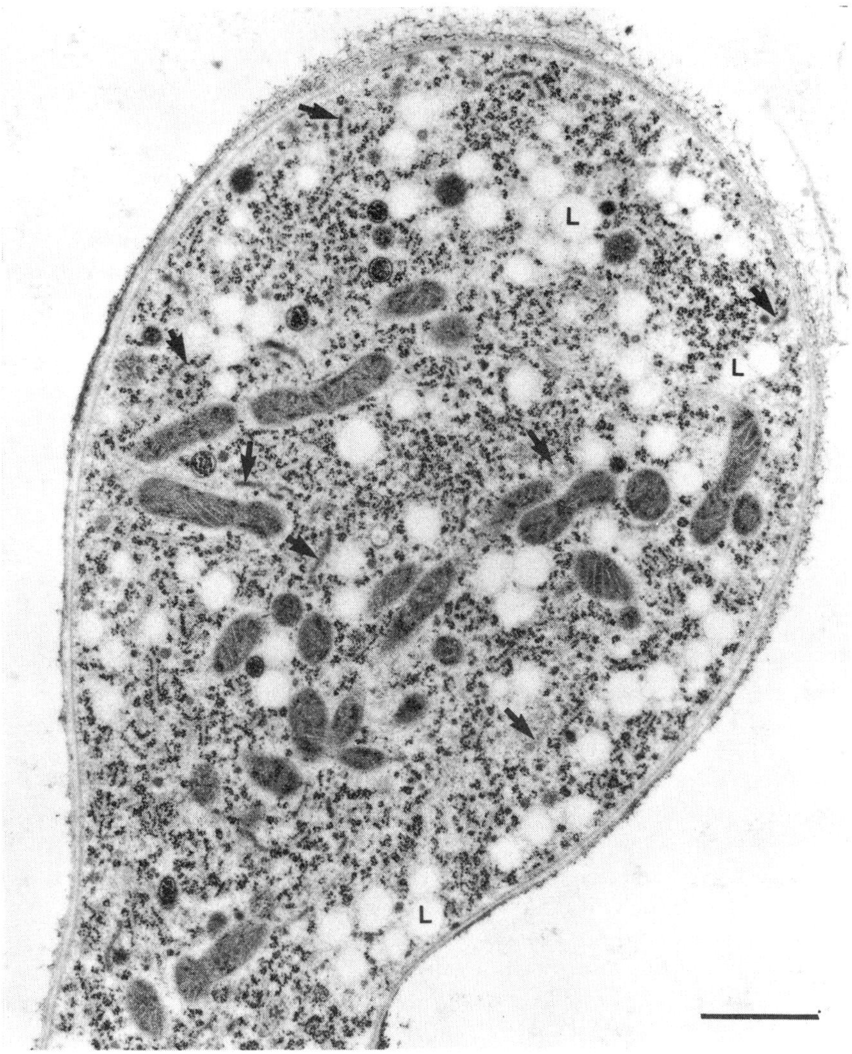

Fig. 1.26 Transmission electron micrograph section of a germinating *Magnaporthe grisea* conidium. The germ tube contains smooth membrane cisternae (arrows) and lipid bodies (L). Bar represents 1 μm. Reproduced from Howard *et al.* (1990), courtesy of Prof. R. J. Howard and Springer-Verlag.

distributed and occur in many ecologically diverse and complex habitats. Since supplies of reduced carbon compounds are essential for vegetative growth and development the position occupied by a particular individual is largely dependent on the availability of suitable substrates at a site. Survival will be influenced by the ease with which appropriate nutrient materials are encountered and can be assimilated. Additionally, other factors both biotic and abiotic, will have a moderating role on fungal development. For example, competition for resources may be imposed by other individuals. Some habitats may be subject to high levels of disturbance and rapid changes in conditions, which impose sudden, although often transient, ecological pressures. Additionally, environmental extremes occurring at a site may result in a reduction in the amount of biomass produced and are therefore regarded as stresses. Since such conditions often operate more or less continuously at a site, species may become adapted or tolerant to these extremes.

Fungi gain access to available nutrient substrates in different ways and are often grouped together accordingly. Three different nutritional modes, saprotrophy, necrotrophy and biotrophy, are recognised. It is important, however, to remember that these are not mutually exclusive. Some species may be able to adopt any of these modes under different conditions, and since many fungi are inherently very variable switches between modes may be made during one life cycle. The term saprotrophy describes the use of non-living, organic materials as resources for growth. Necrotrophy occurs when the living tissues of an organism are first killed by the fungus and are then utilised saprotrophically, whereas in biotrophic mode only the living cells of an organism are used to obtain resources.

As a result of selection pressures in the natural environment a range of strategies for survival have evolved. Strategy theory suggests a continuum between two types of organisms (Pianka, 1970), along which selective forces operate. Strategy theory has recently been applied to saprotrophic fungi (Pugh, 1980; Cooke and Rayner, 1984; Cooke and Whipps, 1987) and predicts that some individuals, R-selected, are likely to occur transiently in an ecosystem and have a high capacity for rapid reproduction. Others, K-selected individuals, are more likely to persist for a longer time, reproducing towards the end of their life cycle and sequestering only a relatively low proportion of their available resources for reproduction. Within this continuum the forces of natural selection have resulted in three primary strategies (Grime, 1977). C-selection for a combative (or competitive) strategy, results in the exploitation of resources under conditions of low stress and low disturbance. C-selected species are likely to be persistent in the environment, probably growing and reproducing relatively slowly, but are also likely to be antagonistic and able to defend nutrient resources. S-selection (stress-tolerant strategy) has resulted in adaptation to conditions of persistent environmental stress and R-selection (ruderal strategy) to conditions of high ecological disturbance where there is low stress and easy access to readily utilised nutrients. R-selected species usually have a short vegetative growth phase, reproducing rapidly. These strategies also form a continuum and where overlap occurs between them secondary strategies may develop which combine features giving rise to

C–R (combative–ruderal), S–R (stress–tolerant–ruderal) and S–C (stress-tolerant–combative) strategies.

A great many fungi adopt a saprotrophic mode of nutrition and since such organisms can be cultured readily much information has been gathered concerning their physiological and ecological adaptations. Saprotrophs are often able to utilise complex materials, such as the highly resistant and recalcitrant structural components of plant tissues, as sources of nutrients. However, it has been suggested that saprotrophy is a relatively primitive state and that necrotrophy evolved with the ability to overcome the resistance of a potential host to invasion. Further specialisation of the fungal invader resulted in the reduction of the degree of physiological damage inflicted on a host species during invasion, the successful diversion of host resources and hense the evolution of biotrophy. Biotrophs are therefore regarded as highly adapted individuals, making use of relatively simple materials (sugars and amino acids) directly from living host cells and often having a very limited host range. An alternative view, suggesting the evolution of nutritional versatility and therefore saprotrophy as the more advanced state, has also been proposed (Cooke and Whipps, 1980).

These modes have been further subdivided to categorise the fungi (Table 1.8) on the basis of their nutritional and ecological behaviour in the natural environment (Lewis, 1974). This scheme considers the dependence of a fungus on a particular mode of nutrition. Fungi may utilise non-living organic materials as resources, by saprotrophy, or may interact with other organisms obtaining nutrients through symbiosis. In this context symbiosis is taken to include parasitism, in which nutrients are taken directly from, and to the detriment of, a host organism (either by necrotrophy or biotrophy) and mutualism, associations from which both partners derive benefit.

Obligate saprotrophs have no capacity for necrotrophy or biotrophy and never breach host defences. A very great many fungi fit into this category and as a consequence this group shows a very diverse range of life-styles. Many of these organisms can be grown readily in culture on chemically defined medium. Others have nutrient requirements, which are not always fully understood, but can often be grown on undefined media. In some cases the ease of culturing has resulted in the accrual of much physiological information concerning species which fall within this category. Most species which inhabit leaf and root surfaces of plants and the surrounding soil are obligate saprotrophs, as are many species which contribute to the rotting and degradation of plant remains in the soil environment.

Facultative necrotrophs have at least some capacity for necrotrophy although they may live saprotrophically for much of their life cycle. Examples of such species are those which can take advantage of a

Table 1.8 The ecological and nutritional behaviour of fungal categories (after Lewis, 1973)

Obligate saprotrophs:	saprophytes with no capacity for biotrophy or necrotrophy
Facultative necrotrophs:	normally saprophytes with ability for necrotrophy or species equally successful as saprotrophs or necrotrophs
Obligate necrotrophs:	necrotrophs with very restricted saprotrophic capacity: specialised pathogens
Facultative biotrophs:	normally biotrophic with limited saprotrophic capacity; includes facultative mycorrhizal fungi and lichens, may have independent existence
Obligate biotrophs:	no saprotrophic capacity

weakened host. *Pythium* and *Rhizoctonia* live in the soil and may cause damping-off diseases and foot rots of small seedlings and immature, or damaged, plants, especially in wet situations. Obligate necrotrophs have only a limited capacity for saprotrophy, with low competitive ability away from a host, and are regarded as specialised pathogens. These may cause a great deal of damage to the host plant and give rise to important plant diseases which cause notable economic losses. Those fungi (e.g. *Fusarium* spp. and *Verticillium* spp.) causing vascular wilt diseases and take-all of cereals and grasses (*Gaeumannomyces graminis*) are considered in this category.

Facultative biotrophs are capable of independent existence in the field although much of their life cycle may be spent in association with another species. The host species may be damaged very little although nutrients are drawn from it. This category includes those fungi which form lichen associations but also have the capacity for free-living and additionally, mycorrhizal species which can exist independently. The obligate biotrophs do not have any capacity for saprotrophy and do not have any free-living phase. These organisms grow and reproduce on the living tissues of hosts. Most are extremely specialised and have very restricted host ranges. Some of the most important examples are the smuts (Ustilaginales), rusts (Uredinales) and powdery mildews (Erysiphales), as well as *Phytophthora infestans* (late blight disease) and *Claviceps purpurea* (ergot). Species which form ectomycorrhizae, vesicular–arbuscular mycorrhizae and some lichen-forming species, also come within this category.

Plants as an environment

2

INTRODUCTION

There are over 250 000 species of vascular plants growing world-wide, in an enormous range of habitats with greatly differing environments from arctic regions to tropical climates, deserts and waterlogged marshes. The distribution of any one species may be very wide and many species are adapted to harsh or unusual climates and situations. The variety of form exhibited is immense, from small annual plants to long-lived woody herbs and trees.

Plants absorb water and inorganic nutrients directly from the soil in which they grow. These are translocated from the roots through the xylem vessels into the shoots and leaves which are held above ground. In the presence of chlorophyll in the leaves, light energy together with carbon dioxide (CO_2) from the air and water from the soil, are transformed into chemical energy by the process of photosynthesis. The products are then transported to areas of the plant where they are utilised for growth and metabolism. It is this ability (autotrophy) to use simple molecules for the manufacture of complex organic nutrients, which distinguishes plants from other organisms. Any disturbance to a plant, either directly from the environment or arising from the activities of other organisms, may serve to disrupt the highly integrated physiological and biochemical processes which constitute normal metabolism and will therefore influence the growth and development of that plant.

As a direct result of the metabolic and synthetic processes that plants carry out, large amounts of organic material are produced and eventually enter the soil environment when the plant dies. Dead plant tissues provide good substrates for the support of microbial growth and metabolic activities. Living tissues also form a nutrient-rich environment for those microbes which are able to breach plant defences.

LIVING TISSUES AS AN ENVIRONMENT

The vegetative structures of plants are extremely variable in their form. In general, however, the roots of a plant act as absorbing organs, for water and inorganic salts, and also anchor the plant to the ground, stabilising any aerial structures which are formed. The main stem grows above the soil surface and will be more or less thickened depending on the amounts of branch and leaf material which must be supported. The leaves are the main sites of photosynthesis, acting as sinks for the water supply and as sources of nutrient molecules. Reproduction occurs either asexually, by vegetative propagation, or sexually by the formation of flowers and, after fertilisation, the production of seeds in which an embryo will be accompanied by a nutrient source to support the early stages of germination and future development.

World-wide, plants encounter many different environmental and ecological conditions and overcome the problems posed in a myriad of different ways. Cells become differentiated to perform specific functions within the tissues of the whole plant. However, it is the integration of the activities of cells and tissues which maintain the identity and integrity of plants. The presence of a vascular system, by which nutrients and water may be carried to any area of the plant, to supply and service all the component cells, must be the most important aspect of plant evolution.

PLANT CELLS

Ultrastructural characteristics

Plant cells are bounded by a wall, which is composed primarily of cellulose, with varying amounts of other complex carbohydrates (polysaccharides). The wall may be more or less thickened and may be composed of several layers. Initially primary walls are laid down and subsequently the secondary walls, which may be multilayered, are formed on the innermost surface. Channels through the cell wall, the plasmodesmata, maintain cytoplasmic connections with neighbouring cells and probably have an important role in cellular communications. The wall is lined by the plasma membrane, a semi-permeable barrier, which regulates the flow of molecules into and out of the cell. Damage to the plasma membrane results in the loss of selective permeability, loss of turgor and eventually cell death. The membrane encloses the cell cytoplasm, containing the organelles involved in metabolism and normal functioning.

Plant cell nuclei are large and spherical, often occupying a high proportion of the total cell volume. Most plant cells also have chloroplasts, which contain chlorophyll and are the sites of photosynthesis. Each is composed of a double membrane system enclosing the stroma. The innermost of these membranes is highly folded forming grana, lipoprotein lamellae arranged in stacks, which are the sites of light energy absorbtion. The reduction of CO_2 to produce carbohydrates, occurs in the stroma. Other important organelles are mitochondria which are often present in quite large numbers. These are bounded by a double membrane, containing the matrix. The inner membrane is folded and invaginated into christae and it is on this that oxidative enzymes are located. Mitochondria, chloroplasts and amyloplasts, which are involved in starch synthesis, are thought to develop from smaller organelles called proplastids, also contained in the plant cell cytoplasm.

Plant cells usually contain one or more vacuoles, which may occupy a very large proportion of the total cell volume. These are bounded by the tonoplast membrane and such vacuoles are sites for the storage of many molecules and are therefore rich pools of nutrients. Other membraneous structures in plant cells are the endoplasmic reticulum, a double membrane system which forms a network, ramifying throughout the cell. Membraneous connections between the endoplasmic reticulum and the outer membranes of mitochondria and nuclei have been observed under the electron microscope. It has been suggested that such connections maintain communications throughout the cell cytoplasm. Large numbers of ribosomes, which are sites of protein synthesis, also occur in plant cell cytoplasm. In some areas ribosomes appear to be located on the endoplasmic reticulum (rough endoplasmic reticulum). Stacks of membrane enclosing cisternae, called dictyosomes, are also formed, from where small vesicles arise, and to which a secretory role has been attributed.

Meristematic development

Plant cells originate by division from the meristems, constituted by groups of undifferentiated cells which retain the capacity to divide. These are regions of high cellular activity, and are usually located at the apical regions of stems and roots. The meristem of flowering plants is a dome-shaped mass of enlarging and dividing cells (Fig. 2.1). The meristematic cells are arranged in layers which can be traced into the differentiated tissues of more mature regions in the stem. This is particularly apparent in roots where these radiating rows of cells grow to form the central vascular tissues, the cortex and surface layers of the young root. Root meristems are enclosed and protected, from the abrasive effects of the soil, by the cells of the root cap. Many of the root cap cells slough off as the root grows but usually provide good protection during early development. At apical regions of shoots, meristems are protected from physical damage from the environment by the newly formed leaf primordia which essentially cover

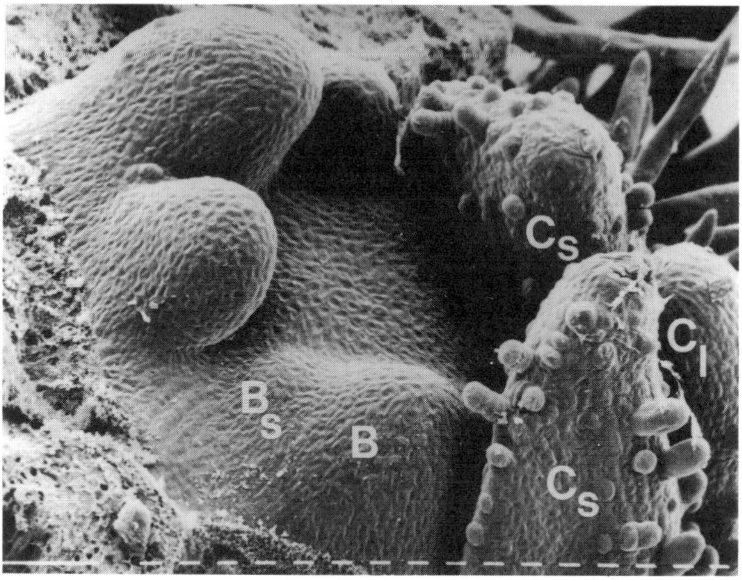

Fig. 2.1 Meristematic region of cocoa plant (scanning electron micrograph). Note the tripartite arrangement of the leaf lamina (B and C_l) and two adjacent stipule primordia (B_S and C_S). B and C_l are primordia in sequence. Bars represent 10 μm. Courtesy of Dr K. Hardwick and Dr R. Machado.

it in successive layers. Additionally, many apices are covered in surface hairs which also afford good protection.

Cellular differentiation

Cells subsequently become differentiated to undertake particular functions. Much of the main body of a plant is composed of fairly unspecialised parenchymatous cells, which are usually rectangular in shape, half as broad as they are long, with relatively thin walls. These form the ground tissues in the root cortex, leaf mesophyll and the pith of stems. Often parenchyma cells contain many chloroplasts and are very active in photosynthesis.

More highly specialised cells make up the vascular tissues of plants. Xylem vessels and the seive tubes of the phloem contribute the main transporting system, together forming the vascular bundles which ramify through the plant. The xylem vessels are the water-conducting cells and create channels through the plant. Xylem cells become elongated and stretched longitudinally as the plant grows, and since the cross walls are perforated or absent, very long tubes are formed. Water passes through the plant in the transpiration stream via these vessels (Boyer, 1985) and so, at times, the cells must withstand quite high pressure. The walls are thickened and lignified. Spirals, rings or a network of lignin is laid down to strengthen the xylem vessels, and as a result these cells also provide the plant with considerable mechanical strength. At maturity the xylem vessels do not have any living contents.

Sieve tubes create a corresponding route for the transport of nutrients through the plant. These are located in the phloem tissue and are rows of vertically aligned cells which maintain contact with their immediate neighbours through the end walls. These end walls become perforated (sieve plates) and cytoplasmic strands pass through between the cells. The living contents of these cells are retained although the nucleus and other organelles have usually degenerated by the time these cells are mature. Each sieve tube retains close contact with an adjacent cell (companion cell) which is usually smaller and has very dense cytoplasmic contents. It is thought that these companion cells are essential to the functioning of the sieve tubes.

Plants also contain many other types of specialised cells which may be encountered by invading microbes. Some will be mentioned in context later and further discussion of these is not appropriate here. A potential pathogen may proliferate through particular tissues of a host plant and bring about physiological influences. It is the arrangement of differentiated cells which makes up tissues and organs of whole plants.

TISSUES AND ORGANS

Roots

The kinds of root system and the types of roots developed by plants are extremely variable in form. The roots of a plant act as an anchor for the aerial shoots and leaves and therefore the overall size and structure of the root system must be adequate as a support. Roots must be sufficiently thickened and rigidified, or extensive ramifying networks must be developed to balance and stabilise the plant. Some root systems grow mainly in the surface regions of the soil whereas others grow deeply into the substrate, exploiting the lower layers. Some roots act as storage organs for nutrient materials or for water reserves in very dry environments.

The generalised arrangement of those tissues which make up the root structure are shown in Fig. 2.2. It can be seen that the vascular tissue is located towards the middle of the root, composing the central stele or vascular cylinder. The outer region, or cortex, is composed of parenchymatous cells surrounded on the outside by the epidermis. In general roots are underground organs, although some species do form aerial roots with quite specialised functions, and as a result root tissues lack chlorophyll. The main function is that of water absorption (Passioura, 1988), from the substrate. The surface area of the root system is large since many narrow, finely

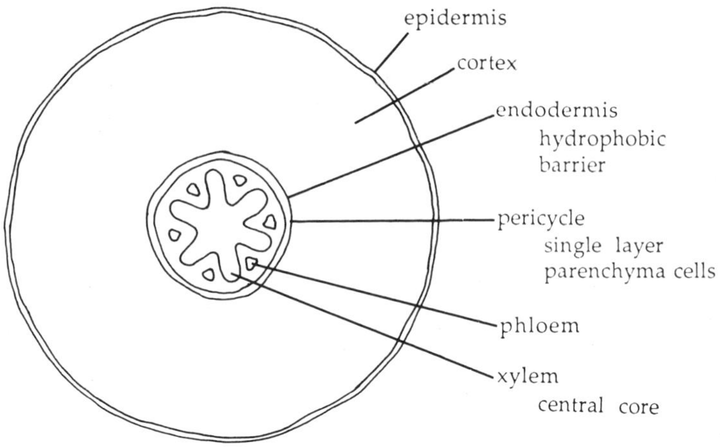

Fig. 2.2 Generalised diagrammatic representation of transverse section through a young root to show the arrangement of the vascular tissue. Note the large region of cortical tissue. The vascular tissue is arranged within the central stele.

divided, adventitious roots are usually formed. These lateral roots arise from deep within the root structure and grow out through the cortex and epidermis into the surrounding soil. The damage to the epidermal layers which results from lateral root development may provide a means of entry for soil-borne pathogens. The surface area is often increased further, just behind the meristematic region, by the formation of root hairs. These are projections from the epidermal cells which increase the lateral projection of the root and hence the volume of surrounding soil from which water may be absorbed. New root hairs are formed as the root extends further into the soil but are usually transient, not persisting throughout the life of the root. The youngest regions of roots are usually the most active in water uptake. Young roots are often surrounded by a layer of mucilage which is attractive to microbes and is a site of much microbial activity.

Stems

The stem of vascular plants is composed of a cylinder of tissue surrounded on the outside by the epidermis, covered by a cuticular layer. Inside this is the cortex, or ground tissue, composed of parenchymatous cells. Vascular tissues are arranged in many different ways, usually with bundles of phloem and xylem cells running through the stem. Angiosperms fall into two groups, the monocotyledonous plants and dicotyledonous plants (in which the embryo has one or two cotyledons, respectively). In monocotyledenous plants vascular bundles are distributed throughout the ground tissue (Fig. 2.3), with the xylem cells arranged towards the centre and the

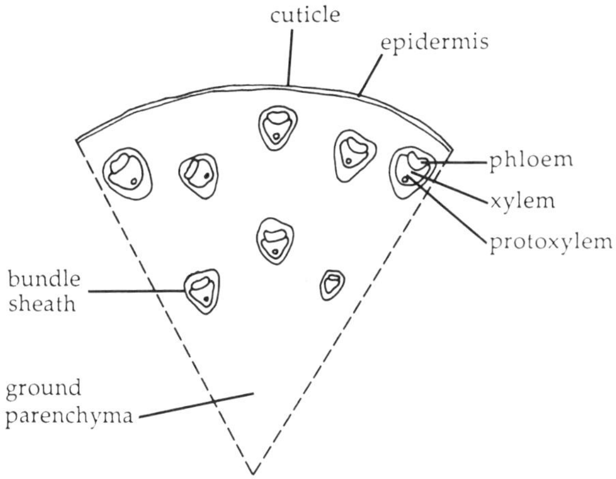

Fig. 2.3 Transverse section through monocotyledonous plant stem. Vascular bundles are scattered throughout the stem parenchyma.

phloem towards the outside (collateral bundles). In dicotyledonous plants the collateral bundles are arranged in a ring (Fig. 2.4), with the xylem vessels differentiating from the centre towards the outside (centrifugal) and the phloem differentiating towards the centre from the outside (centripetal). A layer of undifferentiated cells (cambium) occurs between the two types of tissues and it is from these cells that the secondary xylem and phloem are formed, creating a ring of vascular tissue with an internal core of pith.

Vascular tissues are clearly vital to the functioning of a plant. It can be seen that extreme environments result in adaptations within the vascular system of exposed plants. It has been suggested that the differentiation is controlled by the levels of plant growth regulator (auxin) and is mediated by the auxin gradient which occurs through the plant (Aloni, 1987). Highest concentrations of auxin are found in the growing shoots of plants and roots naturally act as a sink for auxin. If plant growth is curtailed by climatic conditions, or the growing season is restricted by seasonal variations, then small plants are formed with a high density of small vessels in the stems. Relatively short distances from leaves to roots results in a small auxin gradient. Less restricted growth gives rise to larger plants with a low density of wide vessels in the stems and consequently longer distances from leaves to roots creates a high auxin gradient.

Secondary growth in stems takes place by means of the cambium and results in an increase in the stem diameter. In perennial, woody plants activity in the cambium is seasonal, beginning in the spring each year. This gives rise to growth rings in the stem tissues, by which it is possible to deduce the age of the plant. The woody tissues of a stem are composed

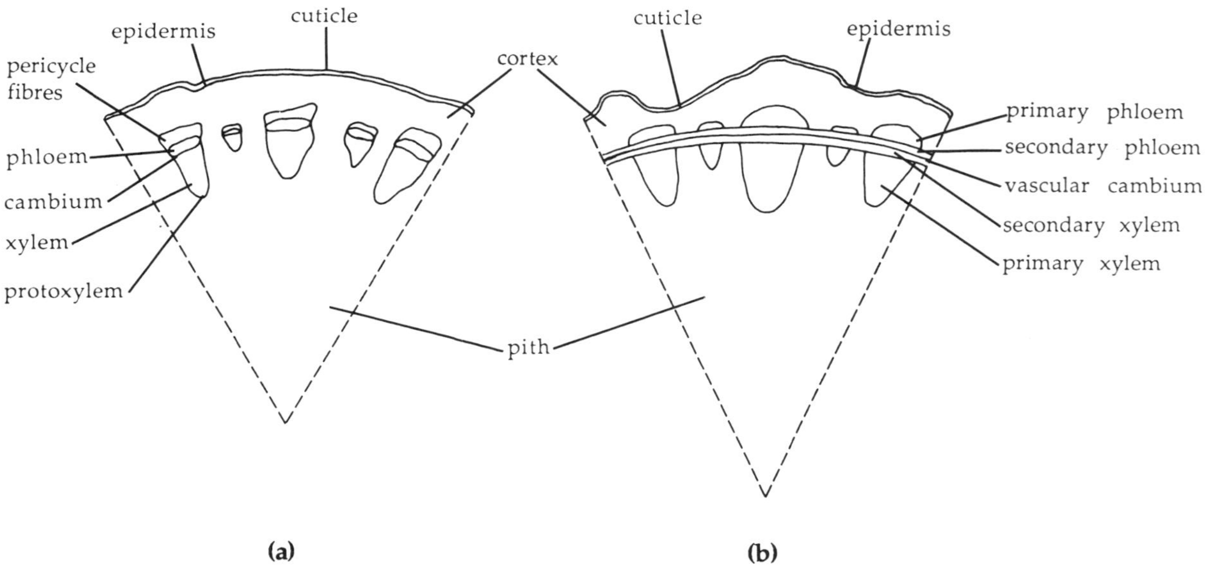

Fig. 2.4 Diagrammatic representation of a transverse section through a dicotyledonous plant stem to show the regular arrangement of vascular bundles around a central pith region. (a) Young stem with primary vascular tissue; (b) older stem showing secondary vascular tissue.

almost entirely of xylem. Secondary xylem has vascular rays which pass horizontally through the wood and tracheids which pass vertically. These components probably have a role in aeration and storage in the woody tissues. Stem tissues are closely packed and relatively dense.

Leaves

Leaves are organs of the plant, held laterally from the stem which are important in the biosynthesis of nutrient materials. A great diversity of leaf shapes and structures occurs amongst the vascular plants, from simple, undivided laminae to compound leaves with leaflets, arising in many different ways. Indeed leaf structure on any one plant may be variable depending on prevailing conditions, and in some species leaf shape changes as the plant matures. Sizes range from very small to several metres in length. The positioning of leaves on the stem (phyllotaxy) is also very varied although spiral arrangements are most common. It is probable that this results in the least shading of any leaf by those above it. This pattern is set up in the apical meristem during the early development of the leaf primordia. In many monocotyledonous plants the bases of leaves act as sites of storage for nutrient reserves, which overwinter and support early growth in spring.

Leaves are generally flattened, to provide a large surface area, and are often only a few cells thick. The epidermis, usually covered by a cuticle of waxy material of variable thickness, is a single layer of cells which do not contain chloroplasts. The mesophyll cells beneath (Fig. 2.5), which form the bulk of the leaf tissue, do have chloroplasts and are the main sites of photosynthesis. The upper mesophyll layer, palisade tissue, is composed of rectangular cells packed vertically in rows. The lower layers, spongy mesophyll, contain cells which are less regular in shape, sometimes branching forms, and which are less regularly, and more loosely packed. The form of the mesophyll tissues can be markedly affected by the amount of incident light received by the leaf. Much of the development of a leaf is attributable to cell expansion (Cosgrove, 1986; Dale, 1988) rather than to cell division, which is largely complete early after the growth of primordia.

Stomata occur on most aerial parts of plants, including stems. However, most are found in the epidermal layers of leaves, and more commonly on the lower epidermis. Some plants have stomata on both surfaces , some only on the lower (abaxial) surface. The stomatal pore is surrounded by two guard cells (Fig. 2.6), which are the only epidermal cells to contain chloroplasts. The immediately adjacent epidermal cells may be larger in size than other epidermal cells and these are known as subsidiary cells. In some cases, especially conifers, the stomata may be highly concealed and held quite deeply within the leaf tissue (Fig. 2.6). Stomata also afford a means of entry to plant tissues for pathogens able to grow through the

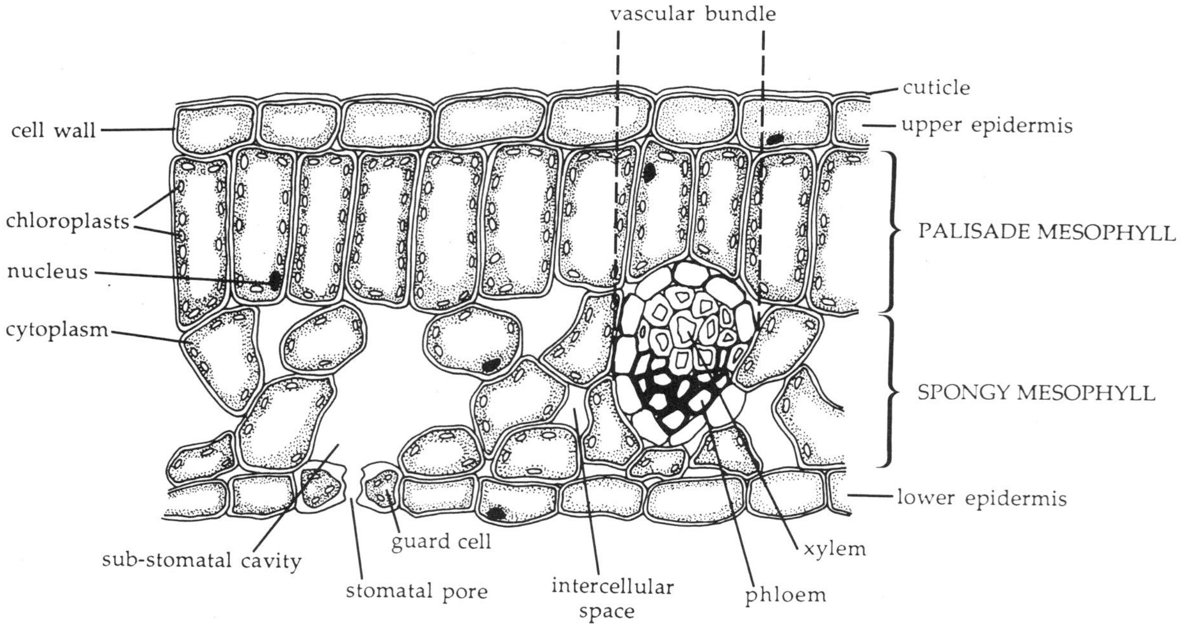
vascular bundle
cuticle
upper epidermis
cell wall
chloroplasts
nucleus
cytoplasm
PALISADE MESOPHYLL
SPONGY MESOPHYLL
lower epidermis
sub-stomatal cavity
stomatal pore
guard cell
intercellular space
phloem
xylem

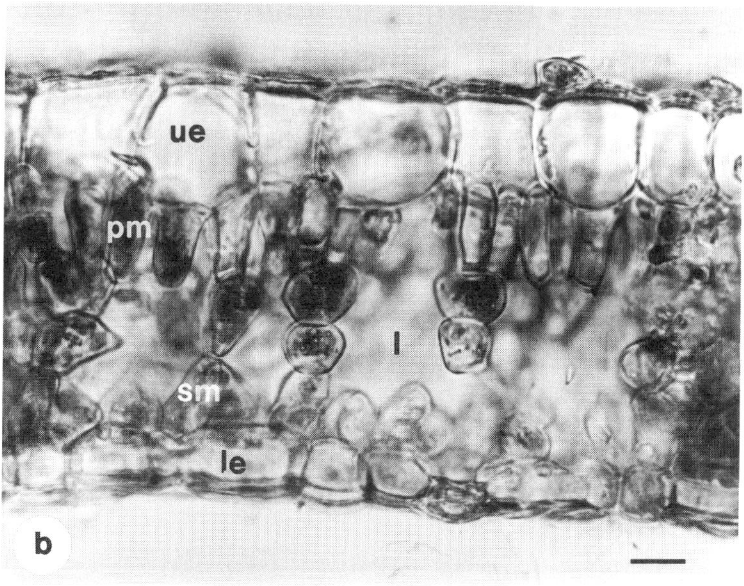

Fig. 2.5 Vertical sections through leaf tissue to show the arrangement of cells and intercellular spaces. (a) Generalised diagrammatic representation; (b) light micrograph of mature cocoa leaf. Large upper epidermal cells (ue); cells of lower epidermis (le); palisade mesophyll cells (pm); spongy mesophyll cells (sm); intercellular space (I). Bar represents 40 μm. Courtesy of Dr K. Hardwick and Dr R. Machado.

pores. Osmotic changes in the leaf influence the turgor of the guard cells; when the cells are turgid the stomatal pore is open, when flaccid, the pore is closed. This opening and closing allows the exchange of CO_2 and the diffusion of water vapour from the leaf tissues. Very large air spaces occur immediately around stomatal pores (substomatal cavity) and also within the structure of leaves, between the mesophyll cells (Fig. 2.5). Air spaces may occupy considerable volumes of the whole leaf. Plant cells are also leaky and many nutrient materials will be present in the intercellular spaces, providing rich substrates for invasive pathogens (Fig. 2.7).

The growth of a leaf is closely related to the formation of the vascular network. Vascular bundles are formed throughout leaf tissues (veins) so that no cells are far from the transport system which provides water and mineral nutrients (xylem) and redistributes newly synthesised carbohydrates (phloem). Vascular bundles may be surrounded by cells which contain large chloroplasts (bundle sheath cells), which are important in the conduction and storage of nutrients. New leaves are usually considered

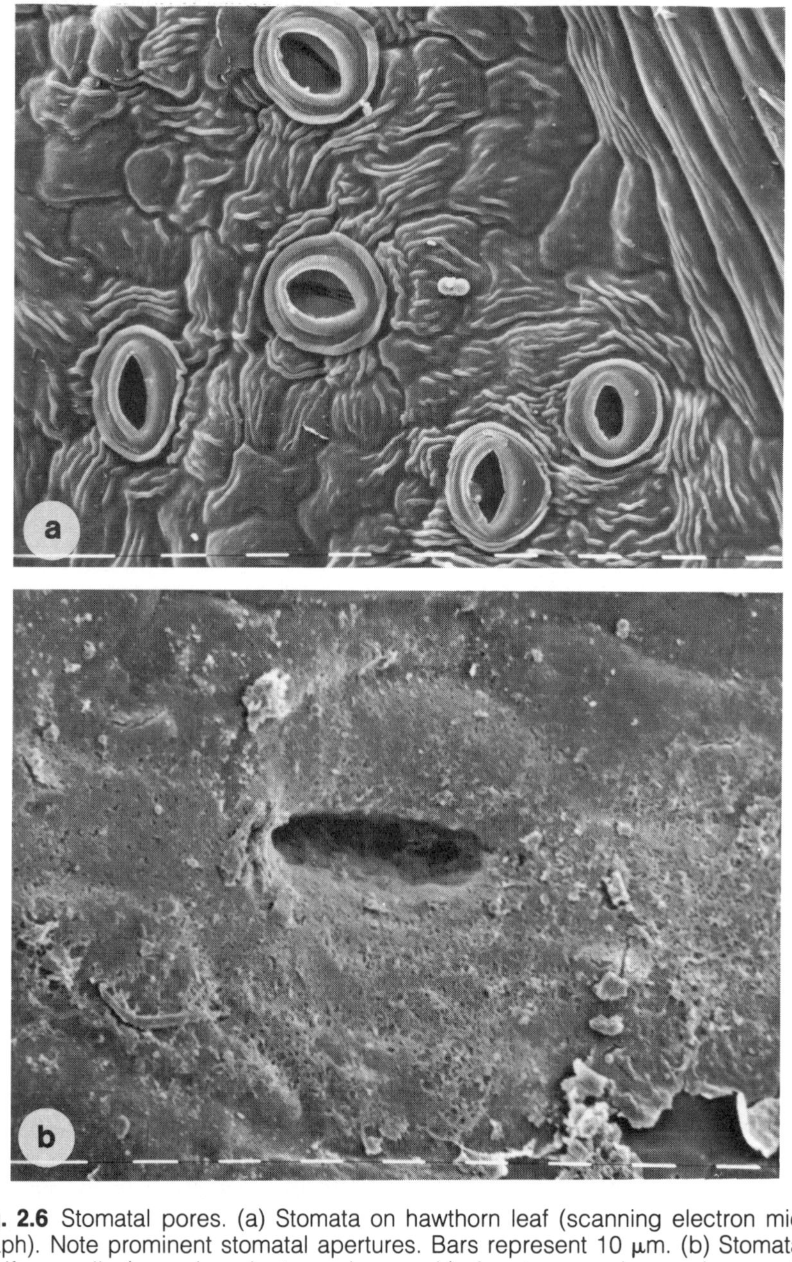

Fig. 2.6 Stomatal pores. (a) Stomata on hawthorn leaf (scanning electron micrograph). Note prominent stomatal apertures. Bars represent 10 μm. (b) Stomata of conifer needle (scanning electron micrograph). Apertures sunken and concealed. Bars represent 6.25 μm. (c) (p.87) Transmission electron micrograph section through celery stomatal pore. Guard cells (G), substomatal cavity (SC). x 5 600. (c) Courtesy of S. Edwards.

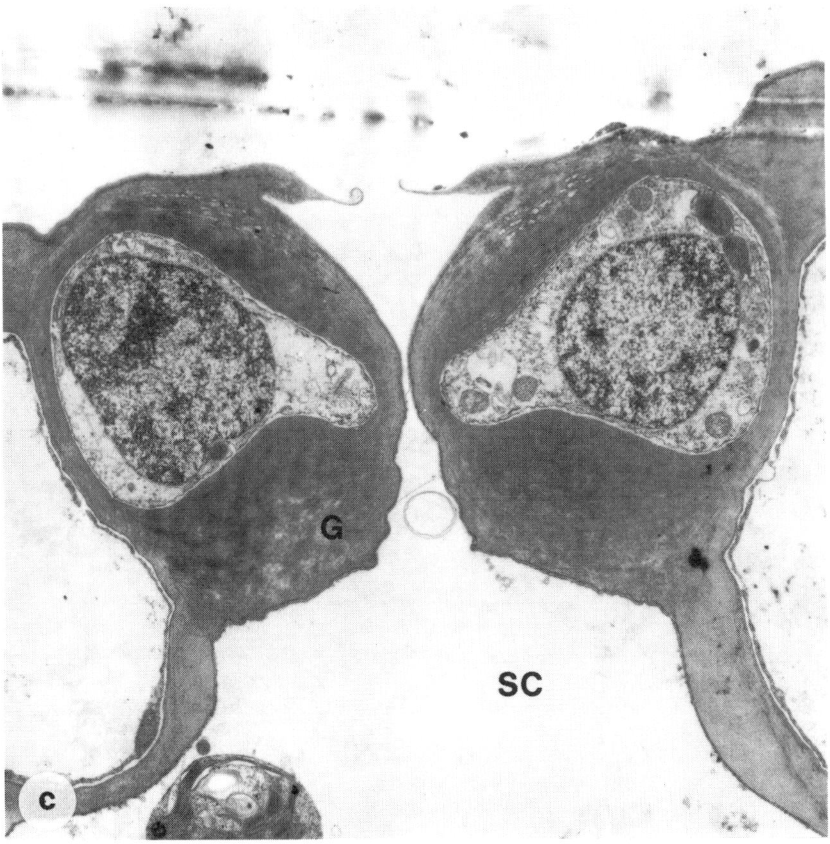

to be a nutrient sink but later, when the cells are fully photosynthetically active, these become an important nutrient source.

PLANT SURFACES

The outer surfaces of living plants provide important ecological niches for a large number of microbial species, and it is in that environment that many fungal–plant interactions occur.

The rhizosphere

The outer surface of the plant root is often designated the rhizoplane. Since much of the immediate surface is composed of dead and/or senescent plant cells it is not always easy to define the position of the interface with the living root material. The region of soil immediately surrounding the root, over which the plant has direct influence, is known as the rhizosphere.

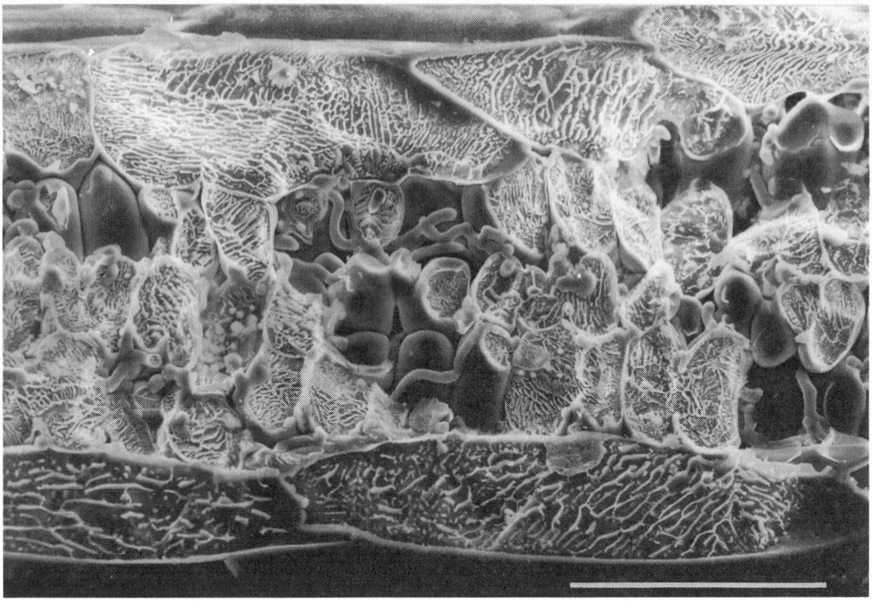

Fig. 2.7 Fungal growth in intercellular spaces of leaf tissue. Scanning electron micrograph of *Puccinia recondita* in winter wheat variety Kanzler. Transverse planed section of a frozen hydrated specimen using a freezing microtome. Bar represents 100 μm. Reproduced courtesy of Dr S. Humphreys and Dr P. G. Ayres.

This zone is also difficult to measure but is usually taken to include the region 1–2 mm from the living root tissue. The characteristics of these regions are dependent on the nature of the soil environment, and also the age, status and health of the plant concerned. In general, however, this is considered to be a relatively low stress environment, buffered from many rapid environmental fluctuations by the surrounding soil.

Exudates from active roots stimulate microbial activity in the soil adjacent to the root. Such exudates have been shown to consist mainly of carbohydrates with some amino acids. These form an important part of microbial nutrition. Numbers of bacteria are usually higher in the rhizosphere compared with the surrounding soil. This difference is most noticeable in nutrient-poor soils where any input from a plant represents a marked increase in nutrient status. In more nutrient-rich soils however, the rhizosphere effect is less noticeable. The activity of microbes in this region has been shown to stimulate roots to produce greater amounts of exudates and more dead material. Less of these are formed by roots maintained under aseptic conditions. Ultimately, after the death of the root, the accompanying microbial flora will contribute to the progress of root decomposition.

In general terms, the influence of the rhizosphere on the proliferation of bacteria is probably greater than on the fungi, especially in terms of numbers of individuals and sizes of populations. However, roots are the regions in which mycorrhizal associations are set up. These constitute extremely important interactions which influence most species of land plants (Chapter 5).

The phylloplane

In comparison the leaf surface, or phylloplane, environment is a high stress habitat. Influences from general climatic changes, local weather conditions, degree of exposure, extent and duration of wetting and drying, heating and cooling, are pronounced and important to the survival of microbes. Fluctuations in these parameters can be very rapid and can be extreme even in moderate environmental conditions.

Leaf surfaces provide inhabiting microbes with nutrient materials from the exudates which are produced, particularly near veins. The nutrient composition of such exudates is influenced by the age and status of the whole plant concerned and also of individual leaves. In very general terms, young active leaves liberate more exudates than older leaves. However, older leaves often have more weathered and damaged surfaces through which microbes can obtain nutrients and sometimes access to deeper-seated leaf tissues. The phylloplane flora acts as an inoculum for decomposition once the leaf has become detached from the plant, and many species inhabiting the phylloplane are also considered to be litter decomposers and soil inhabitants.

The surface structures of leaves have a great influence on the habitat conditions. Leaves may have wrinkled or folded surfaces with many undulations, they may be waxy and smooth or have surface hairs. Such characteristics must mediate microclimatic variations at the surface. Microbes proliferating beneath a weft of surface hairs will be protected from the washing effects of rainfall, whereas those growing on a smooth open surface will have little chance of remaining at that site after a shower. Fungal mycelium is relatively well adapted for growth on leaves; hyphal extension often tends to follow troughs in the topography. Additionally many leaf-inhabiting species produce a mucilagenous sheath which is sticky and by means of which the mycelium adheres to the surface. Fungal pathogens, however, are often not able to survive for long on the plant surface but penetrate leaf tissues rapidly after landing.

EFFECTS OF ENVIRONMENTAL CONDITIONS ON PLANT HEALTH

Plants require light for photosynthesis, a reliable supply of water and minerals from the soil. The absolute requirements for these components varies between species and this must influence the habitats which can be exploited by individuals. Additionally, plants are affected by environmental and general climatic factors, different species having different ecological amplitudes. Relatively few species are able to colonise extreme environments and a greater diversity of species occurs in more equitable habitats. The distribution of species will be influenced by the physiological effects of environmental factors. A species of plant may become dominant at a particular site if it is able to fully exploit that environment and therefore compete successfully with other species. Plants are able to adapt to the environment and may adopt different phenotypes (phenotypic plasticity) depending on the prevailing conditions (Bradshaw, 1965). Adjustments can be made in the growth form and orientation of leaves, morphology of tissues (e.g. development of aerenchyma in flooded plants), and in the biochemical pathways operated.

A combination of factors influences the distribution of species between and within habitats. Plants are constantly challenged by fluctuations in the prevailing conditions and by disruptive influences from other organisms. All such factors will tend to depress the rate of growth and development from the maximum theoretically achievable by that plant. It is interesting to contemplate the extent to which cultivated plants, in apparently well-fertilised, watered situations with limited competition from other plants and relatively controlled predation from pests, are in optimum conditions. It is unlikely that many plants ever reach maximum growth for more than very limited periods of time and some may always suffer deleterious effects of their growth conditions.

MINERAL NUTRITION

Although all plants need essentially the same minerals there are in fact wide quantitative differences in the mineral requirements between species, indeed the lack of some minerals may result in limitations to the field distribution of plants. Mineral deficiencies in plants may be the direct result of the lack of those ions in the substrate or may arise from the plants inability to absorb those ions.

Plants accumulate ions from the soil, against concentration gradients. Ionic concentrations are always higher in the root than in the surrounding soil and as a result these gradients may sometimes be quite steep. Plant root surfaces have a negative electrical charge and therefore cations are able to enter cells down the electrochemical gradient but against the concentration gradient. Anions, however, are transported into roots against electrochemical and concentration gradients. Active uptake is an energy-requiring process. Sodium and potassium pumps probably operate and other ions may also move passively through the membrane maintaining the electrochemical balance.

Mineral availability

The concentrations and availability of minerals in soil are very variable and may be patchy even within a soil, depending on the structure of that soil, associated vegetation, climatic and seasonal variations, the degree to which ions are adsorbed on to soil particles and the extent to which microbial activity mineralises organic complexes.

The pH of a soil largely determines the availability of ions. Most of the major elements (e.g. nitrogen, phosphorus and potassium) are less available in acid soil than in alkaline soil, although the converse is true in other cases (e.g. iron and magnesium). Cations readily bind to organic particles, which are negatively charged. Those ions which are present in the soil solution are readily absorbed by plants. The adsorbtion of ions on to colloids restricts their movement within the soil rendering them largely immobile, although some interchange with the soil solution will occur. Ions are absorbed from the soil immediately surrounding the root which may rapidly result in a zone of depletion. Roots can overcome this to some extent by continued growth into unexploited soil or by the formation of mycorrhiza. Ions will move through the soil by mass flow with water and some movement will occur by diffusion down concentration gradients, but these processes are often very slow and unreliable. For example, diffusion must be extremely limited in a dry soil.

Essential elements

The basic mineral requirements of plants are met from the soil environment. Some are needed in relatively large amounts and are regarded as major nutrients, whereas others are required in lesser or even trace quantities and although important to plant growth, are considered to be minor nutrients. It has been shown that a lack of some minerals reduces plant growth and the yield of biomass produced. Plant vigour and health may be reduced and disruptions to cellular function may cause symptoms of deficiency to develop (Table 2.1). This may increase the susceptibility to attack by some pathogens. Additionally an excess of minerals may also alter plant growth with a similar overall effect.

Major nutrients

Nitrogen is a component of proteins, enzymes, purines and pyrimidines. It is also a vital constituent of chlorophyll and coenzymes. It is an extremely

Table 2.1 Nutrient deficiencies in plants

Major nutrients required in relatively large amounts

Nutrient	**Deficiency symptoms**
Nitrogen	Stunting, chlorosis. Leaves very pale green. Very poor growth, stunting. Plants slow to fruit.
Phosphorus	Leaves dark bluish-green; older leaves red to purple.
Potassium	Shoots thin. Leaves chlorotic, dying back at tips; leaf edges distort. Some necrosis.
Magnesium	Older leaves chlorotic with necrotic spots (appear mottled). Edges of leaves curl up. Leaf drop may occur.
Sulphur	Chlorosis, young leaves pale green to yellow (similar to nitrogen deficiency).
Calcium	Stunted growth, especially roots. Young leaves distorted with brown scorch marks.

Minor nutrients required in trace quantities

Nutrient	**Deficiency symptoms**
Boron	Poor, stunted growth. Root and shoot necrosis.
Copper	Leaves chlorotic and wilted (especially at tips).
Iron	Chlorotic leaves with some green veins, appear mottled. Leaves dry out and may fall.
Zinc	Stunted growth. Shoots die back. Leaf drop common.
Manganese	Chlorosis, except in veins (mottled). Necrotic spots.
Molybdenum	Very poor growth.

important element for plants and is required in relatively high amounts. However, most of the nitrogen content of soils is in organic form and therefore comparatively unavailable to plants. In general, they absorb nitrogen from the soil solution, in fixed form, as nitrate. However, some species in acid environments may utilise ammonium ions. Nitrates are very soluble and are easily leached out, not remaining in the soil for very long and are therefore often added routinely in agricultural situations. The form of nitrogen which is available to plants has been shown to be important in terms of disease resistance. The addition of ammonium fertiliser may render the plants more susceptible to some diseases (root rots, damping-off disease) while the addition of nitrates may favour the development of others (take-all of wheat). In the natural environment the fixation of nitrogen by the activity of soil microbes is vital to the continued supply of nitrates which are used by plants.

Phosphorus is required as a component of nucleic acids, phospholipids and ATP in plants and quite large quantities are needed. It occurs in soil in both organic and inorganic forms. In the organic form phosphorous is unavailable to plants since it is bound into nucleic acids and phospholipids and must be decomposed by microbial activity before it can be taken up by roots. However, plants are able to utilise inorganic phosphate ions, $H_2PO_4^-$ (mainly available in acid conditions) and HPO_4^{2-} (available in neutral to alkaline conditions). These occur in soil as exchangeable ions adsorbed on to colloids which exchange slowly, with the soil solution. The ions are readily taken up by plants but are adsorbed very strongly on soil particles and depletion zones occur rapidly around the roots of growing plants.

Additionally, potassium is required by plants for osmoregulation and as an enzyme cofactor. It is often present in soil in relatively large amounts but most is immobilised in a non-exchangeable form. Some is present in exchangeable form and some in solution, however, and most is available to plants where soil pH is around 7–8.

Minor nutrients

Lesser amounts of other elements are needed for healthy growth. Calcium affects the permeability of plant cell membranes by its influence on the structure of membranes and lipid components. It is also present in cell walls and middle lamellae of plant cells, as calcium pectate. Calcium is not mobile in phloem and is not translocated. At the onset of leaf senescence plants tend to remove nitrogen, potassium and phosphate from their leaves prior to leaf fall but calcium is not removed and is often present in quite high levels in leaf litter. It is interesting that on invasion by a pathogen Ca^{2+} ions accumulate around the point of penetration and are also found in papillae which form on plant cell walls as a resistance response in invaded plants (Kunoch, 1990). Calcium is a major exchange cation in soil although most of the soil content is in non-exchangeable form because much is adsorbed on to the negatively charged surfaces of soil particles which attract the Ca^{2+} ions. In acid conditions more hydrogen ions are bound to soil particles and more Ca^{2+} ions are available to plants.

Sulphur is a constituent of amino acids (methionine and cysteine), proteins and vitamins (biotin and thiamine). In soil most sulphur is present in the organic fraction and it is sulphate ions which are taken up by plants. More sulphur is adsorbed on to particles in soil with low pH, but these ions are replaced by anion exchange and most becomes available near neutrality. Magnesium is an essential component of the chlorophyll molecule and as an enzyme activator. It is also an exchange cation in soil. More is available at high pH but magnesium is less abundant than Ca^{2+}. Iron is involved in the synthesis of chlorophyll and as a component of the cytochromes. In soils most is available in ferrous form. Soil pH is important to iron availability, although this element is not usually limiting to growth. More is present in solution in acid soils whereas it is more insoluble in neutral or alkaline situations.

Other elements are required in trace amounts, often as components of enzyme systems. Manganese has an essential role as enzyme activator in cell respiration. Bivalent ions (Mn^{2+}) are usually available in soil solution or alternatively in exchangeable form on particles. Copper is a component of enzymes (e.g. phenolases and laccase) and often only limited amounts are available in soil solution because copper is strongly adsorbed on to soil particles, forming non-exchangeable complexes with organic matter. Zinc is used in the biosynthesis of auxin and is least available in alkaline soils. Boron is involved in carbohydrate transport and cellular differentiation. Although some boron occurs in solution it is often low in concentration. Most boron is available in acid soils. Molybdenum is essential for the processes of nitrogen metabolism and is most available in alkaline soils.

Mineral toxicity

Large amounts of some minerals in a soil may lead to the inhibition of plant growth and the element may then be regarded as toxic to the plant. In very general terms, an excess of those elements which are required in large amounts is likely to be less damaging to a plant than excess of those elements required in relatively small quantities, which may quickly become toxic (e.g. boron, manganese and zinc). Not all plants are equally sensitive to elevated levels of all minerals. The response may be quite variable and some species may be quite tolerant in particular conditions.

Much of the effect of ions can be attributed to the acidity of the soil. In acid soil salts are generally more soluble and often more available to plants and may therefore reach toxic levels. The effect is normally due to direct cellular injury, or to interference with the uptake of other elements. As a consequence an excess of one element may lead to deficiencies in others. Some plants have evolved strategies to overcome high levels of some elements, such as sodium chloride. Some have mechanisms for avoiding excess salt, by secreting it or by sequestering it in to specialised cells. Others, however, appear to have a requirement for elevated levels and do not grow well in the absence of salt. Salt tolerance is polygenic and not yet well understood (Ramagopal, 1987).

WATER AVAILABILITY

A reliable source of moisture is important to plant health. Most of this is provided directly from the soil although atmospheric moisture also has a role in plant–water relations. The input of water to the soil from rain or dew is obviously vital but the supply to the plant is largely mediated by other environmental conditions. Soil structure governs the water-holding capacity of the substrate. Large particles provide a well-aerated soil but with poor capacity for water retention; however, a soil with small particles has fine pores which can retain high levels of moisture, but will be poorly aerated and at extremes may become waterlogged. Combinations of conditions such as low relative humidity, elevated temperatures, and air movements tend to dry out the surface layers of soil and therefore limit water availability to plant roots. The high tensile strength of water results in its being drawn up through a plant, even over considerable distances, as a result of the cohesive forces which operate. Water moves through the capillaries of the soil into the root apoplast, xylem vessels, and through to the leaf apoplast provided that the column of water is not broken, and that the continuity of that stream is maintained.

Transpiration

Water constitutes up to 90% of the fresh weight of herbaceous plants, but it is the constant stream of water through the plant which is important for normal healthy functioning. Water is a solvent for biologically important molecules and as such has an essential role in plant function. The movement of water through the plant from roots to leaves also brings about the transport of solutes between tissues, distributing carbohydrates, ions, proteins and other cellular metabolites. A good water supply also maintains cells in a turgid and functional state, without which membranes loose integrity and normal functioning is disrupted.

The flow of water through the plant is maintained by the transpiration stream. Water evaporates from the moist surfaces of mesophyll cells into intercellular spaces and is lost to the atmosphere through stomatal pores. A smaller amount of water vapour is also lost directly from the cuticle and from lenticles. The opening and closing of stomata is controlled by the turgidity of the guard cells. During the day a gradient of water potential is set up between the roots and the leaves of a plant, resulting in the transpiration stream to maintain cellular turgidity. In a dry soil, where water flow cannot be maintained, leaves may suffer water stress during the day. Stomata have a very complex role in the regulation of water loss from a leaf. Pores are closed in the dark, but in the light, and under conditions of lowered CO_2 concentration, potassium and chloride accumulate in the guard cells which therefore take in water. This increases the turgor pressure over that of other cells in the epidermis, and the stomatal pores open. In conditions of water stress or elevated temperature, respiration rates are increased which results in higher concentrations of CO_2 in intercellular spaces and hence, in stomatal closure. This conserves water in the leaf.

Water is absorbed through the roots of plants. Although a root system may have a very large surface area variable amounts of the whole root are involved in water uptake at any one time. Additionally, the amounts of above-ground biomass supported by a root system will vary with water availability. Plants adapted to very dry conditions may produce most biomass below ground (up to 90%). However, in adequately watered situations the root:shoot ratio will be lower (25–30%). Many plant roots are adapted for water storage (e.g. bulbs and tubers).

Water stress

A long-term lack of soil moisture may lead to the development of symptoms in the plant. Growth will be reduced, plants become pale green in colour, wilting occurs and, eventually, even death. In some cases plants have evolved morphological adaptations to conserve water, such as surface hairs on leaves, small leaves with reduced surface area for potential water loss, and rolled leaves. The effects of short-term water stress are usually reversible and plants recover, even though the physiological effects may last for several days.

Metabolic processes proceed at a maximum rate when cells are fully turgid and when turgor pressure is low metabolic disturbance occurs. When plants are under water stress the stomata become closed, CO_2 uptake ceases and the rate of photosynthesis is reduced. Physiological processes are then affected but have different tolerance to the degree of water stress experienced. Cellular and leaf growth are very sensitive to water stress and cease once turgor pressure falls since this drives cell expansion. The synthesis of proteins and chlorophyll (Steponkus, 1981) is disrupted by relatively mild water stress, followed by hormone synthesis and CO_2 assimilation. Electron transport through the photosystems (PS I and II; see Chapter 4) is disrupted by water stress (Bradford and Hsiao, 1982) and photophosphorylation is also sensitive to high temperatures. These changes affect the progress of carbon metabolism. Respiration and carbon metabolism are less sensitive and may remain unaffected until moderate to quite severe stress occurs. Water stress is also reported to impare the speed and effectiveness of wound healing and therefore, as a consequence of this, the response to invasion by pathogens must also be detrimentally affected since these processes are very similar in plants. Additionally, under severe water stress drying may displace protein units from membranes and thereby impare normal function.

Waterlogging

High moisture levels in the soil lead to reduced oxygen levels in the substrate and therefore limited supplies for roots. This leads to root stress, asphyxiation, death and eventually rotting. Such anaerobic conditions favour the activities of some microbes and hence the production of high levels of nitrites in the soil, which are toxic to plants. As the root cells loose selective permeability toxic materials can enter, which leads to cell death.

TEMPERATURE

In general, plants are able to grow at temperatures between freezing and 40–45°C, although species have an optimum range between these extremes. The overall temperature at a site will be moderated by climatic and environmental factors, such as the degree of exposure, wind speed, seasonal variations and time of day.

The aerial portions of a plant are subjected to more rapid fluctuations in temperature than are roots, which are buffered and protected by the soil environment. The transfer of heat through soil is relatively slow and the amplitude of variations at depth is small. More deeply rooted regions have the most constant environment. Roots may experience the highest soil temperatures in autumn and a high proportion of root growth occurs at that time. The structure of a soil also mediates the temperature regime. Soils with high water-holding capacity tend to warm up very slowly and are regarded as cold soils in a cool environment. However, in hotter climates a high water content may provide a cooling system for plants and provide protection from the effects of temperature extremes. Near ground level the air temperature is mediated by heat exchange with the soil but above the soil surface wider fluctuations occur and and may influence leaf morphology.

At elevated temperatures the rates of biochemical reactions increase. For example, respiration is slow below 20°C but the rate increases at higher temperatures. The optimum temperature for the biochemical processes of photosynthesis is between 20 and 35°C, although the pathways do operate below this temperature. Above the maximum temperatures scorching may occur and cells are rapidly damaged, particularly if high temperatures occur in combination with high light intensity or low water availability.

There is also evidence that a number of proteins are synthesised very rapidly and *de novo* under conditions of heat stress (Schoffl and Key, 1983), indicating a change in gene expression. However, the functioning of these 'heat-shock' proteins is not well understood although they may act as a protection for the plant.

Low temperatures may, however, be more damaging to plants than the higher levels reached in most climates. Membrane permeability may be altered by low temperatures with the effect that cells retain water, stomata remain open and water stress results (McWilliam, 1983). Chilling also reduces the synthesis of chlorophyll and therefore affects photosynthesis.

Any temperature below 0°C can result in frost damage. The formation of ice crystals within leaves causes particular problems owing to the effects on membranes. Water in the intercellular spaces may freeze at, or just below, 0°C. The solutes inside cells freeze at lower temperatures (−3 to −5°C). A great deal of damage results from the formation of extracellular ice which draws water vapour from the apoplast and has the effect of dehydrating the surrounding cells. Additionally, rapid thawing may compound the problems for plant cells and result in water soaking of tissues.

LIGHT

The amount and quality of the light received by a plant has a profound affect on photosynthesis and therefore influences the growth and metabolism of that individual. Solar radition is utilised in plant processes by virtue of the energisation of pigments. The carotenoids and chlorophylls absorb radiant energy at the blue (450 nm) wavelengths, and chlorophyll at the red (660 nm) wavelengths for photosynthesis. The phytochromes are also energised by 660 nm wavelength light and revert to the inactive state, either in the dark or in light of 730 nm. Ultraviolet radiation can cause damage to proteins and therefore the effects are detrimental. Plants are able to synthesise flavonoid pigments which protect the cells, although much of the energy and those wavelengths have been filtered out by the atmosphere before reaching plants in most situations.

At very light-deficient sites a plant may become etiolated, producing thin, elongated stems with few leaves, and will suffer retarded chlorophyll production rendering the leaves very pale yellow in colour. However, sunlight is rarely limiting or damaging to plants, although sunscorch can occur if high light intensity is combined with elevated temperature or water limitation. Clouds absorb some wavelengths of solar radiation and on a cloudy day relatively diffuse light reaches a plant. However, a higher proportion of that radiation can be used for photosynthesis, than of the direct radiation which reaches a plant when there is little or no cloud cover.

The amounts of light reaching a plant at different times of the year can alter the physiological behaviour of a plant, also affecting its morphology and behaviour. The positioning of leaves on a stem, the shape of those leaves (shade plants develop large leaves to absorb maximum light energy), the timing of flowering and leaf fall are all influenced by day length and the quality of the incident light. Plants do exhibit adaptations to the prevailing lighting conditions. Exposed leaves receiving high light intensity may develop increased cuticular covering or larger numbers of cells in the mesophyll layers. In relatively shaded leaves changes in the positioning of chloroplasts in the cells will enable the maximum exploitation of available energy.

OXYGEN SUPPLY

Low oxygen availability usually occurs in soils which are waterlogged. The air spaces in the soil structure become filled with water and oxygen is then available for plant roots only by diffusion, which is very much slower than through air. With reduced levels of oxygen, $NADH_2$ accumulates in cells because the cytochrome chain can no longer function (Chapter 4). The tricarboxylic acid (TCA) cycle is repressed and the levels of the end products of fermentation (acetaldehyde and ethanol) build up in the cells and may quickly reach toxic levels. Energy yields from these anaerobic processes are also less than from aerobic metabolism and in such conditions plants are rapidly under pressure to increase the rate of metabolism to compensate (Pasteur effect). This quickly gives rise to higher levels of toxic metabolites. Additionally, activities of microbes, which are favoured in such conditions, result in the loss of nitrate from the soil and the accumulation of hydrogen sulphide, soluble Fe^{2+} ions and other toxic chemicals, which puts plants under additional pressure. There is also evidence that low oxygen availability reduces the capacity of plants to respond to wounding by limiting the rate at which the protective compound suberin is laid down in, or on, cells walls.

Some species have morphological and physiological adaptations which allow growth under such anoxic conditions. Oxygen diffusion is encouraged in the plant roots by the development of large air spaces (aerenchyma) within the root cortical tissues. Some plants are able to avoid the Pasteur effect by suppressing alcohol dehydrogenase activity and in some instances plants are able to exclude the toxic products of fermentation.

AIR POLLUTION

Gases produced by industrial processes, burning of fossil fuels, metal smelting, exhausts from car and other internal combustion engines release pollutant gases which affect the growth and distribution of higher plants. The morphological symptoms which can develop as the result of exposure to these pollutants are given in Table 2.2.

Sulphur dioxide (SO_2) enters leaves through open stomata and forms sulphite ions with the moisture in the leaf, which are toxic to the plant. SO_2 uptake may result in increased opening of stomata and hence increased uptake levels. This stimulation of opening can occur even in water stress conditions and consequently the plant may suffer from dessication rather than the direct effects of the gas. Ozone is absorbed in the same way and also disrupts cell membranes, breaking down the plasma membrane in particular by effects on the lipid and protein (Mudd, 1982) components, and having pronounced effects on plants as a result. These gases and also nitrogen dioxide cause severe disruption to cell membrane systems and

Table 2.2 Symptoms caused by toxic pollutants

Pollutant and toxic concentration*	Symptoms†
Ozone (0.1 ppm)	Chlorotic spots on upper leaf surface first, especially at stomata. Leaf drop may occur. Disrupts cell membranes
Sulphur dioxide (1 ppm)	Chlorosis of leaves particularly between veins. Forms acid rain with water
Nitrogen dioxide (2–3 ppm)	Brown spots form between veins. Reduces growth.
Hydrogen fluoride (0.1–0.2 ppm)	Leaf tips and margins necrotic.
Ethylene (0.1 ppm)	Stunting, necrosis and premature senescence.

* The concentrations at which symptom development occurs vary markedly with individual species and with environmental conditions.

† Exposure to combinations of pollutants may change the symptoms displayed and the thresholds which can be tolerated by plants.

inhibit photosynthesis by the effects on thylakoid membranes. Ethylene, chlorine and hydrogen fluoride may also inhibit plant growth and cause premature loss of leaves where exposure is high.

The exposure of plants to combinations of pollutants changes the threshold levels which can be tolerated and also influences the symptoms which are exhibited. Particulate matter such as dust and grit can also be regarded as pollutants and affect plant growth. In some instances the dust may itself be toxic and cause leaf burn when dissolved in rain drops on the lamina surface. Alternatively a buildup on the leaf may result in impared photosynthesis and reduced growth.

THE PLANTS' RESPONSE TO ABIOTIC AND BIOTIC INFLUENCES

REACTION TO STRESS

It can be seen that the health and growth rate of a plant is moderated by a combination of factors which operate and interact within the environment. Any influences, which reduce the growth rate of a plant from the maximum for that individual, place it under strain and can be regarded as stresses. Very severe strain may permanently damage the plant and even cause its death. Moderate or mild strain, on the other hand, may lead to the development of symptoms as the result of physiological disruption, but may not kill the plant. Such changes, for example wilting in response to water stress, or chlorosis in response to damaging levels of pollution, provide the plant pathologist with clues as to the ailment of a plant. However, plants express relatively few symptoms and in some cases physiological changes caused by different stresses may result in the production of very similar symptoms. It may also be easy to diagnose water stress in a plant, for example, but it may be much more difficult to determine the direct cause.

Strains imposed on a plant may occur as the result of environmental factors (abiotic) or due to the activities of other organisms (biotic). In very general terms plants respond in similar ways to these influences because both abiotic and biotic stresses have similar physiological effects on the plant (Ayres, 1984b). If, on removal of a stress the strain is also relieved then it is considered to be an elastic strain. However, if the strain remains then the stress is said to be plastic (Levitt, 1980; Ayres, 1984b). An injury is a plastic strain and may have long-term consequences for that individual at all levels. By virtue of the highly integrated nature of higher plant physiology and metabolism, strains and injuries occurring at subcellular, cellular or tissue levels may rapidly have knock-on effects at other levels in the plant. Any such damage will elicit plant defence mechanisms. It is interesting to note that plant responses to injury and wounding are very similar to the responses to infection by pathogens. Wound ageing also increases the protection afforded by the process of wound closure (Bostock and Stermer, 1989).

The susceptibility of plants to stress changes with the physiological status and with the stage of development of the individual. This may be partly as a direct result of the age and health of the plant or may be attributable to its ability to compensate for the effects. For example, an established plant is more likely to survive effects, particularly water shortage for example, which would cause severe stress, or even death, to a recently germinated seedling. In natural field situations plants are likely to encounter harmful effects from several sources at once, e.g. high temperatures leading to water stress on a hot day. Combinations of strains occurring simultaneously are more likely to have detrimental effects on growth and development and may be additive in their influences. In the case of biotic injury and invasion of a plant, a simultaneous shift in environmental influences may favour either host or pathogen and may therefore either increase or decrease the severity of the disease and the symptoms expressed.

Plants can recover from many stresses and injuries but the duration and severity of these factors will moderate the overall effects. As a result it is likely that the competitive ability of an individual, in the field, will be severely impaired by the additional challenge. Plants are able to adapt to long-term stress conditions too, by changes in their morphology, such as alterations in the root:shoot ratio, or leaf size, and also by adjustments to the chemical composition of membranes and the biochemical pathways utilised.

The effects of stress or injury to a plant are often exerted first on membranes. The plasma membrane has a vital role in the maintenance of selective permeability in cells and disruption of this causes very rapid injury. Increased leakiness of cells and the disturbance of the osmotic balance may result. In addition, membranes are vital components of important cellular organelles such as mitochondria and chloroplasts. In this way disruptions to membrane systems interfere with respiration and photosynthesis and therefore the overall energy balance in a cell. Subsequently, sustained effects may lead to changes in the functioning of biochemical pathways. New or subsidiary pathways may be activated and the production of new enzymes, utilised in plant defence systems, may be induced. The formation of novel compounds is an indication that the expression of genes becomes altered under stress conditions. Recently this area has attracted attention (Sachs and Ho, 1986) with the possibility for engineering plants with built-in, ready-made resistance to detrimental environmental conditions. Increased levels of plant growth regulators (hormones), or any change in the proportions of these compounds, may occur and act as a stimulus for the synthesis of phytoalexins. Phytoalexins are often termed 'stress metabolites' and are formed by plants in response to both abiotic and biotic influences. These factors are often considered to be components of resistance mechanisms, especially in relation to biotic influences, and will be considered in greater depth in Chapters 3 and 4.

NON-HOST RESISTANCE

Although plants suffer many diseases and widespread economic losses occur to crops annually, plants are, in general, resistant to most potential pathogens. Only a relatively few species of fungi (considerably less than 1% of documented species) are pathogenic to vascular plants, and most species of crop plants are successfully invaded by only a very limited number of those pathogens. Even a fungus with a comparatively large host plant range will infect only relatively closely related individuals or those with similar growth habits and life-styles. Most plants are therefore non-host species to most fungal pathogens.

The resistance of a plant to successful infection by a pathogen, for which it is not normally considered to be a host species, is termed non-host resistance. Heath (1986) has suggested that the study of non-host resistance is important to the continued development of our understanding of host–pathogen interactions. Plants are able to exhibit a number of forms of resistance to pathogenic invasion. Some plants may escape disease by avoidance of the pathogen, or may have either general or specific morphological features and adaptations which prevent the pathogen from forcing entry to living tissues. Some may be tolerant of the pathogen and although host tissues are invaded the plant may be able to resist the most damaging effects and continue to survive. There are many potential barriers and defence reactions to pathogen invasion and it is clear that no one factor or combination of factors is responsible for plant resistance. The mechanisms and barriers vary with each host–pathogen combination considered. In some cases resistance may occur as the lack of a factor, normally required by that pathogen for its further development. For example, the presence of a specific nutrient compound which stimulates spore germination may act as a trigger for the initiation of the infection cycle. If such a compound is not present it may therefore act as a barrier to invasion by virtue of its absence (Heath, 1986).

NON-SELF RECOGNITION

It has been suggested that plants possess some means by which self and non-self can be detected and distinguished (Callow, 1984) and therefore the study of such a recognition system may provide a key to the understanding of disease resistance (Matta, 1982; Ingram, 1982). It follows, therefore, that if a pathogenic fungus is recognised by the potential host plant as 'self' then resistance mechanisms will not be initiated and disease may subsequently ensue. However, there is very limited evidence for such a recognition system (Bushnell, 1986).

It is clear that incompatibility mechanisms do operate at the pistil–pollen interface, preventing some crosses and therefore maintaining species boundaries. However, grafting between two higher plant partners is often very successful and vascular elements do redifferentiate across the interface to create a functional pathway between the stock and the grafted scion. It is well recognised that the potential success of such a graft is increased by the closeness of the taxonomic relationship between the partners chosen and intergeneric grafts are most likely to succeed. However, grafts have been made between species of higher plants. In some cases necrotic reactions develop at the graft site and while the reasons for this are often unknown, there is clearly no involvement of a universal non-self recognition response.

Similarly the growth of undifferentiated plant cells (callus) in laboratory culture presents the opportunity to pair different species together, either in close contact or adjacent within a culture vessel (Bushnell, 1986) to test compatibility. In some instances combinations have been shown to grow together without signs of incompatibility, whereas in others the formation of dark necrotic zones indicated incompatibility responses. These responses are not taken to indicate any form of non-self recognition.

Protoplasts have been successfully isolated from many species of higher plants. Such units are liberated from plant material by enzyme digestion. All connections with associated cells are therefore severed and the cell wall is removed to leave an osmotically fragile, membrane-bound, unit of cytoplasm. These units contain all the organelles and cytoplasmic constituents of the original plant material. Fusions between protoplasts from closely taxonomically related plant species have been made successfully and intact hybrid plants have been regenerated from such material. Additionally, crosses between widely taxonomically separated species as well as plant–animal and plant–fungal pairings have been made and such combinations have been shown to co-exist for extended periods (Bushnell, 1986; Lynch *et al.*, 1989). It is therefore concluded that there is no universal recognition system to distinguish self and non-self operating in higher plants.

EFFECTS OF FUNGI ON PLANT HEALTH

Fungi affect the growth of plants in an enormous range of different ways. As the result of symbiotic associations with fungi, for example the formation of mycorrhiza, plants may show increased survival rates, greater biomass production, improved vigour and general health, and better resistance to attack by pathogens. Conversely, infections by fungal pathogens cause reductions in the growth rate and the development of plants, either rapidly or in the long term, which may be so severe as to be fatal. In general, reductions in biomass occur and reductions in yields which have economic importance. These may arise from very specific damage to structural components or to uptake systems, perhaps as the result of toxic activity by the pathogen, or the fungus may act as a sink for nutrient materials, essentially starving the plant of resources.

In some cases very specific tissues are invaded by a fungus and many of the overall effects on host plant physiology then relate to the mechanical and biochemical disruptions which have occurred. Invasion through a plant cuticle results in the rupture of the outer, protective layer and may lay the plant open to other secondary infections too. Intercellular penetration of tissues, whether purely mechanical or aided by enzymic degradation will result in direct effects on the middle lamellae between closely associated plant cells. Intracellular penetration will damage plant cell membranes and increase nutrient leakage into intercellular spaces. Such activities affect the overall functioning of plant tissues. Invasions of meristematic regions may give rise to an altered growth habit in an affected plant. Although relatively few infections occur in this area, increased metabolic activity may result in an already highly active region, and may give rise to the swelling and/or massive proliferation of shoots (e.g. *Crinipellis perniciosa*, causing Witches' broom disease of cocoa; *Peronospora viciae* on peas).

Continuous supplies of rich nutrients can also be gained directly, by the invasion of vascular tissues. Sieve tubes of the phloem carry carbohydrates between regions of production and sites of nutrient use or storage. The levels of nutrients carried are high and few pathogens are able to invade directly into these living cells. However, specialised pathogens are able to invade xylem vessels which transport water and salts from the roots to the leaves of plants. The vessels system of a plant is well protected by highly resistant barriers to invasion but growth through young root tissues enables the eventual colonisation of the xylem by the vascular wilt fungi. From this site the pathogen, and/or its toxic products, may subsequently spread throughout the plant using the xylem pathway.

Any such invasions therefore result in a range of visible or measureable changes, which are considered to be symptoms of the interaction. Many plant diseases are manifested by a characteristic pattern of symptom development which is often referred to as the disease syndrome. Diagnosis may well be complicated, since plants are unable to exhibit many different symptoms. Different causes may give rise to similar symptoms resulting in confusion. Equally a combination of biotic, or biotic and abiotic factors operating simultaneously will also complicate the situation.

DIAGNOSIS OF PLANT DISEASE

Great advantages are offered by experience in the area of plant disease diagnosis. It is obviously important, in the first instance, to distinguish between environmental effects and those of disease. Where microbial pathogens are involved, it is usually possible to detect the organism on the surface of the plant or within the tissues. Detailed examination, with a microscope, may reveal the causal agent, particularly in the case of a fungus which is sporulating. However, if identification within the tissues is not possible, isolation procedures can be employed. In some cases it may be possible to encourage the sporulation of a fungus by the incubation of infected tissues under more favourable conditions, or alternatively the organism may be isolated into axenic culture.

Koch's postulates

In 1882 Robert Koch laid down a series of rules to be followed in the demonstration of the pathogenicity of microbes in medicine. These have been subsequently modified for application to plant pathogens.

1. The microorganism (pathogen) must always be associated with the disease.
2. The pathogen must be isolated from diseased tissue, grown in pure culture and its characteristics described. Alternatively (for obligate

biotrophs), it must be grown on a susceptible host plant and all characteristics of the plant and pathogen recorded.
3. The specific disease must be reproduced when the pathogen, isolated into pure culture, is re-inoculated into healthy host plants.
4. The microorganism must be re-isolated from an inoculated host which has developed the disease and the characteristics must match, exactly, with the previous isolation.

These rules are still widely used in the determination of causal agents although it is now recognised that they are not equally applicable to all pathogens. In the case of obligate biotrophs, which cannot be grown in pure culture, it may be necessary to isolate the unknown organism into healthy plants, rather than on to laboratory medium, and then apply rigorous controls to be certain that a particular organism is the causal agent of a disease. The direct transfer of isolated fungal spores on to a potential host can also be carried out. The sporulation of a fungus is important with regard to the identification of an isolate since many of the taxonomic criteria applied to fungi relate to spore production and morphology.

SYMPTOMS OF PLANT DISEASE

Many fungal pathogens cause the production of characteristic symptoms in host plants. Specific terms have been applied to the physical manifestations of the host–pathogen interactions which result in order to accurately describe the symptoms which are visualised. Many of these have arisen as the result of historical usage. Such terminology has been listed (FBPP, 1973), in attempts to clarify, standardise and clearly define the terms which are used. The definition of such symptoms is often used in surveys carried out to assess the degree of damage which has occurred in a standing crop. In these cases accuracy is important to provide a comparative measure through categorisation. However, the relative importance credited to particular symptoms may be moderated by the economic consequences of their development. Symptoms which affect a whole plant relatively little may be considered as especially important if the marketability of the product (e.g. fruits or grain) is likely to be affected.

As has been pointed out (Wheeler, 1984), it is unlikely that agreement will easily be reached concerning such terminology. The development of symptoms is mediated by the environment and the general health of the potential host at the time of infection. The speed with which the pathogen can influence the physiology of the host, and therefore give rise to symptoms, will depend on the ease with which interactions are initiated and the course which the disease may run. In some cases host plants are quickly overpowered by an invading pathogen (necrotrophs) but in other instances a plant may remain apparently unaffected by the invader (biotrophs) for some time before symptoms develop. Symptoms which develop in different tissues may vary greatly in severity, or apparent severity, and any one disease may then show slightly different characteristics. Additionally, plants under different conditions have variable capacity to compensate for any damage caused by disease.

Reduced growth

In most cases reduced growth is an important symptom of disease, especially in economic terms. This may be manifest as a reduction in the overall yield of a crop, either in terms of total biomass or numbers, size and quality of fruits, grain, seeds, tubers, flowers, etc. Many of these effects arise from disruptions to photosynthetic tissues and mechanisms (increases in respiration rates, nutrient relocation and changes in water relations due to cuticle rupture when the fungus sporulates) as the result of pathogenic invasion.

Rust diseases

Rust diseases occur as devastating infections which have caused great economic disasters throughout the world. Infections commonly occur on stems, leaf sheaths and blades of an enormous range of plants, and are caused by members of the Uredinales (Basidiomycotina). These species are ecologically obligate parasites (although some have now been cultured on laboratory media) with relatively complex life cycles. No fruiting bodies typical of Basidiomycetes develop during the life cycle but these species produce a range of different spore forms (up to five different types). Rust diseases of wheat and other cereals have particularly important implications for world economy.

Puccinia graminis f. sp. *tritici* infects wheat and barley. This is a heteroecious species, with a disease cycle (Fig. 2.8) alternating between a primary host species (cereal) and a secondary host species (barberry; *Berberis vulgaris*). Some rusts are autoecious, in which case all spore types are formed on one host. Uredospores, which are asexual spores, are formed on the cereal host erupting in clusters through the epidermis of leaves. These spores are reddish in colour and give the characteristic rusted appearance to infected plants. Reinfection of the host by uredospores, which invade host tissues and obtain nutrients by means of haustoria (see also Chapter 3), may occur many times. As the host plants reach maturity dikaryotic teliospores are produced. These do not germinate immediately and the fungus often overwinters in this form. Teliospores germinate to produce a short septate hypha, with four compartments which each function as a basidium. Each contains one hapolid nucleus as the result of meiotic

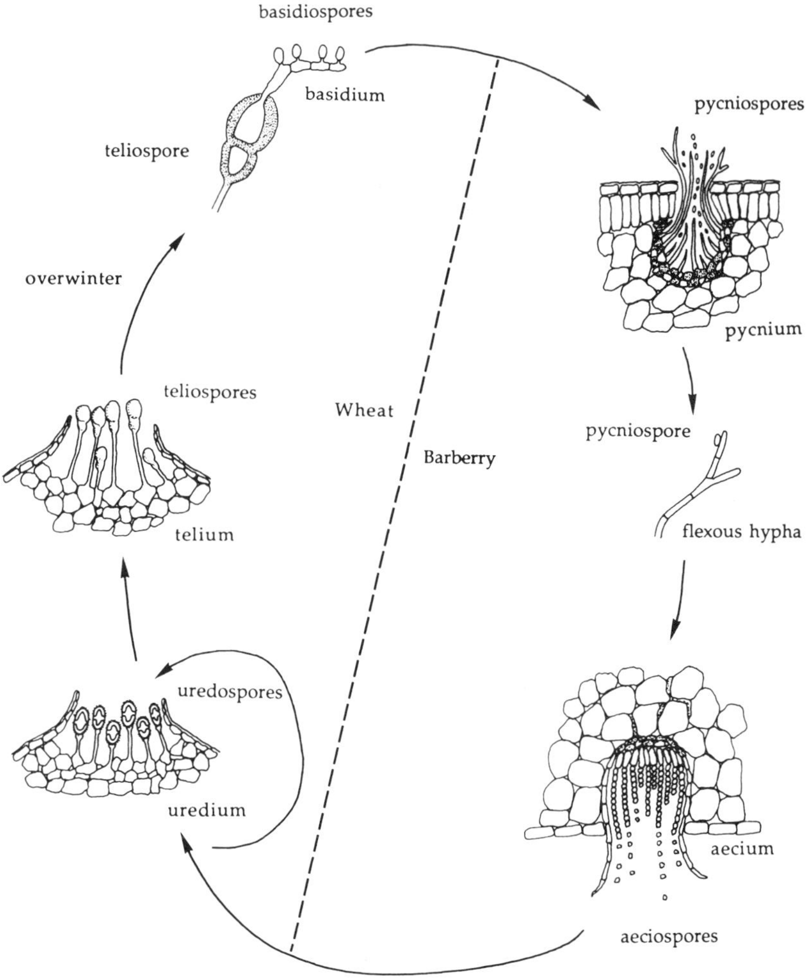

Fig. 2.8 Disease cycle of *Puccinia graminis* f. sp. *tritici*, which causes stem rust of wheat and infects barberry as a secondary host.

division and gives rise to a monokaryotic basidiospore. These infect only the secondary host. Subsequently pycnia, containing pycniospores, are formed on the upper surface of the barberry leaves. These are released and dispersed by insects. *P. graminis* is heterothallic and further development depends on fusion between a pycniospore and a flexous hypha, which occur around pycnia, of the opposite mating type. A dikaryon is formed which colonises the leaf giving rise to aecia on the lower surface. Dikaryotic aeciospores are dispersed by wind and infect only the cereal host. Removal of the secondary host gives good control and reduces the incidence of the disease. Systemic fungicides for rusts are under testing.

Many physiological races have been identified which can very rapidly cause widespread and devastating epidemics. Resistant cultivars are now used extensively for cultivation. Other rusts are also important pathogens although some have simpler life cycles, e.g. *Puccinia coronata* infects oats; *P. sorghi* causes corn rust; *Phragmidium* spp. cause rust on roses; several species of *Uromyces* infect broad beans and peas and *Melampsora lini* causes rust of flax.

Mildews

(a) Powdery mildew

Powdery mildew infections are easily recognised, and affect a very broad range of plant species including cereals and other crops, trees and ornamentals. The causal agents are obligate parasites which have not been grown in laboratory culture. Species such as *Microsphaera alni* (trees, rhododendrons and lilac) and *Spaerotheca macularis* (strawberry and soft fruits) cause notable losses.

Erysiphe graminis (Ascomycotina) is particularly important on cereals and grasses, causing extensive infections in humid conditions. The upper surfaces of infected leaves show a greyish coloration, caused by the presence of a superficial, but often extensive, layer of mycelium. Short conidiophores are formed from this mycelium, bearing long chains of powdery conidia, from which the disease gets its name. These conidia are dispersed easily by wind and may travel long distances. Conidia germinate and rapidly infect leaves, by penetrating directly through the cuticle. Haustoria, which act as feeding structures, are formed in epidermal cells and these provide the external mycelium with nutrients. No further invasion into plant tissues occurs. In unfavourable conditions cleistothecia, containing ascospores (sexual spores) are produced and the fungus may overwinter in this form.

Powdery mildews spread both easily and rapidly by virtue of the light, airborne spores (Fig. 2.9) and are extremely common. Agriculturally, multiline cultivars are now used which carry race-specific resistance and

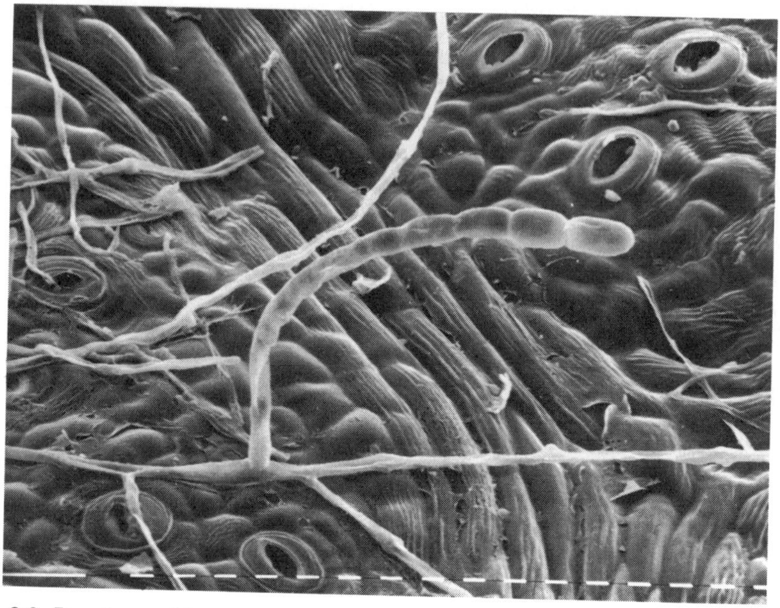

Fig. 2.9 Powdery mildew mycelium and chain of powdery conidia on hawthorn leaf surface. Bars represent 6.25 μm.

also field resistance, in efforts to limit the disease. Chemical control is possible using the pyrimidine fungicide ethirimol, but is not easy to implement in field situations and variable environmental conditions.

(b) Downy mildew

Downy mildew diseases are essentially foliage blights, caused by members of the Peronosporaceae which give rise to the development of discoloured zones on leaf surfaces. White to grey growth of the fungus appears on the under surface of infected leaves, which gives the disease its name. Branched sporangiophores usually protrude from stomata and bear sporangia. These either germinate directly or release zoospores which are motile in surface films of water. Germination often occurs at night owing to the extreme sensitivity of the spores to high temperatures, desiccation and light. Sexual reproduction is by means of oospores but is a relatively rare event.

The disease develops and spreads very rapidly after infection has occurred. Some crop plants have suffered very important economic losses, e.g. grapes (*Plasmopara viticola*), tobacco (*Peronospora tabacina*) and grasses are often affected (*Sclerospora graminicola*). Downy mildew of lettuce is a very common disease caused by *Bremia lactucae*, which flourishes under

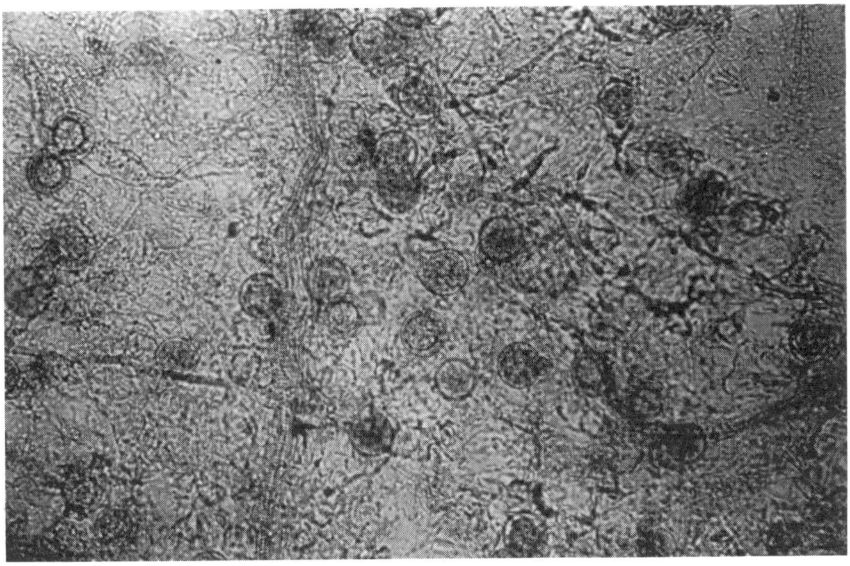

Fig. 2.10 Oöspores of *Bremia lactucae* formed within lettuce cotyledons infected by two isolates of opposite sexual compatibility type (B_1 and B_2). Reproduced courtesy of Dr I. R. Crute, HRI, East Malling.

cool, moist conditions (Fig. 2.10). Infections are often superficial but these species can also cause systemic disease.

Necrosis

The term necrosis is used to describe the death of infected plant tissues. During disease development necrosis may become extensive and rapidly lead to the death of the whole plant, or may be restricted to more isolated zones. Necrotic spots appear on infected plant leaves as the result of localised cell death. A range of diseases cause the formation of necrotic regions, e.g. damping-off diseases, leaf spots, blights, scabs and anthracnose.

Blights

The term blight refers to plant diseases which cause blackening of leaves and other plant organs and has often been applied particularly to infections which spread very rapidly. Probably the most famous of the blight diseases is that caused by *Phytophthora infestans* (Mastigomycotina) on potatoes and other Solanaceae. Late blight of potato caused widespread and devastating famine in Ireland between 1845 and 1850, and is still very common in cool, moist climates.

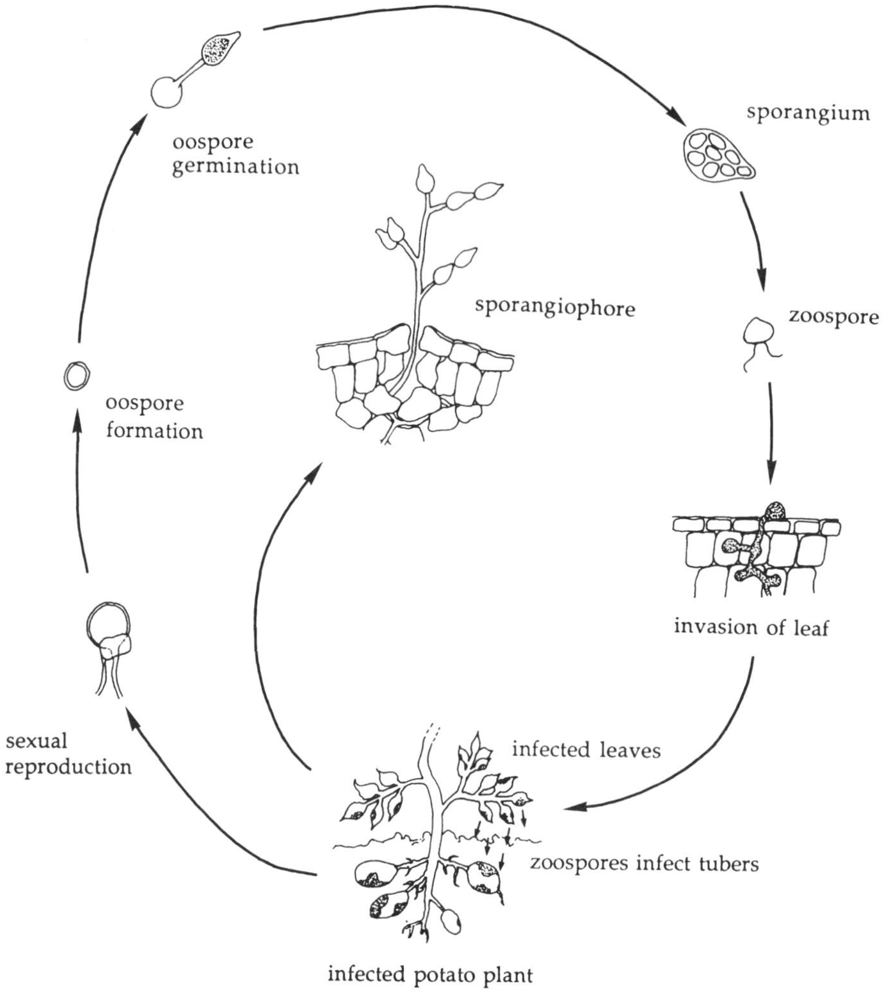

Fig. 2.11 Disease cycle of *Phytophthora infestans* which causes late blight of potato.

Phytophthora late blight kills the foliage and stems of potential crop plants in the field (Fig. 2.11). Initially haustoria are formed in host cells and the fungus lives biotrophically but later, extensive necrotic lesions occur on leaves. Sporangia form around these regions, releasing zoospores which are washed down into the soil. The zoospores are motile in the water film in moist soil. These encyst on the host surface (Fig. 2.12), attaching by means of extruded mucilage. Underground tubers may be attacked in the field and any partially infected tubers continue to rot during storage. This

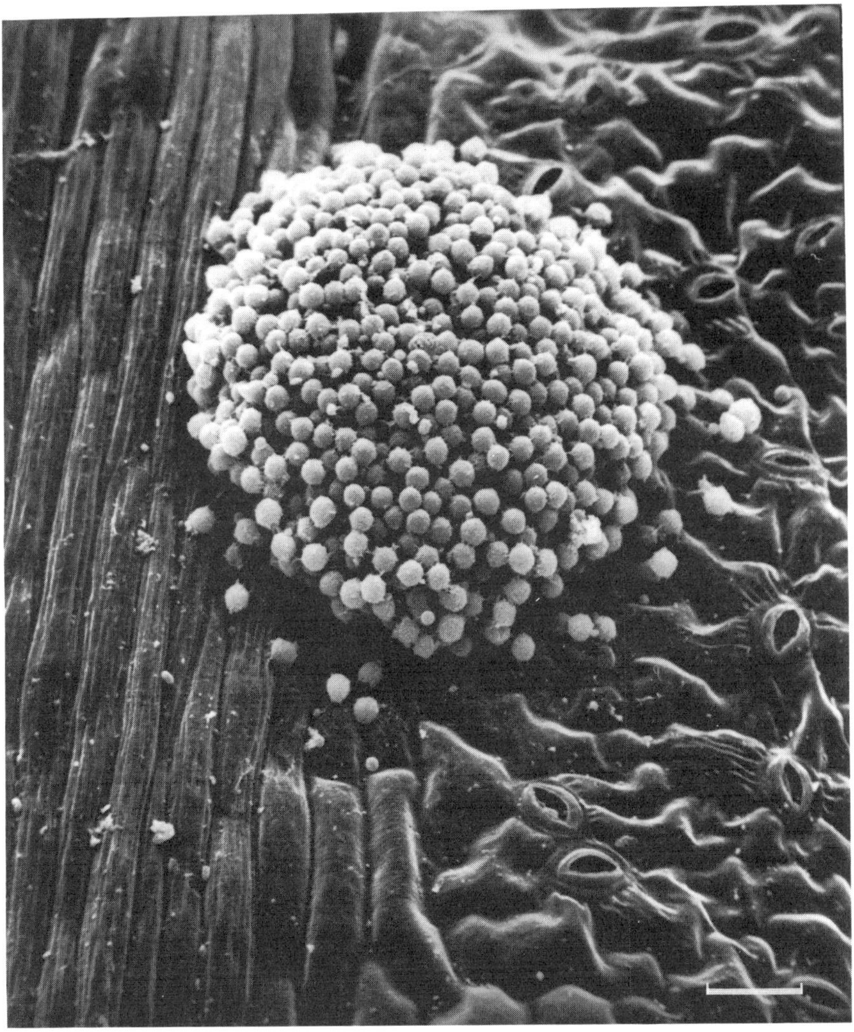

Fig. 2.12 Aggregation of *Phytophthora infestans* zoospores on host plant surface. Bar represents 50 μm. Reproduced courtesy of Dr D. Shaw.

species also overwinters as spores (Fig. 2.11), or as mycelium in infected tubers which may act as a reservoir for infection if allowed to remain in the soil.

Sexual reproduction of *Phytophthora infestans* rarely occurs in the field and requires the presence of two different mating types A_1 and A_2. Some other species are self-fertile. There are many species of *Phytophthora*, some of which have extremely wide host ranges and are particularly destructive and are, therefore, important plant pathogens. Many species cause foot and root rots after the formation of necrotic lesions, especially in damp, cool, poorly drained soils. Young seedlings are most susceptible to infection, rapidly becoming weakened and lodged.

Damping-off disease

Damping-off disease can affect plants of most kinds but is most rapidly damaging to ungerminated seeds, young seedlings and cuttings. Initial infections appear as water-soaked lesions which darken in colour. Affected cells swell, but subsequently collapse which may rapidly lead to the death of the whole plant. Such lesions usually occur at, or just below, the soil line. The diameter of the stems of small seedlings is often reduced and the mechanical strength severely impaired so that the weight of the whole plant cannot be supported (post-emergence damping-off). Older plants may develop necrotic lesions around the whole stem and growth may be significantly retarded.

Pythium species, in particular *P. ultimum*, *P. aphanidermatum* and *P. debaryanum* are common agents of pre- and post-emergence damping-off disease. Such species are widely distributed in soils world-wide and are able to live saprotrophically on dead plant materials. However, invasion and growth within plant tissues is rapid and sporangia are formed on the infected host plant tissues. These sporangia either germinate directly or produce a vesicle from which motile zoospores are released. Zoospores subsequently encyst, germinate and reinfect host tissues. Penetration into the host is by means of mechanical pressure aided by the production of pectolytic and proteolytic enzymes, which weaken the middle lamellae between cells, eventually resulting in extensive tissue maceration. If vascular tissues are invaded then death of the seedling is extremely rapid. Good soil drainage and clean agricultural practices can help enormously in controlling these diseases.

Scabs

The term scab is used to describe superficial necrotic lesions that are formed as a result of host invasion, together with a roughening of the outer layers. In some infections these lesions are accompanied by the formation of cork under the epidermis or by the associated accrual of fungal mycelium. The epidermis eventually ruptures to release fungal spores. *Spongospora subterranea* (Plasmodiophoromycetes) infects roots and tubers of potato and affects the marketability of the crop.

Scab disease of apple and other fruits is caused by *Venturia inaequalis* (Ascomycete). Infections destroy the market value of crops and occur readily in cool, moist climates. Cracks in young shoots near the necrotic zones allow the entry of secondary invaders too. This species produces both conidia and ascospores, which are equally infective. The fungus overwinters in dead leaves. Usually ascospores reinfect leaves or fruits early after formation in the spring.

Anthracnose

Anthracnose is the common name for plant diseases characterised by very dark, sunken lesions, containing spores. The name is derived from the Greek word meaning coal. *Glomerella* and *Colletotrichum* species produce anthracnose of annual plants and affect fruits, such as apples, peppers and tomatoes. *C. lindemuthianum* is an important pathogen of beans (*Phaseolus vulgaris; Vicia faba*) causing anthracnose of leaves and pods. These hemibiotrophic species are members of the Deuteromycotina and live biotrophically for some time, after which necrotic lesions are formed. The necrosis may extend to whole leaves and cause deformations of pods in severe cases. Light-coloured conidia can be seen in the central regions of lesions at maturity, particularly under dry conditions.

Leaf spots

Small necrotic spots often occur on plant leaves and may be attributable to many different causes. These lesions effectively reduce the photosynthetic capacity of the leaves, and when the fungus sporulates water relations can be disturbed owing to cuticle rupture. Stem and leaf lesions are caused by a number of fungi, in particular *Septoria* species (Deuteromycotina). *Septoria apiicola* is the causal agent of leaf spot of celery. To the grower this disease has economic importance, affecting the marketability of the crop but probably causes little problem for the host plant unless a very heavy infection occurs. Early infections affect the outer leaves of the plant and these may gradually die back. Intercellular mycelium gives rise to the production of pycnidia, containing pycnidiospores, on either surface of the leaf. Subsequently spores are spread by rain splash. More devastating effects occur later in the host plants growing season, favoured by moist, warm conditions. Few control measures beyond careful practices are available; fungicide treatment of seed has some effect and spraying can be helpful if applied at the right time.

Eyespot diseases of cereals (especially wheat and barley) are caused by *Pseudocercosporella herpotrichoides* (Deuteromycotina). Infections develop most readily in moist, cool conditions, after spore penetration through stomata. Typical lesions are eliptical regions with a darkened centre and a light brown surrounding necrotic zone. Mycelium is associated with the centremost regions. Later in infections dry conidia are formed at these sites and are dispersed by rainsplash. The lesions occur low down on the plant stem, just above the soil surface. The formation of several lesions eventually weakens the stem, by damage to the strengthening, vasular tissues and causes infected plants to lodge, ruining the growing crop.

Cankers

Necrotic lesions which appear sunken into plant tissues are usually termed cankers. These areas may be surrounded by an overgrowth of plant tissue which occurs as a response to the wounding effect. Plant tissues may be damaged down to the vascular tissues and such infections may be fatal to badly infected plants. The damaged necrotic region is often blackened and plant tissues may become dry and brittle. Such lesions are frequently caused by Ascomycete fungi and it is in these tissues that perithecia-bearing fungal spores are formed. Such infections are particularly important on fruit trees and include, e.g. *Nectria galligena* on apple, pear, elm and oak, *Endothia parasitica* on chestnut, *Ceratocystis fimbriata* on cocoa, coffee and rubber.

Chlorosis

A number of diseases lead to the yellowing of leaves in the infected plant. This symptom is often attributable to an iron deficiency which leads to a reduction in the synthesis of chlorophyll in leaves, rather than pathogenic invasion. However, chlorotic spots on leaves may also be associated with the effects of fungal pathogens on chloroplasts. Reductions in the size or number of chloroplasts, or of the amounts of chlorophyll contained per chloroplast may lead to yellowing. Chlorosis often accompanies the formation of necrotic spots.

Deformations and stunting

A range of pathogenic fungal infections give rise to alterations in the morphological characteristics of host plant tissues and affect the normal development of that plant. The mechanisms by which such changes are brought about are largely unknown, although the involvement of growth factors is often implicated. A range of symptoms come into this category. Hypoplasia is the term given to the general stunting or dwarfing of a plant which results from underdevelopment. Reduced cell division may be the primary mechanism for this effect. Hyperplasia is the term given to the excessive growth of tissues, usually as the result of increased cell division, and hypertrophy describes excessive growth due to the enlargement of individual cells. It is likely that these symptoms are brought about by changes in the levels of growth regulators and inhibitors in infected tissues (Chapter 4).

Leaf curl

A range of plant deformities are caused by species of *Taphrina* (Ascomycotina). *Taphrina deformans* causes leaf curl on peach and nectarine which results in leaf fall and economic losses to crops. In cool, moist climates both leaves and twigs are invaded. Young branches become swollen and blossoms fail to develop normally. Infected leaves develop a crumpled appearance, becoming highly distorted and often curling downwards. It can be seen that cells of the palisade mesophyll grow more rapidly than those of the spongy mesophyll and the leaf does not develop in the usual way. The involvement of the plant growth regulator indole-acetic acid (IAA) is implicated in the production of this effect. Eventually ascospores are produced in leaf tissues and asci break through the cuticular layer of infected leaves, from where the spores are dispersed.

Witches' brooms

Infections occurring near meristematic tissue may give rise to the loss of apical dominance in affected plants. This, in conjunction with changes in the levels of growth regulators may give rise to a great proliferation of shoots, traditionally described as Witches' brooms. *Taphrina* species (Ascomycotina) cause Witches' brooms on trees, such as birch (*T. betulina*) and alder (*T. epiphylla*). *Crinipellis perniciosa* (Basidiomycotina) causes enormous losses in South American cocoa plantations annually, owing to the development of brooms on cocoa trees. Affected plants show massive shoot proliferation and greatly reduced pod formation as a result of infection. It has often been postulated that the effect is due to increases in levels of growth regulators but this has still to be confirmed.

Epinasty

Epinasty is the term given to the downward growth of the petiole of infected plants. This symptom may appear akin to the wilting syndrome of young shoots as a response to water shortage but in this instance the tissues remain fully hydrated and turgid. This is one symptom of vascular wilt disease caused by *Fusarium oxysporum* and may be the result of the growth-regulating activity of ethylene produced in infected plants.

Club root disease

Club root disease of crucifers causes world-wide losses of cabbages and cauliflowers, especially in acid soils. The causal agent is a member of the Myxomycetes, *Plasmodiophora brassicae* (Plasmodiophoromycetes).

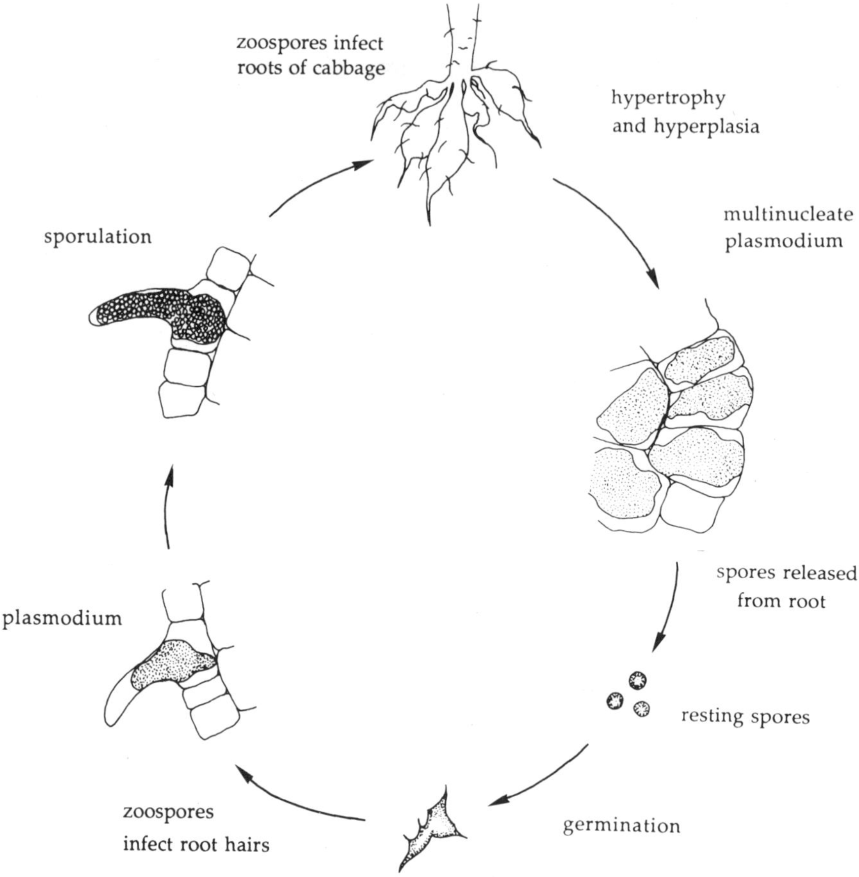

Fig. 2.13 Disease cycle of *Plasmodiophora brassicae* which causes club root disease of crucifers.

Infected plants become generally stunted and above-ground portions wilt readily. Host plant penetration occurs through the root hairs, after which the roots swell and become club shaped (Fig. 2.13). Infections may be limited to the main roots of a plant or, in more severe cases, may extend to include the whole root system (Fig. 2.14).

Highly persistent resting spores which survive in contaminated soils release a motile primary zoospore (biflagellate) which infects the root hairs of susceptible host species. A plasmodium is then formed within the plant tissues which subsequently forms zoosporangia, each liberating secondary zoospores. These zoospores pair and fuse together to form a diploid zygote which reinvades cortical root tissues. In the host plant a plasmodium is formed which penetrates the cortical tissues and stimulates

Fig. 2.14 Galling of kohl rabi (*Brassica oleracea*) symptomatic of club root disease caused by *Plasmodiophora brassicae*. Reproduced courtesy of Dr I. R. Crute, HRI, East Malling.

both cell enlargement (hypertrophy) and abnormally rapid cell division (hyperplasia). After meiosis, resting spores are formed which return to the soil environment after decay of the plant tissues. The main effect of this pathogen on the plant is to interfere with root function but in addition, the outer layers of the root do not develop normally and the possibility of secondary invasion is high. The presence of the plasmodia in the root leads to an increase in the ploidy of the infected cells and to increases of a hundred-fold in levels of auxin and cytokinin growth regulators.

Galls and warts

Galls are localised swellings which occur on host plants and have an altered morphology from normal host tissue at that site. Many galls are formed as the result of a variety of causes, for example insect action, but some are produced by fungal infections. The fungus *Albugo candida* (Oömycete) causes white blister disease of Brussels sprouts. Leaves become senescent and show green island effects associated with the fungal pustules (Fig. 2.15a, b). Lateral buds ('sprouts') often show morphological malformations (Fig. 2.15c). *Synchytrium endobioticum* (Chytridiomycetes) causes black wart disease of potato. In this case the fungus is an obligate biotroph and zoospores invade meristematic cells of tuber tissue. Fungal activity stimulates cell division around sites of resting spore formation and large warty outgrowths are formed on the tuber surface. Such activity is not usually fatal to a plant but may cause economic losses in commercial situations.

Wilt

The wilting of plants in the field may have a number of causes quite unrelated to any disease response. Drought conditions result in the loss of turgor in leaves and shoots and may result in generalised wilting. Unless the water supply is particularly restricted or prolonged this wilting symptom is reversible. Wilt caused by pathogenic invasion is, however, irreversible and may often progress to affect the whole plant relatively quickly. Such wilting may also occur following root damage or death. Additionally, vascular blockage, which occurs during vascular wilt disease, eventually results in irreversible wilting.

Vascular wilt disease

Vascular wilt diseases (caused by members of the Deuteromycotina) are particularly destructive and very rapid in their effects. The most important

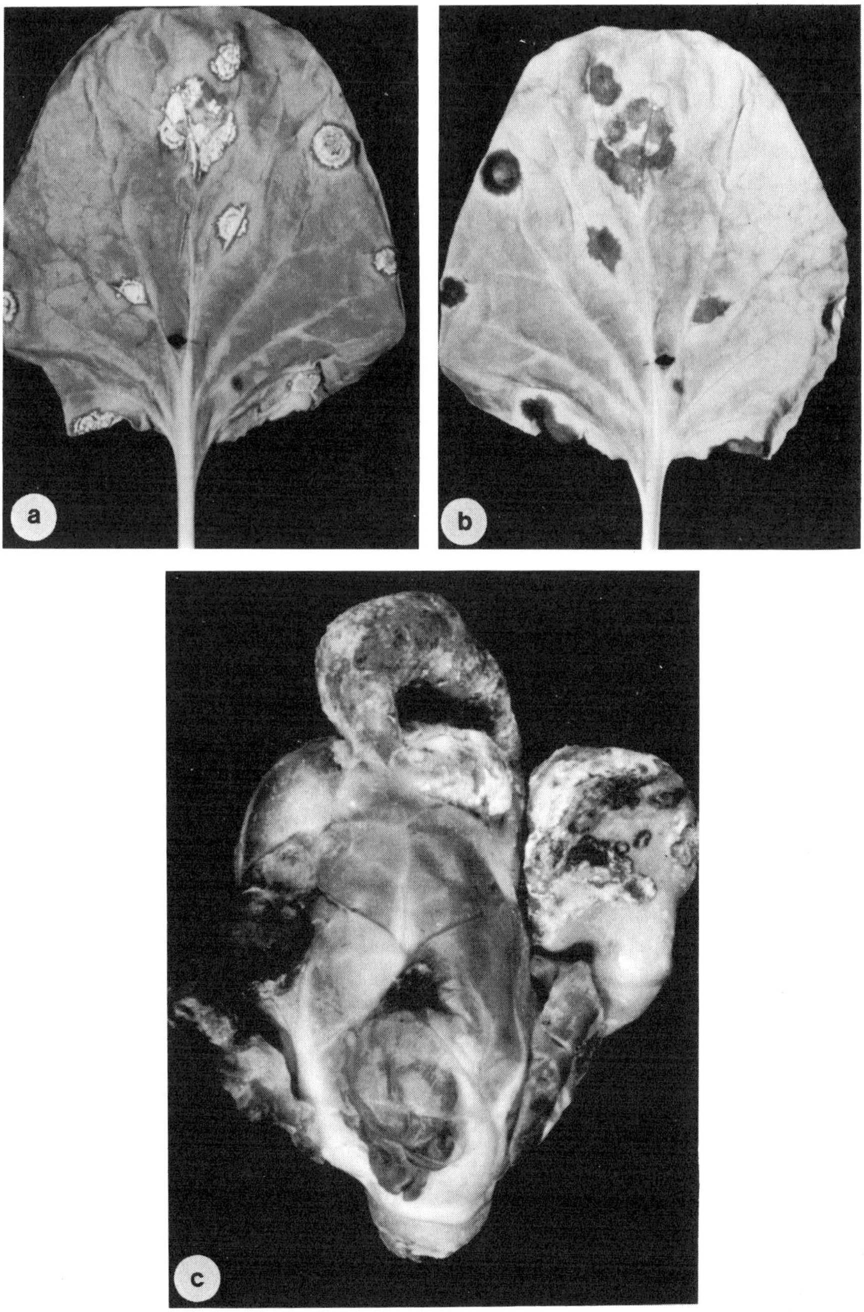

Fig. 2.15 White blister disease of Brussels sprout infected by *Albugo candida*. (a, b) Upper and lower surfaces, respectively, of senescent leaf showing green island effect. (c) Morphological malformation of lateral bud (sprout). Reproduced courtesy of Dr I. R. Crute, HRI, East Malling.

examples are *Fusarium* spp. and *Verticillium* spp. *Fusarium oxysporum* is a very important pathogen of vegetables and flowers, particularly affecting tomato, onion and cabbage. Many different physiological races of this fungus occur, which infect different host plant cultivars. *Verticillium albo-atrum* infects ornamentals, flowers, fruit trees and some crop species, such as tomato, potato, cotton and peanuts.

Invasion into plant tissues occurs through the root system often aided by entry through damaged tissues. Further penetration occurs after the production of pectic enzymes which attack the middle lamellae between plant cells and this allows the fungus to penetrate into deeply seated living tissues. In this way the fungus spreads into the vessel elements from where further dispersal through the host plant is easily achieved as mycelium or conidia. At this stage the lower leaves on the plant loose turgidity, droop and become chlorotic. The vascular elements rapidly become blocked by fungal mycelium and conidia, or by gums and tyloses produced by the host plant as a resistance mechanism. Later in the course of the disease the fungus releases secondary metabolites which act as toxins to host plant cells, disrupting cell permeability and hastening plant death. After the death of the host plant the fungus may sporulate on that dead material. These fungi are able to produce sclerotia to survive in the soil environment and are effective saprophytes, overwintering as very highly resistant spores or as mycelium on dead host plant tissues.

Dutch Elm disease

This disease has had very important consequences for the tree populations in most temperate regions. It is highly destructive and has caused devastating and extremely rapid losses of trees. The disease is caused by the fungus *Ceratocystis ulmi* (Ascomycotina) but is spread by an insect vector which transports the spores between infected and healthy host plants. The elm bark beetle (*Scolytus multistriatus; Scolytus scolytus; Hyurgopinus rufipes*) is responsible for this transmission. As a preference bark beetles lay their eggs in damaged trees and the larvae tunnel gallery systems under the bark. The fungus sporulates in these galleries and consequently when adult beetles emerge in spring they carry mycelium and spores to new host plants. The fungus spreads through the host plant via the vascular system as mycelium or spores. Eventually vessels become blocked and the characteristic syndrome of wilting and chlorosis in new branches develops and leads to death in those tissues. This disease is difficult to control and although systemic fungicides can be used treatments are not easy or reliable and are expensive. In general the main control measure is the felling of diseased trees which is a successful measure in preventing further spread.

Fig. 2.16 Wheat ears infected with ergot (*Claviceps purpurea*). Sclerotia are visible (arrowed). © Crown copyright.

Invasion of specific organs and the disruption of host plant reproduction

Some fungi have specialised growth habits and infect only a particular tissue type on a susceptible host plant. Examples of this strategy include some economically important diseases in which the reproductive system of the host plant is most affected by fungal activity.

Ergot of cereals and grasses

Ergot is a world-wide disease which may cause crop losses when severe infections occur. However, another important consequence of this disease arises from feeding contaminated grain to animals or humans. Any contamination of grain exceeding 0.2–0.3% by weight of fungal material (sclerotia) is sufficient to cause ergot poisoning. Animals are affected by the ergot alkaloids which are produced by the fungus as secondary metabolites. These cause circulatory problems and drastically reduce the reproductive potential of livestock.

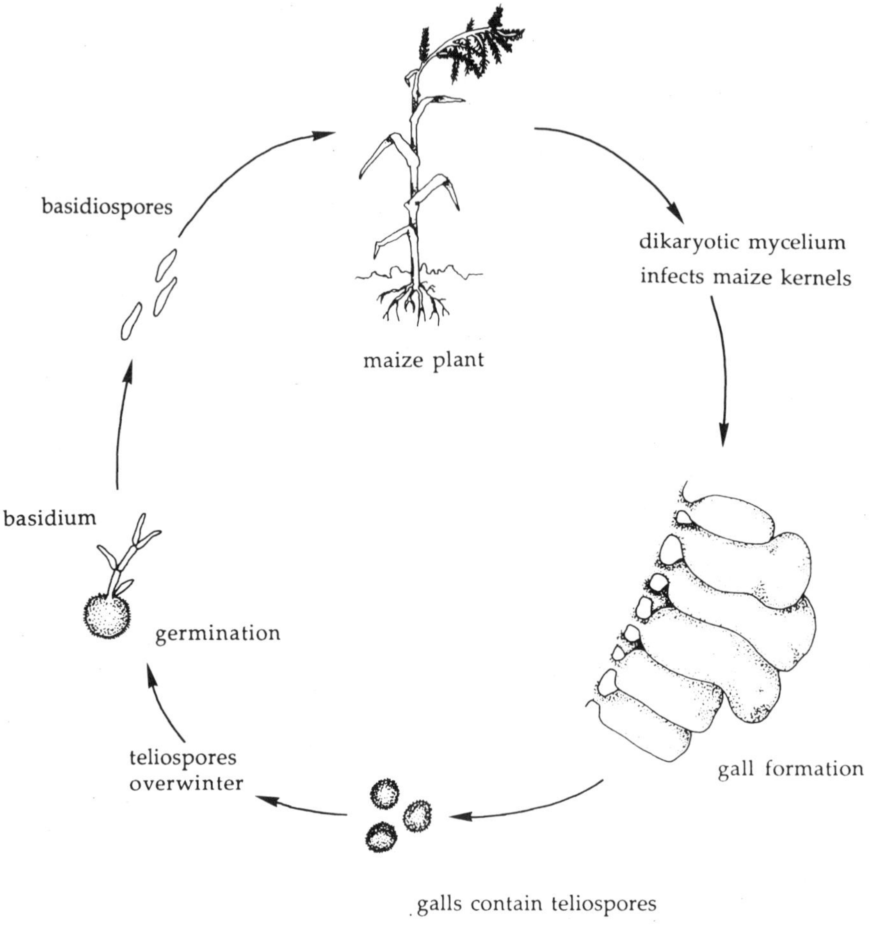

Fig. 2.17 Disease cycle of *Ustilago maydis* which causes smut of maize.

Ergot is caused by the fungus *Claviceps purpurea* (Ascomycotina), which overwinters as sclerotia on host plants (Fig. 2.16). In spring the sclerotia germinate and give rise to a number of stalked structures, each with a head bearing many perithecia. These contain asci and ascospores which are extruded and dispersed, either by insects or the wind, to developing host plant flowers. The fungus penetrates into the ovaries via the stigma. From this tissue the fungus liberates conidia in sticky fluid which is carried, by insects, to other flowers by which means other host plants are reinfected. Subsequently the fungus in the ovary tissue hardens to form a new sclerotium, replacing the grain which would normally have been formed by the plant. These fall to the ground and overwinter to germinate and reinfect new plants in the next growing season.

Smuts

Smut diseases are caused by Basidiomycete fungi (Ustilaginales), of which infections of grain and cereal crops are the most important in economic terms, although other plants are also affected (e.g. onions and sugarcane). Infections cause general stunting of crop plants as well as having more direct effects on reproduction. Such infections are caused by a large number of fungi, including *Ustilago nuda* (barley smut); *U. tritici* (smut of wheat); *U. hordei* (barley and oats); *Tilletia caries* (covered smut of wheat); *Urocystis cepulae* (onion smut) and *Sphacelotheca reiliana* (head smut of maize).

Smuts may infect seedlings and grow through the plants until the inflorescence is reached or alternatively may infect ovaries of grain crops more directly through the flowers (Fig. 2.17). Galls are produced on the host plant by smuts. These contain dikaryotic mycelium which gives rise to teliospores by which the fungus overwinters. In spring these teliospores germinate to produce a short hypha which functions as a basidium and gives rise directly to basidiospores. After basidiospore germination, fusion of compatible haploid hyphae gives rise to dikaryotic mycelium which can infect susceptible host plants. The presence of the fungus stimulates host plant cell division and enlargement, and galls packed with mycelium are formed, often in reproductive tissues.

CONTROL OF PLANT DISEASE

The development of plant disease depends upon a series of dynamic interactions between host plants and potential pathogens at cell, tissue and whole plant level. The outcome of such encounters will be affected by the initial health and status of the plants concerned and will be moderated by the environment. Owing to the wide range of variables involved the course of disease development is therefore often erratic, highly variable and unpredictable.

The occurrence of epidemics has caused very substantial economic losses and imposed serious food shortages throughout the world. For example, potato blight disease (caused by the fungus *Phytophthora infestans*) spread through Southern Ireland in the mid 1840s and devastated the potato crop which was, at that time, the staple food of the predominantly poor population. As a result that population was reduced by 25% due to emigration, or starvation and death, in just a few years. An epidemic such as this might have changed the course of history but, although usually less catastrophic, very serious losses also occur annually, with major impacts on agriculture and the environment. Dutch Elm disease, for example, has imposed changes on the landscape over much of Northern Europe by the devastation of trees which will take many hundreds of years to replace.

Such economic and human considerations, together with the application of modern agricultural practices, including the large-scale cultivation of plants in close proximity, has necessitated the development of both preventative and direct control methods. Serious diseases are often persistent in an area and may spread very rapidly. A good understanding of the infection cycles and life-styles of the pathogens concerned, in interaction with their hosts, is important in the identification of suitable combative measures. The recognition of the mode of infection and the mechanisms by which disease becomes established help to pin-point the possible sites and times of pathogen vulnerability leading to the development of strategies by which the disease cycle may be interrupted. In general terms the steps which are taken include avoidance of disease, direct control measures, usually involving chemical treatments, and biological control. In many instances combinations of methods are most effective and integrated control systems are most useful.

EVASION OF DISEASE

If it is possible to prevent host and pathogen coming into contact then disease will not develop. In most instances this is not possible but some measures can be used to limit the potential contact or to protect seedlings at the most vulnerable stages of development. In some cases simple means can be used most effectively to avoid disease or at least to aid in the limitation of disease establishment.

Quarantine

Strictly enforced Government legislation now prevents movements of diseased plant materials between countries in efforts to exclude non-indigenous pathogens and highly infective strains from previously un-invaded regions. Such controls operate on a global scale. The importation of plant materials, plant products and soils is restricted and as a result the spread of some diseases has been reduced. The introduction of a novel pathogen to an area may result in the massive development of infection, if it occurs, because the indigenous and hitherto unchallenged plant population is likely to have little inherent resistance.

Crop rotation

If a crop can be grown in an isolated area clean from potential pathogens then disease is unlikely to develop. In practice this would be considered a very unusual situation, particularly in agriculturally active regions. However, the rotation planting of immune and susceptible crops in a particular area is an extremely successful, easy and cheap way to limit many plant diseases. This is a widely used crop protection measure. The technique is particularly suitable for soil-borne fungi but in instances where resting spores are very long-lived little success will be gained. The method has proved especially useful for the control of club root of crucifers (*Plasmodiophora brassicae*), and take-all of cereals (*Gaeumannomyces graminis*).

Crop husbandry

An understanding of plant crop management together with the physiological mechanisms of disease production are essential to ensure successful husbandry in this context. Appropriate treatment of crops will ensure good plant health and minimise the influx of disease. The optimisation of various parameters, including for example, planting of seed at a suitable depth, degree of soil compaction, planting distances between rows and individuals within rows and the provision of suitable fertiliser levels, will maximise plant health and the rate of development. If it is possible to protect the seedlings through stages when they are most vulnerable to pathogen invasion a healthy crop is more likely to result. In some instances it is possible to control some environmental factors relatively easily, such as drainage and soil pH, and to great effect in counteracting the activities of selected fungal pathogens. For example, waterlogging increases the potential infection of young plants by some fungal species, e.g. *Pythium*, and *Plasmodiophora brassicae*, which causes clubroot disease, is most likely to thrive and be infective in acid soils.

It may be possible to avoid disease by the eradication of an alternate host. *Puccinia graminis* f. sp. *tritici* infects wheat and barberry alternately, so that the removal of barberry has proved to be an efficient control measure, by the interruption of the infection cycle and hence the protection of the wheat crop.

The addition of high levels of nitrogen fertiliser have been shown to increase the severity of disease in some crops. For example, high nitrate nitrogen levels increase the severity of vascular wilt diseases (*Verticillium albo-atrum*), and rust and mildew infections. The form of nitrogen and other mineral fertilisers may also have marked effects on the outcome of pathogenic invasion and increase the susceptibility of some plant species.

Hygiene

Avoidance of disease is achieved in the absence of the pathogen and therefore a useful protection measure is to maintain plants in conditions as

Table 2.3 Non-systemic fungicides

Fungicide	Usage
Thiram	Seed treatment for vascular wilt disease; celery leaf spot. Treatment against *Botrytis*, rust diseases and as a turf treatment.
Maneb	Potato blight; rusts of cereals; black spot of roses; also used as post-harvest treatment for soft fruits.
Zineb	Downy mildews, tomato blight and root rot; also used to protect mushroom crops.
Captan	*Botrytis* on lettuce and soft fruit; apple scab; tomato stem rot.
Captafol	Potato blight (*Phytophthora infestans*).
Dinocap	Powdery mildew diseases.

free from infected material as possible. At planting, clean, disease-free seed should preferably be used. It is interesting and an important observation, that even a very low proportion of infected seed material may rapidly give rise to disease outbreaks and therefore it is clear that this is an important step in disease prevention. Parent plants should be grown in clean conditions for seed production. Much seed is now treated with suitable chemicals for the control of pathogen proliferation. Hot water treatments (50°C) have been used successfully to remove cereal smut (*Ustilago*) infections from seed. Additionally, tubers, bulbs and cuttings require treatments to protect plants until they are well established.

Good hygiene in fields has proved to be essential for crop protection. The removal of potential sources of infection such as dead plant remains or previously infected material, which may act as sources of infection, is essential. The careful pruning (rouging) of infected plants is an especially useful treatment for long-lived plants, trees and shrubs, to remove sources of infection. This is a labour-intensive means of control but is a useful system providing that the rate of disease spread is not too rapid. If the spread is particularly fast then such treatment may not be cost-effective.

In greenhouse situations it may be practical to sterilise the soil to be used in cultivation programmes. In some instances steam can be piped through the soil in perforated lines beneath polythene sheeting and give good results. However, it is important to ensure even heating throughout the soil for treatment to be effective, particularly in the removal of vascular wilt fungi (*Fusarium* spp. and *Verticillium* spp.). If the soil temperature is raised too high the properties of the soil may be altered and levels of some salts released following such treatments may have toxic effects on the subsequent plant crop. In relatively small-scale concerns chemical fumigants may also be used to clear residual pathogens and potential inoculum for any new crop.

DIRECT CHEMICAL CONTROL

Chemical treatments have been very successful in the control and eradication of potential fungal pathogens, particularly with fruit and vegetable crops. Large amounts of fungicides are used annually to great effect, especially when used in combination with practices for disease avoidance. Some chemical compounds are applied to the aerial parts of plants and remain at those specific sites of action, protecting the plant from invasion. Success with such treatments is reliant on good timing of the application, to ensure maximum effectiveness, and changing conditions may necessitate subsequent applications later in the year (Table 2.3).

Other fungicides are systemic in their action (Table 2.4). These may be applied to the soil around plants or to the leaves, from where they are absorbed and translocated through the plant. Absorbtion through the roots is often highly efficient and although absorbtion through the plant cuticle may be less reliable this is also a useful system for the prevention or treatment of many diseases. A systemic fungicide becomes distributed

Table 2.4 Systemic fungicides

Fungicide	Usage
Antibiotics	
Streptomycin	Downy mildew on hops.
Cycloheximide	Wheat stem rust.
Kasugamycin	Rice blast disease (*Pyricularia*).
Benzimidazoles	
Benomyl	Very wide usage. Powdery mildews, apple scab, leaf spots, eyespot of cereals.
Thiabendazole	Post-harvest treatment.
Pyrimidines	
Ethirimol	Powdery mildews, especially *Erysiphe graminis* on barley.
Morpholines	
Tridemorph	Cereal mildews; leaf spots.
Others	
Metalaxyl	Treatment of soil-borne Oömycetes, often as seed treatment.
Chloroneb	Treatment of Oömycetes.

around the plant effectively but may not always reach new growth which occurs after the treatment.

The eradication of a fungal pathogen is complicated by the mycelial growth habit and may become very difficult once disease is established. Even very small portions of hyphae are able to grow and generate a mycelial colony so that it is important that any treatment is fully effective on all fragments. A surface treatment may kill external mycelial growth but any infective structures within the host tissue may survive and become re-established. Hyphae ramify thoroughly through plant tissues and so it is difficult to ensure suitable treatment. Fungal spores are also highly resistant, some may be resting spores which are particularly resilient. Spore walls are generally multilayered and thickened and are not readily absorbtive. However, at spore germination when the outer layers of the spore wall are breeched the young extending germ tubes are particularly vulnerable.

Many fungicides have therefore been developed to protect the plant from invasion in order to try to prevent establishment in internal tissues. Protective chemicals must obviously be applied prior to the arrival of the pathogen and are required in a form which will persist sufficiently to be effective at the appropriate sites. Many formulations have been tested, considering the physical and chemical properties of the compounds used. Applications as gels, pastes, wettable powders, sprays and dips have been used and much research continues to establish cost-effective treatments.

Inorganic fungicides

Copper

Copper treatment is one of the oldest and most famous treatments for the protection of plants. Bordeaux mixture, a combination of copper sulphate and calcium hydroxide (lime), was used in the 1880s as a foliar spray to treat powdery mildew disease in the vineyards of France. However, copper ions in solution are also toxic to plants and insoluble copper salts are now used. Some plants are particularly sensitive, especially in wet weather. For example, the leaves of some apple varieties become scorched and photosynthesis is then impaired. The formulation of Bordeaux mixture may be changed by the addition of extra lime to protect plants from the toxic action but today this treatment is not used. Cuprous oxide (Cu_2O) is employed as a seed treatment and cuprous oxychloride ($CuCl_2.3Cu(OH)_2$) as a foliar spray or dip treatment. Some copper derivatives are only used as timber treatments because of their extreme phytotoxicity.

Copper-based treatments are effective against a wide spectrum of fungal pathogens. Their action is fungistatic and prevents spore germination. However, if the spores are well washed after treatment the copper is removed and the spores may subsequently germinate. It is thought that the effects on mycelium are as a result of reaction with sulphydryl groups of amino acids, which then denature proteins and enzymes.

Sulphur

Elemental sulphur is useful for the control of a wide range of fungal diseases, particularly powdery mildews (Erysiphales), leaf blights, rusts and apple scab. It is mostly used as a fine powder or as a paste. Some plants, such as tomatoes, and apple varieties, are sensitive to sulphur under dry conditions when leaf burn occurs, with loss of chlorophyll (chlorosis) and premature senescence of leaves and fruit drop. The action is thought to be fungitoxic and sulphur is suggested to interfere with electron transport through the cytochromes. Hydrogen sulphide is formed which denatures essential proteins.

Mercury

Compounds containing mercury are effective against fungal pathogens. Mercurous chloride (calomel; Hg_2Cl_2) has been used as a grass treatment and is especially suitable for the eradication of clubroot disease (*Plasmodiophora*). Mercuric chloride ($HgCl_2$) has also been used in aqueous solution as a dip or seed treatment. Phenyl mercury chloride is an effective foliar spray for apple canker (*Nectria galligena*). However, these compounds are extremely toxic to mammals and are therefore not generally in use today for such environmental reasons.

Non-systemic organic fungicides

Dithiocarbamates

The dithiocarbamates, derivatives of dithiocarbamic acid, are the most widely used of the fungicidal preparations which are commercially available today. The fungitoxic action is derived from the inactivation of sulphydryl groups in amino acids, which inhibits the production of proteins and important enzymes in the pathogen.

Thiram (tetramethylthiuram disulphate; Fig. 2.18) is used mainly as a seed dressing against vascular wilts and *Botrytis*. It is used as a very successful seed soak (24 h at 30°C) to protect against deep-seated, seed-borne pathogens, such as leaf spot of celery crops and as a control for *Alternaria* diseases of carrots. Maneb and zineb (Fig. 2.18), which are zinc and manganese salts of dithiocarbamate, are used widely for the treatment of vegetable crops, either as seed treatments or often as post-harvest dips. These are particularly useful for the control of some rust diseases and powdery mildews. However, the possible effects of dithiocarbamates on mammals currently give rise to concern.

Phthalimides

This group of fungicides is also very widely used and is agriculturally important. Captan (Fig. 2.18) is a very effective, broad spectrum, insoluble fungicide. It is used for the treatment of apple scab, tomato blight and *Botrytis* diseases. It is taken up into fungi very readily and the metabolism of these compounds results in interference with the production of amino acids and proteins in the cells. Captafol is a chemically similar compound

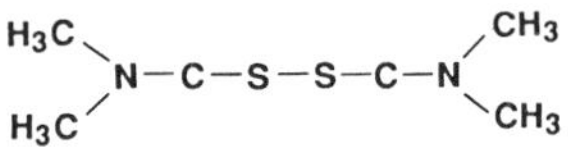

Thiram

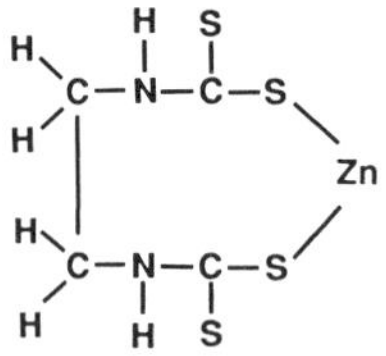

Zineb

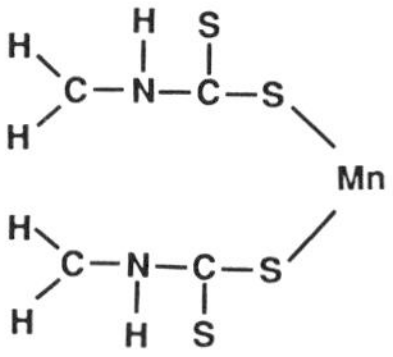

Maneb

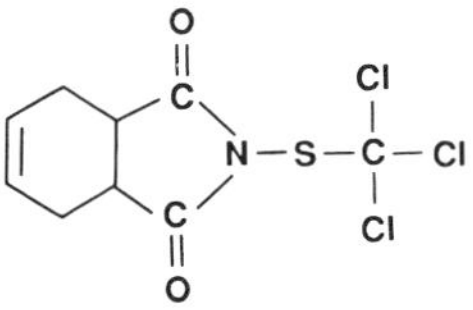

Captan

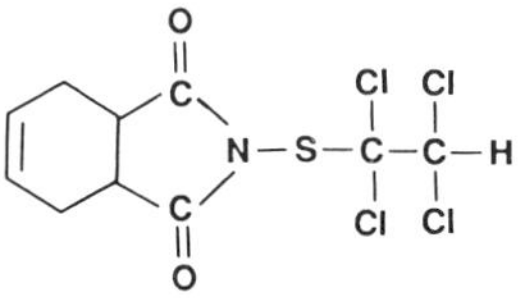

Captafol

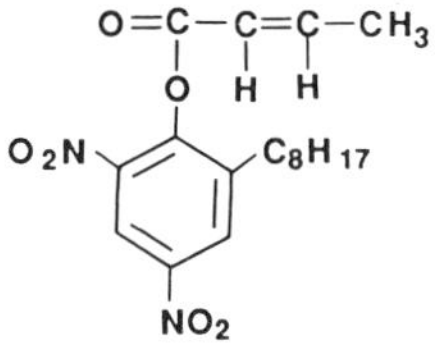

Dinocap

Fig. 2.18 Chemical structures of some non-systemic fungicides.

which is used widely for the control of potato blight and peach leaf curl (*Taphrina deformans*). Folpet is another similar compound but has a more phytotoxic action.

Dinitrophenols

The dinitrophenols are used as fungicides and also as insecticides and herbicides. These compounds uncouple oxidation from phosphorylation in mitochondria and are therefore highly efficient fungicides. Dinocap is used particularly to control powdery mildew diseases (*Erysiphales*), especially on apples and soft fruit. It is often used, in wettable powder form, as a spray in fruit growing areas.

Systemic fungicides

Antibiotics

Streptomycin which is produced by the actinomycete *Streptomyces griseus* is soluble in water and is readily absorbed by plant roots. The main fungicidal action is interference with mitochondrial protein synthesis. It is used for the treatment of downy mildew disease of hops, although it lowers the rate of chlorophyll synthesis and may therefore be slightly phytotoxic under some conditions.

Other antibiotics produced by species of Streptomycetes are effective as fungicides but are also phytotoxic and are, therefore, used only in limited applications. Cycloheximide, which inhibits protein synthesis, is used to control stem rust on wheat. Kasugamycin is used against rice blast disease, very effectively, since this has lower phytotoxicity than other antibiotics.

Benzimidazoles

Benomyl (Fig. 2.19) is a very useful, broad spectrum, antifungal agent which has been used for the successful control of many pathogenic fungi. It is used as a foliar spray as a protectant from powdery mildews, as a seed treatment for wheat smut, as a dip for bulbs to protect against *Fusarium* spp., and *Sclerotinia*, and sometimes as a soil treatment against vascular wilts. The toxic action is by interference with nuclear division and blockage of further mycelial growth. The formation of β-tubulin is prevented so that the microtubules, normally present when chromosomes separate, do not form and mitosis is therefore blocked. Thiabendazole, thiophanate and thiophanate-methyl are also used as very effective protectants and treatments for fungal infections.

The benzimidazoles have proved commercially effective in disease control but resistance has been identified in *Botrytis cinerea*, *Venturia inaequalis*

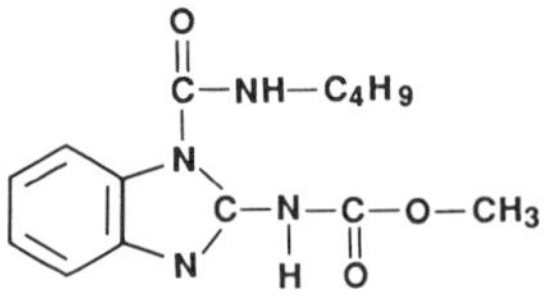

Benomyl

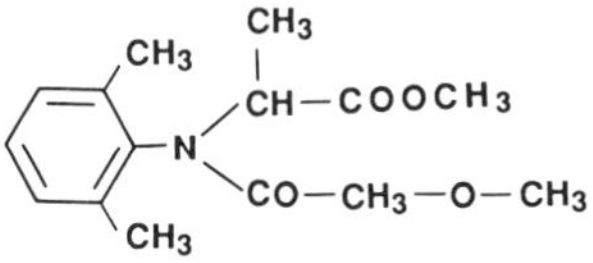

Metalaxyl

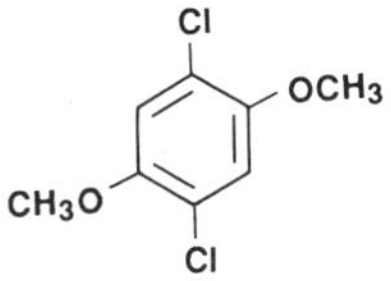

Chloroneb

Fig. 2.19 Chemical structures of some systemic fungicides.

and *Pseudocercosporella herpotrichoides* in field situations. Benzimidazole resistance is very stable and is probably a major gene resistance so that the effectiveness of treatments can be easily lost in agricultural situations.

Pyrimidines

This group of fungicides is most widely used for the control of powdery mildews. Ethirimol is highly effective against *Erysiphe graminis* infections on barley. It is absorbed readily through plant roots and is translocated through to leaf margins. The toxic action is probably by the interference with nucleotide synthesis. This group also includes the fungicides triarimol and dimethimirol.

Morpholines

The most important fungicides in this group are tridemorph and dodemorph and these are used against powdery mildew infections and leaf spot diseases. These compounds interfere with sterol biosynthesis (ergosterol) in fungi and as a result membrane function is disrupted.

Other systemic fungicides

Metalaxyl (Fig. 2.19) is a useful fungicide (an alanine derivative) which is effective against Oömycete fungi and can be used in the treatment of potato blight (*Phytophthora infestans*). It is also used as a protective treatment against a number of fruit diseases.

Chloroneb is also efficient for the treatment of soil-borne Oömycete fungi. The toxic action is limited to *Rhizoctonia* and *Pythium* but it is effective as a seed treatment in preventing seedling blights. It is persistent in soil and protection to the root system is therefore long lasting.

Fungicide resistance

The use of systemic fungicides has increased the occurrence of resistance in pathogen populations. This has occurred because the toxic action is very site-specific in the pathogen and resistant strains have arisen by the selection of resistant individuals or as the result of single mutations. Resistance is less likely to arise if compounds which are less site-specific, or multisite fungicides, are used. Some of the more important pathogens, which are difficult to control, have developed resistance to some systemic fungicides, e.g. *Botrytis, Verticillium, Fusarium, Colletotrichum* and *Ustilago*.

In physiological terms, resistance to fungicides may develop in a variety of ways, and it is obviously important to understand the mechanism of the toxic action in order to understand resistance. The pathogen may become less permeable to the compound concerned. The compound may become bound, to cell or wall materials, and therefore not exhibit toxic activity in the fungus. There may be changes in the metabolism of the compound in the fungus so that toxic intermediates are not formed and the fungicidal action is lost. If the action is by virtue of a metabolic block in a biochemical pathway, then the fungus may become able to bypass that block and circumvent the action of the fungicide. Alternatively, if the action is to block the activity of a particular enzyme then the fungus may produce more of that enzyme and compensate for the fungicidal action.

Resistance to some fungicides can be aquired by changes in major genes within a pathogen. This type of resistance can develop very rapidly and be very stable and will therefore seriously affect the use of these chemicals, repeatedly, in the field. Resistance to other compounds may be multi- or polygenic and may develop only over a long period of exposure to those chemicals (Georgopoulos, 1987).

Knowledge of resistance to systemic fungicides and its development is important. It is now realised that such chemicals should be used in combination and that chemical treatments can be alternated, effectively minimising the possibilities of resistance developing in a particular area. Repeated applications of the same fungicides, especially in relatively quick succession, provide the fungus with a perfect opportunity to establish resistance. It is now apparent that a more likely approach to plant disease control in the future will be through programmes of Integrated Disease Management (IDM), the use of a number of fungicides in conjunction with resistant cultivars (Baldwin and Rathmell, 1988). The diversification of the control systems employed is obviously beneficial and the use of limited quantities of fungicides is important in both economic and environmental terms.

Disease assessment and fungicide testing

There are a number of criteria which define a successful fungicide for plant protection. It is important that the degree of disease control which is provided is sufficient to form an economically viable treatment. It is also important that the control can be consistently achieved with minimal loss of fungicidal activity under changing environmental conditions. The treatments should be easy to apply and as non-hazardous as possible to humans and animals. The toxic effect should not extend to the crop plant concerned, to other local plants or to wild life. A farmer will also require cost-effective results.

The criteria for safety and for product efficiency have resulted in many, often lengthy, assessment programmes to test and screen new products and new formulations. Large numbers of plants are involved, and the effects are often traced over several growing seasons. It is necessary to assess the extent and the impact of disease in order to estimate the degree of control afforded by any fungicide under test. In small-scale experiments, under very controlled conditions (e.g. in a greenhouse), it may be possible and useful to measure the effects of a pathogen on the growth of infected plants directly. However, on an agricultural scale it is more usual to use subjective means for the assessment of disease severity and potential crop losses.

Assessment keys are often employed to estimate the extent of disease for parts of plants, such as leaves or organs, or for whole plants. Drawings are used to indicate the percentage infected area for a particular crop species, so that assessors have a common base from which to work. Such keys have been constructed to provide reference scales so that comparisons between crops at different sites may be compared from year to year.

In terms of fungicide testing it may also be necessary to isolate the pathogen concerned and to create an inoculum with which to infect treated crop plants. Each host/pathogen combination will present different problems in this context (Hickey, 1986) and each must undergo specific trials. The production of a viable inoculum may present some difficulties if the pathogen cannot be cultured, and induced to sporulate in laboratory medium. It is also important that subsequent inoculations are designed to mimic likely field conditions. Small-scale trials are often superseded by larger experiments and a great deal of testing is required. It is important, particularly for large-scale field trials, that maximum information is gained from the experiments carried out and the planning of such work is vital (Nelson, 1986). Much of the data analysis can now be efficiently handled by computer systems.

BIOLOGICAL CONTROL

The term biological control will be used in this context to describe the use of living organisms or their products, to combat the damaging activities of other organisms, in this case potential pests or pathogens of plants. Pests do have natural enemies and biological control systems are designed to manipulate and enhance these in order to reduce the pest populations and to limit their activities.

This can be achieved in a number of ways. Many pathogenic fungi are poor competitors and may be quickly excluded from a site, such as a leaf surface, if species which are more antagonistic are present. Some such combative fungi are highly aggressive and produce toxic metabolites which quickly affect less competitive individuals. Additionally, some fungal species are able to parasitise and directly attack insects, nematode pests, or indeed, other fungal pathogens. The use of various inoculation systems to encourage these interactions has been shown to enhance the effectiveness of such natural biological control.

Biological control systems are preferred to the use of chemicals and in recent years a great deal of research activity has been directed towards the development of efficient and reliable systems. In environmental terms the effects, and particularly the longer term consequences of biological control, are much less damaging than the routine use of pesticides or fungicides. Treatments can be economic and cost-effective providing that good control can be established, particularly when repeated chemical spraying is required during the growing season. Biological systems have great potential in the control of soil-borne microbes which are particularly difficult to treat by spraying alone. Any reduction in the application of chemical pesticides is welcome. Additionally, since the pest and antagonist are developing within a natural situation, there will be co-evolution between them and the potential for the development of stable resistance to the biological control agent is much reduced from that of chemical treatments (Briese, 1986).

It is interesting to note, however, that relatively few instances of biological control have been effectively implemented in commercial field situations to date. Much research has led to the development of efficient systems on laboratory or greenhouse scales where the environment is highly controlled and predictable. However, once trials are scaled up the extreme variability and unpredictability of natural field sites can lead to problems. In theory, biological control, once established in a balanced situation, could be self-perpetuating but in practice such systems are more often used as part of an integrated control programme (IPM: Integrated Pest Management) with pesticides and fungicides as supplementary treatments (Way, 1986; Burge, 1988) in much reduced quantities.

Fungi as biological control agents

Mechanisms of fungal biological control

A number of the natural characteristics of the life-styles of fungi confer qualities which make them potentially useful biological control agents against a variety of pests and pathogens of plants.

(a) Competitive ability

As mentioned above, the competitive activities of some fungi render them highly antagonistic and ideal as potential combative organisms. In theory, at least, increased levels of such species introduced to leaf surfaces would lower the potential rates of infection from other pathogenic fungal species, which tend to be less competitive and aggressive.

Antagonistic activity is also apparent in the soil environment, in close proximity to root tissues. The fungus *Phialophora graminicola* has been shown to develop on root surfaces of cereals, making use of nutrients from the senescing outer cortical cells. This fungus competes with the causal agent of take-all disease of cereals, *Gaeumannomyces graminis*, and in this way effectively reduces the incidence of the disease (Kirk and Deacon, 1987).

(b) Antibiosis

Antibiosis is defined as the inhibition of the growth of a microbe by substances produced and liberated by another microbe. The term most usually refers to antibiotic activity. However, whilst it is relatively easy to prove that an organism produces antibiotics in culture it is difficult to ascertain whether similar production occurs under natural conditions (Williams and Vickers, 1986; Fravel, 1988), and even more difficult to establish a role for these compounds in competition within natural environments. Little antibiotic activity has been detected in the soil environment and it has been suggested that these compounds are degraded or adsorbed on to soil particles. Additionally, a strain newly isolated from the environment may demonstrate antibiotic activity although this ability may be rapidly lost on subsequent subculture. However, the production of antibiotics by *Trichoderma* species has been implicated in the biological control of *Pythium* species on pea seeds (Lifshitz *et al.*, 1986), particularly since the antagonistic action was very rapid. Antibiotic production by *Pseudomonas fluorescens* has also been cited in inhibition of *Gaeumannomyces graminis* var. *tritici* and *Pythium ultimum* (Fravel, 1988).

(c) Mycoparasitism

The term used to describe the direct parasitism of one parasite (usually a primary parasite) by another is hyperparasitism. Fungal preparations are now used and marketed commercially for control of insect pests and nematodes, particularly in controlled, greenhouse conditions (see below).

Fungi which derive most or all their nutrients from another fungus are termed mycoparasites. All the major fungal taxonomic groups contain mycoparasitic species. Biotrophic mycoparasites may have relatively long-term associations with living cells of the invaded species; however, necrotrophic mycoparasites often kill the target fungal-host cells prior to penetration and invasion (Whipps *et al.*, 1988). Interactions between hyphae may be initiated and target hyphae accurately located from a distance. The mycoparasite may coil around the potential host hyphae after contact has been made and in some cases appressoria-like structures are formed prior to host penetration. Subsequently cytoplasmic granulation and degradation may occur in host cells and mycoparasitic species of *Trichoderma* and *Pythium* have been shown to produce wall-lysing enzymes which aid penetration of the target species (Whipps *et al.*, 1988; Lewis *et al.*, 1990).

Some mycoparasitic species are adapted to the exploitation of fungal spores, either asexual or sexual resting spores. Exploitation of this ability, particularly the mycoparasitism of sclerotia, would be of great agricultural and horticultural interest since these structures are extremely long-lived and very difficult to eradicate from soil. The hyphomycete fungus *Sporidesmium sclerotivorum* destroys sclerotia of *Sclerotinia* spp. and *Botrytis cinerea*. This mycoparasite forms haustoria within cells of the infected sclerotia and subsequently grows out as mycelium, giving rise to spore formation and proliferating into the soil from where infections of other sclerotia can occur. In addition, the fungus is most active in lower soils where moisture levels are high. This coincides with the region in which sclerotia are most persistent. It has been shown, in field conditions, that long-lasting biological control of *Sclerotinia minor* on lettuce plants can be economically attained by inoculating soils with spores from *S. sclerotivorum* (Adams, 1990).

Natural mycoparasitism occurs on plant surfaces and the exploitation of this type of combative activity has been suggested as a useful alternative to fungicide treatment (Sundheim, 1986), particularly in the case of biotrophic plant pathogens. Although instances of success and potential exploitation have been reported it may be difficult to obtain consistent results due to interactions arising between the primary host, the pathogen and the hyperparasite. In theoretical terms there are few problems in the use of such systems but there are more practical and ecological considerations, particularly if the species to be inoculated has a wide spectrum of antagonistic activity and a high reproductive capacity.

The use of fungi for biological control

Fungi have been tested as biological control agents for various plant pests and pathogens with varying degrees of success. All these potential areas of application are now the focus for continuing research and development.

(a) Insect control

Insects cause severe crop losses throughout the world and although the use of pesticides can provide good control the development of resistance to such treatments is cause for concern. Entomogenous fungi, species which attack and parasitise insects, have attracted a great deal of attention as potential biocontrol agents. The host ranges of some such fungi may be very limited but many others have a wide host range.

Conidia of entomogenous fungi which reach the outer surface of an insect host, germinate and penetrate the integument by a combination of enzymic degradation and mechanical pressure. Many species enter directly into the insect but some species (e.g. *Metarhizium anisopliae*) form appressoria prior to penetration. Inside the host the fungus grows as small yeast-like bodies which multiply by budding and circulate around the insect. These structures compete with the host for nutrients and also produce toxic metabolites which are important factors in determining the death of the insect. Following host death the fungus proliferates and eventually bursts through the insect cuticle to sporulate and conidia are liberated from the body, often wind dispersed or washed away by rain splash.

Fungal entomopathogens have been most useful as insecticides in glasshouse environments (Zimmermann, 1986; Gillespie and Moorhouse, 1990). Fungal spores can be isolated from infected insects for testing. *Metarhizium* species have been used with effect to combat various insect pathogens of plants and are particularly useful in the treatment of pests which infest soil. For example, *M. anisopliae* has been used to limit the numbers of surviving black vine weevil larvae in controlled environments. *Verticillium lecanii* has been used to control red spider mite and aphids in glasshouses and *Beauveria bassiana* has been used to combat the Colorado potato beetle. The long-term effects of such treatments have proved to be most useful although the initial stages of establishment by most of these biological insecticides are rather slow.

Toxins play an important role in the killing of host species and it has been suggested that these compounds may be formulated as insecticides. However, in most cases the toxins act only on specific hosts, acting particularly within the insect gut, and therefore making delivery to the site of action difficult. Of more potential importance are the toxic metabolites produced by endophytic fungi which live entirely within plant tissues without causing the development of any symptoms. Compounds liberated by these fungi render plant tissues toxic to insects, particularly affecting larval stages and effectively disrupting the reproductive potential of insect pests which feed on such infected tissues (Carroll, 1986; Clay, 1989). This aspect of fungal endophytes is considered in further detail in Chapter 5.

(b) Control of plant parasitic nematodes

Nematophagous fungi produce a range of different specialised structures, called traps, for the capture of nematodes in soil. Some form sticky knobs which attach to the nematode (*Dactylella ellipsospora*), while others form networks of cells in rings which snap around the nematode body to ensnare it (*Arthrobotrys dactyloides*). From these adhesions the fungus grows into the body of the nematode to derive nutrients from it and, once these have been depleted, to sporulate through the host surface.

Nematodes can be particularly damaging to plant crops and their reproductive potential is high, populations often increasing several hundred-fold in a season. Although the potential application of nematophagous fungi has been recognised (Kerry, 1990), *Paecilomyces lilacinus* is the only species to have been tested in field trials. Biological control agents which can effectively eradicate both adults and eggs are required, preferrably with a host range which is not too restrictive. It is most likely that nematode control by a fungal agent might effectively be combined with other chemical treatments (nematicides) to allow more rapid rotation of crops in fields prone to infestation.

(c) Control of fungi pathogenic to plants

Plant pathogenic fungi are subject to mycoparasitism. Some such mycoparasites have been tested under controlled conditions to ascertain their potential as biological control agents. There are some practical problems in the deployment of these, but theoretically such fungi are a very attractive alternative to fungicide treatments.

The control of powdery mildews has been tested under specified conditions (Sundheim, 1986). When applied as conidial suspensions, *Ampelomyces quisqualis* has been shown to penetrate and parasitise the powdery mildew *Sphaerotheca fuliginea* on greenhouse-grown cucumber plants. This mycoparasitism may be useful for the long-term control of a number of species of the Erysiphaceae in the field.

A number of mycoparasites of rust fungi have also been tested. *Verticillium lecani* has been shown to control bean rust disease adequately in controlled environments but is less successful in the field. A number of other species (e.g. *Aphanocladium album* and *Cladosporium* spp.) have also been considered as biological control agents, but in many cases further characterisation of all the factors which influence the host–parasite interaction is required.

Species of *Trichoderma* which are able to grow rapidly and readily have been used in trials for the control of phylloplane fungi (Tronsmo, 1986). *Trichoderma* spp. that liberate cell wall lysing enzymes as part of the antagonistic activity are highly competitive for nutrients. *T. viride* has been used to combat *Chondrostereum purpureum* infections on timber.

(d) Control of weed plants (mycoherbicides)

Some preparations of fungal pathogens of plants have recently been marketed for the control of weed plants. Following some successful applications and the identification of other host–pathogen combinations it seems likely that further treatments will become available in the next few years (Templeton and Heiny, 1990), although some research is still required. The potential of toxins isolated from culture filtrates from pathogenic species has also to be investigated further in this context.

Colletotrichum gloeosporoides f. sp. *aeschynomene* has been licenced in the USA for control of Northern joint vetch in areas of soybean cultivation. The application of the dried conidia in the marketed product (Collego) gives a very rapid response and a very high level of control. *Phytophthora palmivora* has been used to successfully control milkweed vine (*Morrenia odorata*) among citrus plants. However, the potential marketability of this product is limited because the viability of the preparations to be applied is very short.

Work carried out with bracken (*Pteridium aquilinum*), a serious weed which is difficult and costly to eradicate from woodland areas, has pinpointed (Burge, 1988) some difficulties with mycoherbicide systems. The natural pathogen *Phoma aquilina* can attack some bracken stands but in other cases the natural defense systems of the plant prevent massive fungal invasion. In other instances it is the shelf-life of the preparation which renders a potential control agent unsuitable for marketing.

(e) The role of mycorrhizas

Plant roots often form symbiotic relationships, known as mycorrhizas, with fungal species. These associations improve the general health of the plants concerned by increasing the uptake of essential nutrients into the host plants, improving water relations and increasing their natural resistance to pathogens. These relationships are considered in greater detail in Chapter 5.

Mycorrhizas, particularly those in which a sheath of fungal material forms around the root surface (ectomycorrhizas), act as a physical barrier which protects roots from invasion by soil-borne pathogens. These ectomycorrhizas also promote the establishment and increase the survival of tree and shrub seedlings and stimulate growth, particularly in harsh environments. Inoculation systems have been used to encourage the formation of such relationships. Indeed the establishment of some trees (e.g. pines) is dependent on the formation of mycorrhizas. It is particularly important that mycorrhizal associations develop soon after seedling germination, since it is at this stage that individuals are most vulnerable to pathogen invasion and the most useful fungal species are those which are aggressive and establish functional associations quickly.

Field trials have been carried out at sites in the USA which have highlighted the great potential of mycorrhizas in the promotion of increased biomass production and in resistance to pathogen attacks. Marx and Cordell (1990) report trials with *Pisolithus tinctorius* and *Pinus* species which have been successful.

The enhanced health of plants infected with vesicular-arbuscular mycorrhizas (endomycorrhizas) also increases disease resistance. No outer protective sheath of fungal tissue is present in this type of relationship and so this action is likely to have a more physiological basis. These fungi cannot be grown in pure culture and, as yet, the potential marketability of treatments cannot be realised. Future prospects are under investigation and may prove to be good for controlled or small-scale inoculations (Gianinazzi *et al.*, 1990), especially where soils known to contain fungal propagules can be used.

Commercial application of fungal biological control agents

Biological control can be achieved by the enhancement of natural host–pathogen interactions or by the encouragement of other less usual combinations by the provision of suitable conditions. Many apparent successes have been reported in laboratory trials but the use of biological control systems is limited, especially in field situations, and not yet fully commercial. A range of factors which must be considered in relation to the development of a potential biocontrol agent are listed in Table 2.5.

Any biocontrol agent must fulfil a number of important criteria. The potential fungus must have good antagonistic ability. It must be able to compete aggressively for nutrients and space and to colonise the rhizosphere or phylloplane environment rapidly after inoculation. The speed of establishment is particularly important for the protection of young seedlings. Tolerance of the prevailing environmental conditions is also vital to the proliferation, long-term survival and dominance of the potential agent. It is, therefore, important that tests be carried out under natural conditions of temperature, moisture, pH, ultraviolet light, etc., and also in the relevant soil, taking into account soil texture, levels of humic material, degree of compaction and also any indigenous microbial population which may also be antagonistic. Although many trials have been successful in small-scale highly controlled environments bulk application in field conditions often leads to the breakdown of the control system due to unexpected detrimental variables (Powell and Faull, 1990) or is uneconomic due to practical considerations.

It has been shown in tests with fungal isolates that in most cases inoculations may be made effectively, on to plant surfaces or soils, using either vegetative mycelium or spores. Theoretically the production of either of these can be carried out relatively easily on a large scale, so

Table 2.5 Criteria to be considered in the commercial development of a biological control agent

Safety	Microbe and/or toxic products must have low toxicity to man and animals. No adverse effects on crop plants to be treated.
Field persistence	Must persist in field conditions until good level of control achieved. Aggressive species colonising quickly may be advantageous. (If to be used as part of an integrated control programme must be resistant to any chemicals used.)
Environmental criteria	Organism should be indigenous and suited to the prevailing environmental conditions (soil type, temperature, pH, etc.). Should not be genetically engineered.
Stability	Organism and/or formulation should have a practical shelf-life under equitable storage conditions. Formulation should be easily stored and distributed. Reproducible results should be obtained in use.
Formulation	Should be easy to apply on a large scale, without the need for specialist or expensive equipment.
Production	Production systems should be conventional and applicable on a suitably large scale for the potential market.
Cost	Must be a cost-effective treatment for crops. Should compare favourably with alternative control systems (e.g. chemical control agents).

that the preparation of suitable inoculum is not usually a problem. In general, production of large amounts of vegetative biomass can be achieved using liquid cultures. Spore production is more usually carried out on a smaller scale utilizing solid or semi-solid medium. Greater difficulties may be encountered in guaranteeing the shelf-life of any preparation. It is important that both viability and the natural qualities of the fungus are maintained throughout the distribution time and under standard storage conditions. The product must also be supplied in a suitable form for application to the crop. Many formulations are distributed as dry powders. These contain spores mixed with carrier materials to increase the bulk, for ease of application, and wetting agents to ensure efficient dispersion on the material to be treated. Liquid formulations are also attractive since these can be very easily applied (Powell and Faull, 1990). Some agents can be efficiently applied in pellet form or as gels. For commercial purposes it is vital that the preparation is reliable and reproducible in its action, safe to use and with a minimum requirement for specialist equipment. Initially at the research and development level, a great deal of expenditure is required for testing and formulating potential agents into marketable products at a reasonable cost to the farmer.

Once a potential agent has been identified it becomes essential to preserve the features which make it valuable. With the development of strain improvement programmes and the application of molecular biological techniques it may become possible to manipulate the qualities of natural biological control agents to make these a more versatile and commercially viable option for disease control.

Suppressive soils

Conditions which are known to favour the development and infective ability of some soil-borne pathogens are well known and the development of plant disease under such conditions can be reliably predicted. The massive proliferation of some pathogens occurs in such conducive soil conditions. However, in some soils, termed suppressive soils, the development and infectivity of these species is much less. The reason for this is not always clear and may be the result of interactions between both biotic and abiotic factors.

Antagonistic microbes such as *Trichoderma* and *Penicillium* spp. are often found in suppressive soils. The activities of *Trichoderma* spp. are inhibitory to a range of fungi. *Trichoderma harzianum* mycoparasitises the mycelium of some other fungal species in soils, e.g. *Rhizoctonia* and *Sclerotinia*, and inhibits the growth of others, e.g. *Pythium* and *Fusarium*. The fungus *Gliocladium virens* mycoparasitises mycelium of *Sclerotinia sclerotiorum*. Inoculation of soils, to increase the levels of such antagonistic species has been shown to be a useful control measure for pathogens. Some suppressive soils mediate against pathogen development by virtue of abiotic factors too and in such cases the mixing of soil types may be advantageous.

Integrated Plant Protection

At present biological control systems appear to have great potential for the control of plant disease and pest management although this has not yet been realised, particularly on a fully commercial or agricultural basis. However, the realisation of the need to reduce the levels of pesticides and fungicides released into the environment has provided greater impetus for further research and development in this area. Most certainly there is currently both a need and a willingness to limit the levels of chemical control which are in operation but there is also a general lack of suitable reliable alternative control measures.

Major problems for biological control arise from the breakdown of control, particularly when applied on a large scale in the natural environment. Treatments in glasshouse or nursery situations tend to be more stable and persistent. However, the integration of chemical with biological control measures at a site may prove most cost-effective in both economic and environmental terms. Additionally, the use of plants with partial resistance combined with the application of suitable biological control agents may provide good plant protection, with the possibility of chemical applications, in greatly reduced levels, if control is not adequate in a particular situation. It is likely that future developments will be in the area of 'integrated plant protection'.

Fungal–plant confrontation 3

INTRODUCTION

Fungi are invasive heterotrophs and are well adapted for the penetration and efficient utilisation of solid substrates. A spore may carry with it some stored nutrient reserves which will allow the limited outgrowth of germ tubes, given a source of moisture and appropriate physiological conditions. However, continued growth, to form mycelium, must be fuelled by the uptake of nutrients from the external environment. This is particularly critical for those species (obligate biotrophs, e.g. rusts, Uredinales; smuts, Ustilaginales; powdery mildews, Erysiphaceae; etc.) which are not capable of independent saprophytic growth and rely solely on host species for nutrient supplies. A range of diverse, extracellular enzymes are released from growing hyphae and these break down complex polymers locally. As nutrients become exhausted growing apices extend and proliferate further into the surrounding substrate. This penetrating, invasive life-style is clearly an advantage for efficient colonisation. However, if the substrate is a living organism then the situation is more complex.

Any encounter between a fungal pathogen and a potential host plant will immediately give rise to interactions between the two individuals, beginning when a spore settles on the plant surface or when hyphal contact is made with roots in the soil, but before any physiological contact is established between them. The sequence of events which follow involve an integrated series of mechanisms which depend directly on the properties and capabilities of both the organisms involved. A range of consequences may subsequently ensue, from unsuccessful host penetration, to the alteration of host structure and function or even death of the plant.

Those fungi that are capable of host plant invasion and infection are often grouped in terms of the general means by which they derive nutrients from the host. Biotrophic fungi exploit the resources provided by a living host. These fungi cause minimal damage on invasion and spread only slowly through the infected plant. However, they often spread as spores and even a relatively small inoculum may give rise to many entry points on the host. After invasion, physiological contact is set up but the host cell membranes are not penetrated and host cells do not die. Within this group of fungi are the specialised biotrophs (rusts, powdery mildews and downy mildews) which have the ability to form haustoria, structures thought to be associated with nutrient absorption.

Nectrotrophic fungi on the other hand are very destructive to the host and often kill cells before invasion. They do not maintain contact with a living host but kill cells in advance of their spread, often by the liberation of toxins and degrading enzymes. After initial invasion the growth of a necrotroph may remain very localised within host tissue, particularly if the host is still living. When death has occurred in more deeply seated tissues the fungus can then reproduce and spread. Hemibiotrophic fungi may grow and initially invade a host plant as biotrophs, even forming haustoria in host cells, but later become necrotrophic and live saprophytically on dead host tissues.

The means by which fungi invade and interact with potential hosts are numerous, diverse and complex, often involving a range of integrated mechanisms.

FUNGAL INVASION

INITIAL CONTACT

Prior to fungal penetration of a plant many factors influence the sequence of events from germ tube emergence to attachment, adhesion, appressorium development and growth on the plant surface. These may relate directly to endogenous metabolic components of the spores themselves or may be exogenous factors such as influences from the environment, competition from other microbes or factors relating to the host plant such as leaf age, cultivar type and physiological condition.

Spores often reach the host plant surface in a dormant state. Spore dormancy may be broken by chemical signals from the host plant, e.g. flavour compounds and fragrances (French, 1985). Leaf exudates may theoretically be stimulatory to the germination of spores by virtue of the increased nutrient levels which they may provide, but equally moisture films may contain inhibitory substances leached from the leaf surface. In the soil environment too, compounds are liberated from root surfaces which have been shown to act as selective depressors of spore germination. For example, in *Fusarium* wilt of peas, exposure to exudates from pea plants reduced the germination of spores from races of *Fusarium oxysporum* f. sp. *pisi*, to which the host variety was resistant (Buxton, 1957).

Within the rhizosphere, the region immediately surrounding the plant root and which is influenced by root activity, there are often high levels of competition between microbes. Indeed the plants themselves apparently encourage such competition. Plant root exudates, usually richer in nutrient components than the surrounding soil, increase the proliferation of competing microbes. Many pathogenic species are poor competitors and this may limit the levels of infection by potential pathogens. Water films in the soil also allow the chemical attraction of motile zoospores and encourage encystment prior to germination. The formation of the cyst wall in *Phytophthora* species, which is essentially a newly formed glycoprotein coat, results in attachment to the host surface.

On leaf surfaces there may be less competition than in the rhizosphere and external biotic factors are less important influences on spore germination. However, the phylloplane is an extreme stress environment, subject to high levels of desiccation alternating with extreme wetting and exposure to ultra-violet light. In general, spore germination requires a supply of free water, and equitable temperatures. Saprophytic species germinate and grow extensively on leaf surfaces but do not apparently respond to topography or attempt to penetrate leaves. Spores of pathogenic species, however, will germinate becoming firmly attached to the leaf surface at outgrowth, and are often induced to form morphologically differentiated infection structures. Pathogen spores will often germinate in water droplets alone, although this may not occur on leaf surfaces owing to inhibition by the bacterial flora. Failure of germination owing to the lack of specific nutrient requirements has also been recorded, e.g. mutants of *Venturia inaequalis*, which cause applescab, have been isolated, which display very specific requirements for vitamins, bases and amino acids (Keitt and Boone, 1954).

After dispersal uredospores of the rust *Puccinia graminis* are capable of very rapid germination but are also very environmentally sensitive. The recognition of such environmental stimuli induces the morphological changes associated with leaf surface adhesion and germination. Spores will also influence the environment themselves. Endogenous self-inhibitors prevent spore germination at high densities in rust fungi. Factors which influence these processes have been studied at length in a series of elegant experiments with the rust fungi (Uredinales).

Uredospores, one of five spore types in the rust fungi, *Uromyces* spp. and *Puccinia* spp., have an essential role in the development of epidemics. Lucas and Knights (1987), in a review of this work, proposed that the photodormancy of spores synchronises germination with periods of darkness. Since the leaf surface is hostile during the day, spores germinating at night will run less risk of desiccation. The formation of dew also helps to break dormancy and hydration of spores is linked to photosensitivity. Ultraviolet light is a major factor in spore mortality.

Rust spores utilise endogenous nutrient reserves for outgrowth. Both

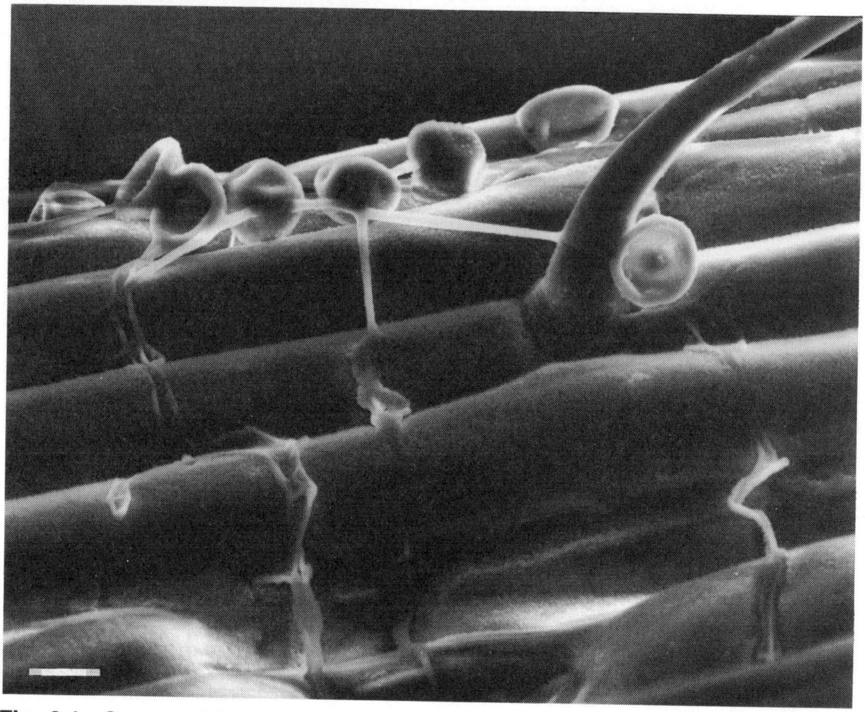

Fig. 3.1 Spores of the brown rust fungus *Puccinia recondita* germinated on the surface of winter wheat. At the bottom of the photograph two infection hyphae can be seen which have penetrated a stomatal pore. Bar represents 20μm. Courtesy of Drs S. Humphreys and P. G. Ayres.

lipids and carbohydrates are used and thereby the processes of novel wall synthesis and protein synthesis are catered for at this stage. The formation of infection structures requires other changes, such as DNA synthesis and nuclear division in addition to the synthesis of novel proteins. Hydrated spores germinate fast but are much more responsive to environmental stimuli than dehydrated spores, which implies that physiological changes increase sensitivity (Lucas and Knights, 1987). Self-inhibitors of germination may have a role in the co-ordination of the lysis and synthesis of cell wall components in addition to inhibiting germ plug dissolution. The germ plug is present as a distinct region of the uredospore wall, at the site of germ tube emergence, which dissolves at germination.

Prior to penetration in wheat stem rust and many other rusts uredospore germ tubes become very specifically orientated on leaves and grow across the surface at right angles to veins until they reach stomata (Fig. 3.1). It has been suggested that this orientation (90° to surface ridges) must maximise the chances of a germ tube reaching a stomatal pore (Dickinson, 1949). The frequency and height of the ridges on leaf surfaces also affects the branching and differentiation of germ tubes (Dickinson, 1949). It has been shown that the germ tubes respond to topographical stimuli and that unless adhesion to the host surface is satisfactorily achieved (Mendgen *et al.*, 1988) the next stage in the infection process does not occur normally. Wynn (1976) studied waxless mutants of corn and found that adhesion of germ tubes was reduced. Insufficient footing, on such hydrophilic surfaces, often gave rise to the production of infection pegs which were wrongly orientated and the resulting infection structures formed aberrantly (Wynn and Staples, 1981), often without any host penetration occurring. Wynn (1976) also used plastics to copy leaf surface topography and showed that germ tubes were induced to form appressoria above the images of stomata. The contact between the tip of a germ tube and the lip of a stomatal guard cell induced the formation of infection structures. This sort of differentiation of infection structures in response to surface topography has been termed 'thigmodifferentiation'. Once appressorium development is complete, then infection structures form in response to chemical stimuli, which has been termed 'chemodifferentiation'(Staples and Macko, 1984). The development of infection structures is shown diagrammatically in Fig. 3.2.

MECHANISMS OF PENETRATION

Penetration of the host plant by an invading fungus gives rise to the potential establishment of physiological contact between the two organisms. A range of both physical and chemical means of penetration have been described but it is important to remember that most usually a combination of mechanisms is effected by fungi. Additionally, any one species may well affect different modes of penetration under different conditions. The penetrating, invasive life-style of a fungus is certainly an advantage. However, the underlying physiological mechanisms and control of penetration are poorly understood at present.

Physical penetration

Natural openings

Plants exhibit several types of natural openings which are relatively unprotected and therefore vulnerable to invasion by fine fungal hyphae. Stomata are probably the most common route for pathogen entry to plants. These are openings, occurring most usually on the undersurface of leaves, which tend to open during daylight hours and close at night. In some instances, germ tubes or zoospores (e.g. *Phytophthora* spp. and *Plasmopara* spp.) may

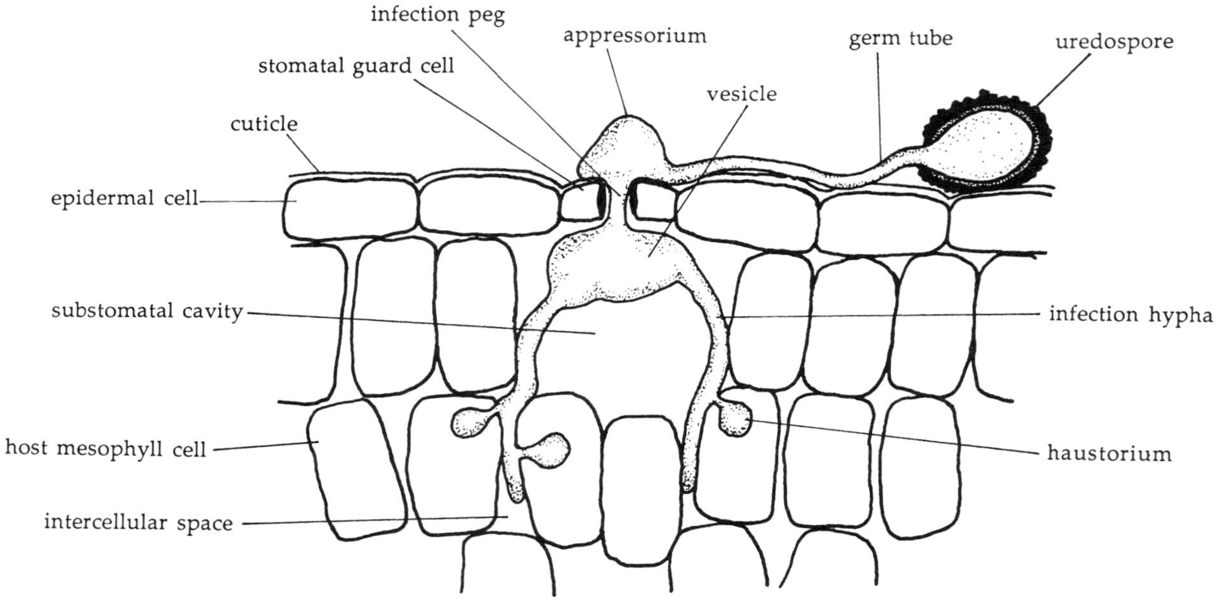

Fig. 3.2 Diagrammatic representation of infection structures produced by rust fungi during the invasion of host plant leaves.

be chemically attracted towards stomata prior to penetration. However, germinated spores of some species which do penetrate leaves via stomata may often grow past, or even over, stomatal pores without any attraction or attempt at penetration occurring, e.g. *Septoria apiicola*, which causes leaf spot disease of celery (Fig. 3.4a, b). Some hyphae are able to enter through closed stomata so that the extent to which stomatal closure may act as protection against invasion is not clear. The penetration of a leaf via a stomata may, or may not, involve the formation of an appressorium. Substomatal vesicles are often formed following penetration but prior to further host tissue colonisation.

Hydrathodes are natural, open pores on the surface of leaf margins which are connected to veins and often exude droplets of liquid (guttation fluid) containing a range of nutrients which can act as attractants to growing germ tubes. Lenticels are open pores on woody stems through which hyphae may easily penetrate, e.g. *Penicillium expansum* causes brown rot of apples, entering through lenticels in apple fruit.

Wounds

Wounds in a plant surface are ideal penetration sites for fungal hyphae. Damage may result from surface breakage, environmental influences, animal and insect activities, etc., and also by natural processes of the plant, e.g. leaf scars, sites of lateral root emergence, etc. Normal abscission layers are wounds which are often not completely sealed when leaves become removed, perhaps by wind action. These areas may exude nutrient materials since a gap exists through to vascular tissues; indeed at leaf fall, suction at the site of abscising xylem may result in suck back, acting as an injection, at the abscission point. *Nectria galligena*, which causes apple canker, achieves entry to the host plant by this route. Sites of lateral root emergence are vulnerable to invasion owing to exudates from young roots which attract microbes. Some pathogenic fungi may enter a host plant through lesions caused by a prior invader. *Venturia inaequalis* causes scabs on apple fruits which crack and increase the susceptibility of the host to invasion. Insects may produce small punctures in a plant surface, or cause excessive defoliation rendering the plant vulnerable. *Ceratocystis ulmi*, which causes Dutch Elm disease, enters the host via wounds caused by the Elm bark beetle *Scolytus scolytus*. Wounds created as a result of grafting, pruning, farming and harvesting methods may also give rise to plant infections and have an important role in epidemiology.

Direct penetration

It is clear that adhesion of potential invaders to plant surfaces is important. Some fungal species produce mucilagenous sheaths covering hyphae,

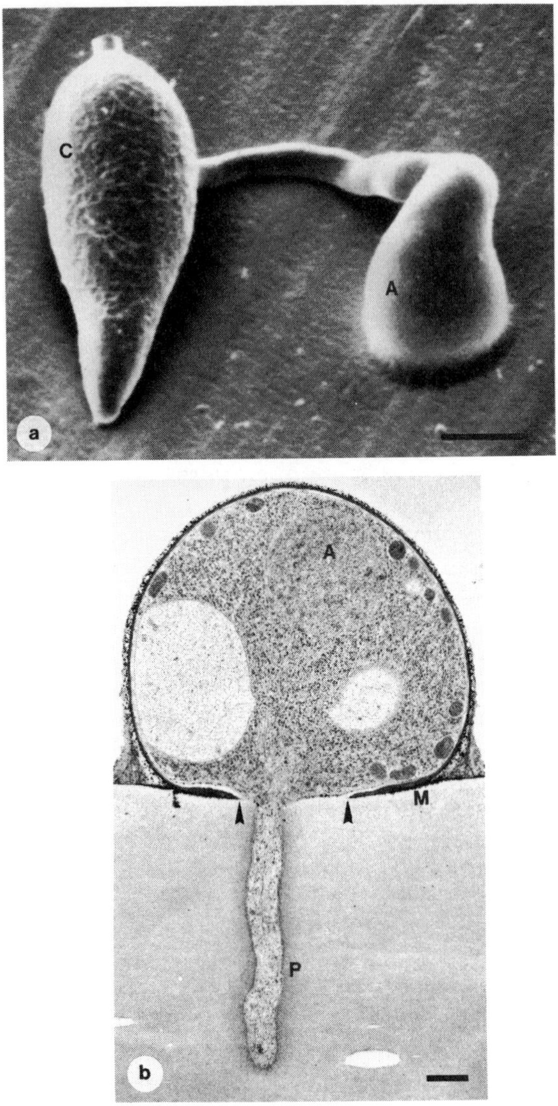

Fig. 3.3 Electron micrographs of the ascomycete *Magnaporthe grisea* which causes rice blast disease. (a) Scanning electron micrograph showing a mature appressorium (A) and conidium (C). Bar represents 5 μm. (b) Transmission electron micrograph (*in vitro*) showing penetration peg (P) extending into cellophane membrane substrate. Appressorium (A) closely adhered to the substrate by spore tip mucilage (M). Penetration peg has arisen from new inner wall layers overlying the appressorium pore (arrowed); probing with wheat germ agglutinin-colloidal gold showed the chitinous wall was not uniformly labelled. Bar represents 1 μm. Reproduced from Howard *et al.* (1990), courtesy of Prof. R. J. Howard and Springer-Verlag.

or other adhesive substances, to ensure close contact (Fig. 3.3a, b). Polysaccharides, glycoproteins, hexosamine polymers and xylans have been implicated, since these become sticky when wet. Bishop and Cooper (1983) have shown that *Verticillium albo-atrum* (tomato roots) and *Fusarium oxysporum* f. sp. *lycopersici* (pea roots) adhere to the plant surface via an extracellular matrix and have suggested that this also contains wall degrading enzymes. After contact between a germ tube and the plant surface, the direct penetration of plant cells must require a combination of mechanical force, which would need to be considerable, and enzymatic pre-softening of the cuticle. Direct penetration is achieved, often beneath a bulbous appressorium, usually by a very fine hypha (penetration peg). This invading hypha regains its normal size after the cuticle has been penetrated. Some species are capable of penetrating through stomata and/or directly through the cuticle surface (Fig. 3.4 b, c, d).

Chemical penetration mechanisms

Once in close proximity to the plant cell wall, many pathogenic fungi are able to launch biochemical attacks on plant tissues and cells to aid and extend penetration. Many complex interactions occur resulting in cellular damage and the development of disease symptoms. Two classes of chemical weapons will be considered here. Firstly, high molecular weight cell wall degrading enzymes which soften and often rupture epidermal cells on contact, leading to tissue maceration, and secondly, low molecular weight toxins which kill cells in advance of hyphal invasion into host tissues. In some instances the production of such chemical weaponry may be essential for pathogenicity and if the ability for production is lost then that individual may become non-pathogenic. However, by no means all pathogens produce such chemicals to assist invasion. It is also interesting to note that although the mechanisms of some enzyme and toxin action have been well documented, very few have been directly isolated from, and linked with, sites of fungal penetration and disease development.

Enzymic penetration of plant cell walls

The plant cell wall is a complex, three-dimensional barrier composed of integrated cross-linked polymers. It is a dynamic, growing and constantly changing structure for which the constituents and their relative proportions may vary during growth. Much information has been collected concerning the types of polymers involved and the major components of them but it must be noted that some of this information may reflect the results of the extraction methods used in identification processes. The molecular and biochemical means by which these polymers are degraded

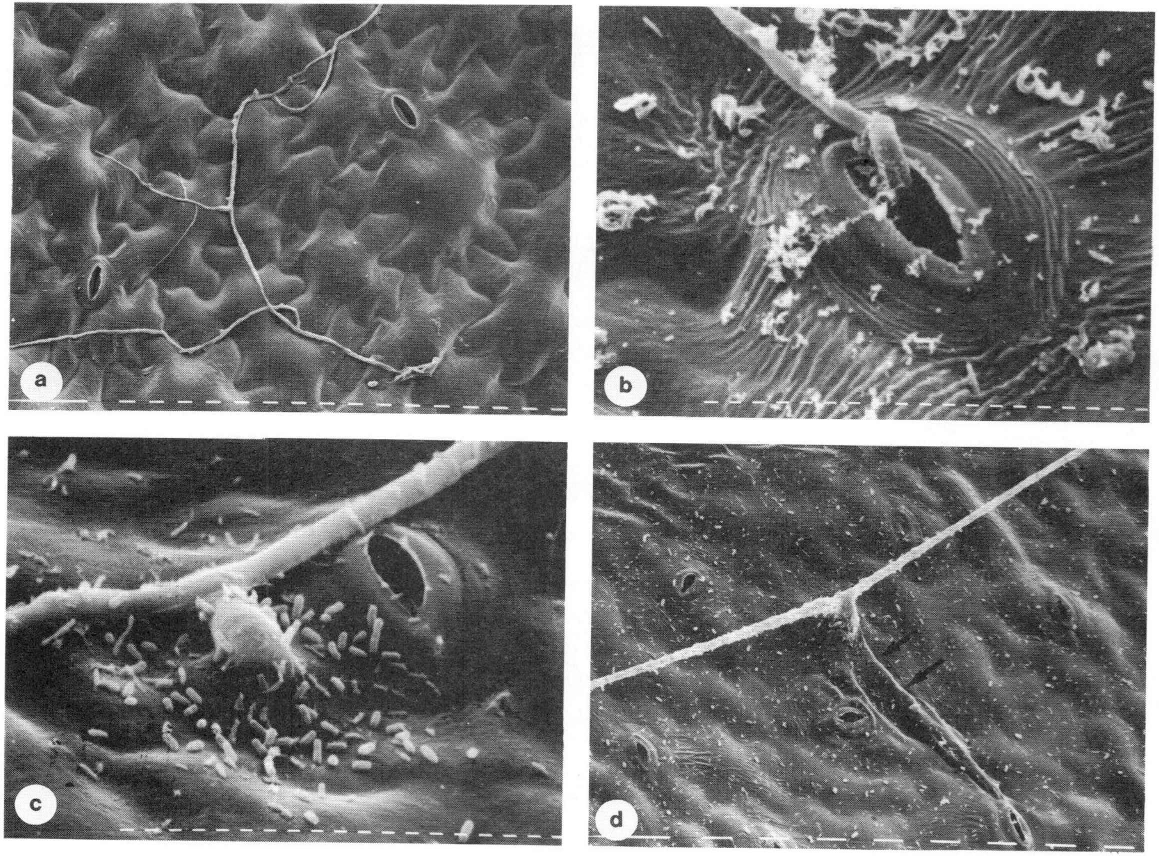

Fig. 3.4 Scanning electron micrographs of *Septoria apiicola* (causal agent of leaf spot disease of celery). (a) Hyphal growth on celery leaf surface. (b) Penetration of celery leaf through stomatal opening. (c) Direct penetration of celery leaf surface. (d) Celery leaf surface showing subcuticular hyphal growth following direct penetration (arrowed). Bars represent (a, d) 10

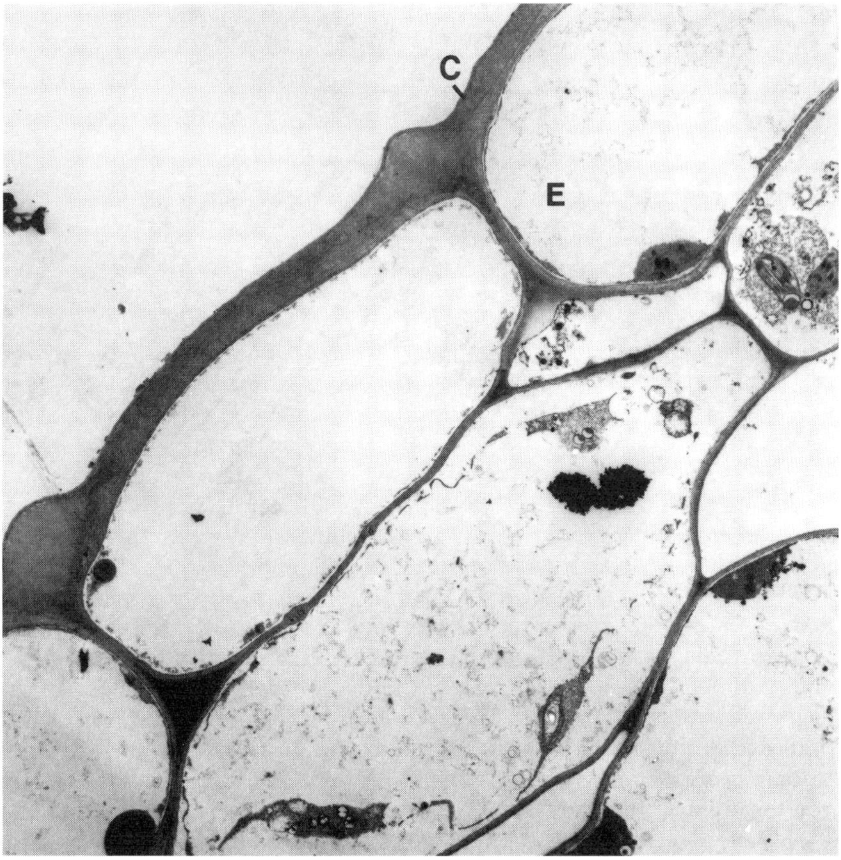

Fig. 3.5 Electron micrograph which shows the variable thickness of the celery leaf cuticle. C, cuticle; E, epidermal cell. x1350 (courtesy of S. Edwards).

is still being pieced together and, in spite of improved modern analytical methods, information in some areas is sparse.

(a) Cuticle penetration

The outermost surfaces of the plant are encrusted with waxes as a hydrophobic layer sheathing the outer cuticle. The cuticle is extremely resistant to attack by microorganisms and is therefore a very important barrier offering great protection to the plant. Cuticular wax is laid down throughout leaf expansion and to some extent, throughout the active life of the plant. The thickness is variable over the leaf surface (Fig. 3.5) and influenced by the stage of development, light intensity received and

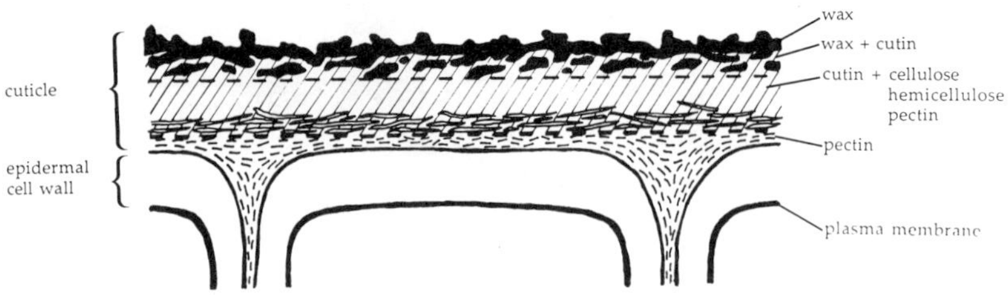

Fig. 3.6 Diagrammatic representation of the cuticular layer of plant leaves.

duration of exposure. The cuticle is multilayered and separated from the plant cell wall by pectic compounds (Fig. 3.6). Potential fungal invaders encounter this outer layer first and usually penetrate by mechanical force. However, *Fusarium solani* f. sp. *pisi* isolate T-8 has been shown (Woloshuck and Kolattukudy, 1984) to liberate cutinase in response to the presence of cutin in the culture medium. Indeed a variety of plant pathogenic fungi produce cutinase (Lin and Kolattukudy, 1980). Additionally, cutinase has been identified at infection sites on pea plants, using ferritin-conjugated rabbit antibodies specific for cutinase (Shayk *et al.*, 1977).

Kolattukudy (1985) has proposed a mechanism for the control of cutinase production in spores. It is suggested that basal levels of cutinase, inherent within fungal spores, release monomers on contact with the host cuticle. This stimulates the switching on of the fungal cutinase gene, resulting in the production of enhanced levels of cutinase which would be released from the germinating fungus. After penetration of the cuticle barrier cutinase production would cease. Mutants lacking cutinase activity are seen to be non-pathogenic (Yoder and Turgeon, 1985a). Additionally, changes in the cuticle may act as a stimulus for morphological differentiation (e.g. production of haustoria) and may aid pathogen adhesion (Kunoch *et al.*, 1990). It has been proposed (Kolattukudy, 1985; Kolattukudy and Crawford, 1987; Kolattukudy *et al.*, 1989) that the ability to produce cutinases is correlated with virulence, although in general the roles of individual cell wall degrading enzymes in infection and disease development are not yet clear.

(b) Cell wall degradation

(i) *Composition of plant cell wall components.* The plant cell wall is a constantly changing structure, the constituents of which vary during growth and development. The main components are listed in Table 3.1 to briefly explain the types of polymer found in plant cell walls and their probable function within the structure of the wall.

In general, plant cell walls (Fig. 3.7) are composed of an outer, or primary wall, which contains cellulose microfibrils widely dispersed in a matrix of

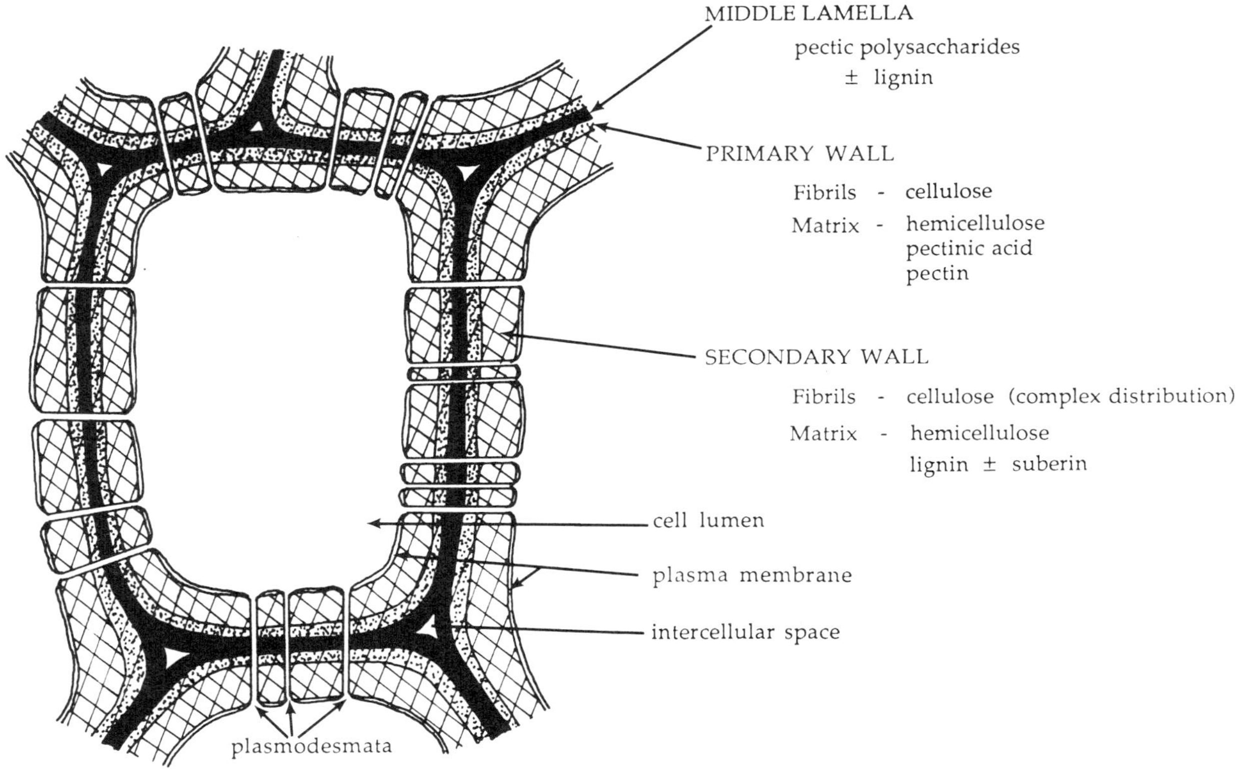

Fig. 3.7 Diagrammatic representation of the plant cell wall.

Table 3.1: Major plant cell wall components

Component	Location	Major residues and function
Cuticle wax	Outer layer Content variable	Primary alcohols, acids and their esters. Secondary alcohols, ketones, paraffin hydrocarbons. Hydrophobic outer surface.
Cutin	Outer layer Content variable	Hydroxy fatty acids. C_{16}–C_{18} family of monomers.
Suberin	Outer layer Root endodermis	Fatty acids, fatty alcohols. Phenol content higher than cutin. C_{20} – C_{26} family of monomers. Chemically and functionally similar to cutin. Often produced in response to wounding.
Cellulose	Primary and secondary walls	Major structural polysaccharide. Long chains β-(1, 4)-linked glucan. Less polymerised in primary wall (random distribution). Highly polymerised in secondary wall (complex distribution).
Hemiculluloses	Primary and secondary walls	Xyloglucan – hydrogen bonds to cellulose. Glucuronoarabinoxylan – hydrogen bonds to cellulose. Matrix material, packs around cellulose fibrils. Provides degree of flexibility.
Pectic polysaccharides	Middle lamella	Homogalacturonan (α-(1,4)-galacturonosyl units). Rhamnogalacturonan I. Rhamnogalacturonan II. Arabinan. Galactan. Matrix materials, cement of middle lamellae. Amorphous gel-forming capacity important.
Hydroxyproline-rich glycoproteins	Primary and secondary walls	Hydroxyproline, also produced as wounding response. Arabinogalactans. Lectins. Structural role. Also providing some elasticity by virtue of 'extensin' contained in pectin/hemicellulose matrix gel which allows slippage of cellulose micro-fibrils.
Lignin	Middle lamellae Secondary thickening	Phenyl propane units. Strengthens and thickens walls. Increasing amounts in differentiated tissues.
Callose	Localised depositions	Unbranched β-(1,3)-glucan. Produced as a wounding response. Also produced as a stress response with lignin, protein and phospholipid.

hemicellulose. Internal to that is the secondary wall which is laid down as the cell ages and may eventually be a thicker layer than the primary wall. The secondary wall is composed of greater amounts of cellulose microfibrils which are distributed with a more complex arrangement than in the primary wall. These fibrils are laid down in a hemicellulose matrix which also contains some glycoproteins. These glycoproteins are often rich in the unusual amino acid hydroxyproline, and it is these components which provide some elasticity to the wall. Plant cells are separated and stuck together by the middle lamella which is composed of pectic compounds (homgalacturonans and rhamnogalacturonans) which have the capacity to form gels. Plant cell protoplasts are interconnected by plasmodesmata.

During tissue development and differentiation secondary walls become thickened and rigidified by the deposition of lignin. Lignin is laid down first in the middle lamella and the wall of xylem cells but can become a major component in woody tissues. Lignins with different monomeric composition are found in different plant groups. Increased lignin synthesis is often associated with resistance mechanisms in plants.

In addition to these components there are other polymers, often laid down in response to wounding or stress, which make up smaller proportions of the cell wall. Individual wall polymers and the components which make up structural layers are often bonded together, with some linkages between wall layers. However, some components are not linked to other polymers, e.g. pectic materials, but are bound together by calcium. The chelating ability and bonding of Ca^{2+} ions confers rigidity on the wall (Cooper, 1984).

Cells of different plant species, cells of different ages and stage of development and cells in different tissues will all have slightly different wall compositions. It has been shown that monocotyledonous plant cell walls differ in absolute chemical terms from those of dicotyledonous plants. However, in spite of these differences the walls are arranged in a very similar way, although there may be some differences in binding between components (Darvill *et al.*, 1980)

(ii) *Degradation of plant cell wall components.* Information concerning the degradation of individual polymers and bound, structural layers of walls, has been pieced together often using purified enzymes and isolated wall material.

Pectic polymers are important structural components and many fungi liberate pectic enzymes which have an important place in wall degradation. The dissolution of pectic polymers is brought about by pectinases or pectolytic enzymes, which solubilise pectin chains. Hydrolytic polygalacturonases break the linkages between galacturonan molecules and pectin lyases remove water from the linkage to split the chains which results in cell wall maceration. Both these types of enzymes occur as endopectinases which break pectin chains at random sites along the lengths of chains, or as exopectinases which break the terminal linkage and release galacturonan units. The subsequent uptake of these monomers into the invading fungus induces the enhanced synthesis and release of enhanced levels of pectic enzymes from hyphae. These enzymes are key factors in soft rot diseases (e.g. apple rot diseases caused by *Penicillium expansum* and *Sclerotinia fructigena*; apple fruits are particularly susceptible), which cause infected tissues to become soft and water soaked.

The breakdown of hemicelluloses requires a mixture of enzymes, collectively known as hemicellulases. Xylanases degrade xylan oligosaccharides, which represent the major component of hemicellulose in plants, and the activities of glucanases and galactanases are also important. It is not clear exactly how much degradation is brought about by these enzymes or the degree to which they contribute to pathogenesis. A nutritional role is often attributed to these enzymes rather than a contributory role in the progress of plant invasion or disease.

Cellulose is degraded by a series of enzyme reactions. Firstly endo-β-(1,4)-D-glucanase (C_1) cleaves the cross-links (glucosyl bonds) between glucan chains, resulting in unbroken, single β-(1,4)-glucan chains. β-(1-4)-D-glucan cellobiohydrolase (C_2) degrades β-(1,4)-glucan (cellulose chains) to the disaccharide cellobiose, which is then converted to glucose by β-(1,4)-glucosidase activity. The endo-β-(1,4)-D-glucanase creates sites at which the β-(1-4)-D-glucan cellobiohydrolase can act. The degradation of cellulose occurs synergistically (White, 1982). Cellulase production is induced by cellobiose, which also acts as a repressor at high levels. Some pathogenic fungi produce all three types of cellulases, e.g. *Trichoderma* spp. Additionally, much cellulose degradation is carried out by secondary invaders, mainly wood degrading species. Cellulases are also important in wilt diseases (caused by *Fusarium oxysporum* and *Verticillium albo-atrum*), where large molecules are released from cellulose in the host plant and cause blockages in the host plant vascular system.

(c) The role of cell wall degrading enzymes in pathogenicity

Cell wall degrading enzymes do appear to be important in the penetration and spread of the fungus within host tissues. However, the extent to which production of such enzymes determines pathogenicity is not altogether clear. It can be seen, by microscopical examination, that fungal hyphae penetrate plant cell walls in a variety of ways which do implicate the involvement of enzymes (Fig. 3.8a). Some species penetrate directly through the plant cell wall causing minimal structural disruption (e.g. rust, mildew and smut). Intercellular degradation associated with an invading hypha may be limited (e.g. *Monilinia fructigena* and

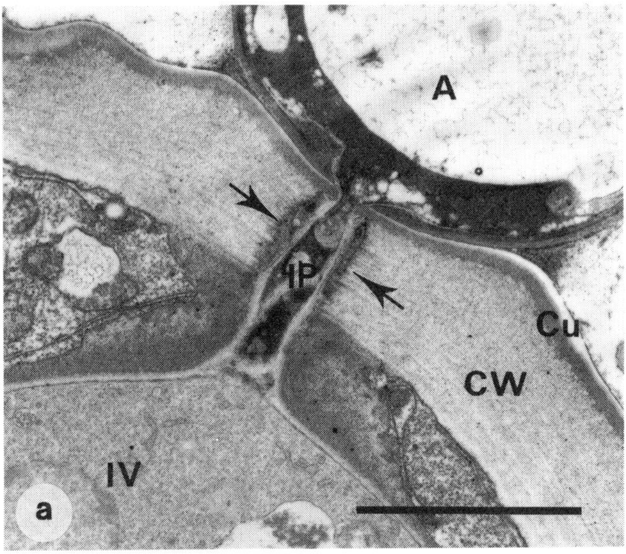

Fig. 3.8 Infection of bean by *Colletotrichum lindemuthianum*. (a) Initial infection: infection peg (IP) emerges from beneath appressorium (A) on plant surface and penetrates host cuticle (Cu) and cell wall (CW) to produce infection vesicle (IV) within cell lumen. Note CW adjacent to IP is electron-opaque (arrows), indicating localised enzymic dissolution. (b) (p.164) Cell-to-cell penetration during early biotrophic phase of infection. Large diameter primary hypha (PH) has penetrated cell wall (CW) between two cortical cells, filling the intervening intercellular space. Note localised wall dissolution and constriction of hypha at points of penetration (arrows). (c) (p.165) Extensive cell wall (CW) dissolution during late necrotrophic phase of infection: small diameter secondary hyphae (SH) develop both inside cell wall and inside cell lumen. Degraded wall around hyphae is swollen and electron lucent. Bars represent (a) 2.5 μm; (b) 5 μm; (c) 10 μm. Reproduced from O'Connell *et al.* (1985), courtesy of Dr R. J. O'Connell and Academic Press, Inc.

Septoria apiicola) or more extensive (e.g. *Botrytis fabae*) and accompanied by localised wall swelling (*Colletotrichum lindemuthianum*), particulary if pectic enzymes are involved (Fig. 3. 8b, c). Some hyphae penetrate thickened secondary walls of plant cells either by means of natural gaps, such as pits in xylem tissues (e.g. vascular wilt fungi) or directly (*Gaeumannomyces graminis*). The degree of cellular injury which results is not always correlated with the degree of structural disruption.

Circumstantial evidence is often used to imply the involvement of cell wall degrading enzymes in infection processes. Cooper (1984) has drawn attention to the need for critical criteria to be considered in experimental investigations. The ability of a fungus to secrete cell wall degrading enzymes *in vitro* allows for characterisation of the enzyme(s) but does not prove any involvement in disease. Similarly, the degradation of isolated

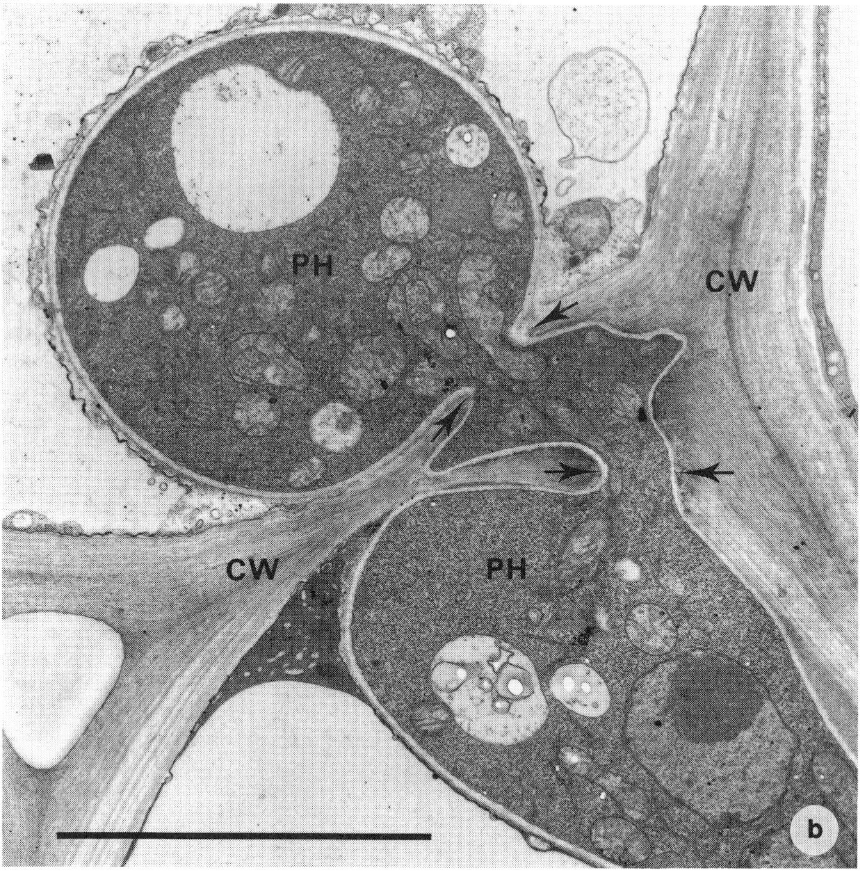

Fig. 3.8(b)

host cell walls or tissues by fungal culture filtrates or purified enzymes does not prove a role in infection. The detection of cell wall degrading enzymes in infected tissues just before, or during symptom development implies involvement in disease but such detection is often difficult owing to the low concentrations which are present and interference from host enzymes. The loss of host cell wall polysaccharide from infected tissues implies cell wall degrading enzyme activity. Ultrastructural examination, using for example, loss of affinity for specific stains, or loss of birefringence of cellulose can provide evidence of the nature of wall degradation. The ultrastructural localisation of cutinases (Kolattukudy, 1985) has been achieved using labelled antibodies, but critical, controlled experiments have only been carried out for a few diseases.

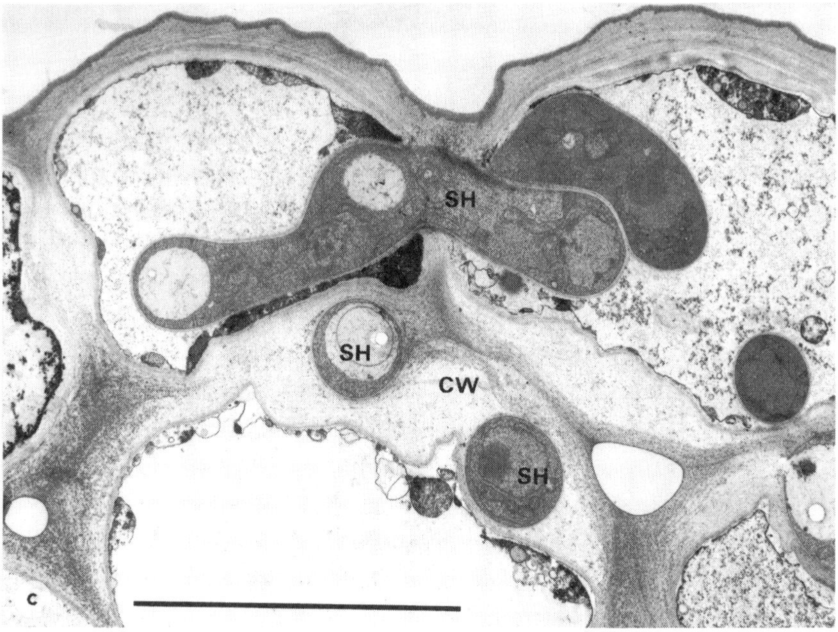

Fig. 3.8(c)

One of the intriguing features of fungal cell wall degrading enzymes is the diversity of molecular forms which are produced. A fungus may secrete several forms of an enzyme which attack the same substrate (Keon *et al.*, 1987). These may differ in molecular weight and isoelectric point and the forms therefore have different electrophoretic properties. The endopolygalacturonases produced by *Verticillium albo-atrum*, *Monilinia* spp. and *C. lindemuthianum* are examples. Such multiplicity of form has been suggested to provide flexibility for potential pathogens (Byrde, 1979). A high isoelectric point (pI value) results in a positive charge at physiological pH values and ionic binding to vegetatively charged cell walls may account for the limited diffusion of such enzymes, and therefore for relatively localised degradation. A low molecular weight might allow an enzyme to penetrate pores in plant cell walls (Cooper, 1984).

It appears that cell wall degrading enzymes may be liberated by a fungus in sequence. Polygalacturonases were liberated first, *in vitro*, when *V. albo-atrum* was grown on extracted cell walls and were also the first enzymes detected *in vivo* (Cooper, 1987). Pectic polymers are important components of the middle lamella of plant tissues and are encountered early after penetration by an invading hypha. Polygalacturonases have been detected in ungerminated *Botrytis* spores which implies involvement in the early stages of infection. Wall damage often precedes the advance of an invading fungus with polygalacturonases selectively degrading the middle lamellae in penetrated regions (e.g. *Rhizoctonia solani* infections). This results in localised maceration of the tissue. The action of polygalacturonases may also be a prerequisite for degradation by other cell wall degrading enzymes. It is likely that very little wall degradation is needed before damage to the plant cell protoplasts occurs. Host cytoplasm rapidly becomes disrupted and ions leak out from the protoplast which cannot then be maintained in a turgid state by a weakened cell wall. Effects of cell wall degrading enzymes on plant cell plasma membranes are also implicated by protoplast damage.

The involvement of hemicellulases in the progress of tissue disruption remains under dispute. Activities are apparent in invaded tissues as infection proceeds but the liberation of these enzymes may have a nutritional role for the invading fungus rather than being a factor directly determining pathogenicity. Cellulases are also liberated much later in infection. However, cellulose degradation may have a more important role in the rapid progress of some chronic diseases, e.g. *Gaeumannomyces graminis* which causes 'take-all' of cereals and *Pseudocercosporella herpotrichoides* which causes eyespot of wheat.

Plant cell walls have been considered as part of the cells lysozome system. Dissolution of the wall by enzyme activity might release molecules which were previously bound to the wall and which may have potentially physiologically damaging effects, e.g. glucose oxidase and peroxidase. This would imply therefore that wall degradation might have an indirect damaging effect on host cell physiology, but has not yet been proven (Keon *et al.*, 1987).

Massive enzymic degradation of host plant tissue may not always be advantageous to an invader. Endopolygalacturonidases have been shown to trigger responses from host cells, activating resistance mechanisms (Cooper, 1987). A more limited degree of tissue disruption might be more advantageous for the establishment of a pathogen within host tissue. Biotrophs which exploit living host cells generally achieve penetration by very localised wall degradation. Softening of wall polymers combined with some (often considerable) mechanical force, which is generated by internal turgor pressure, brings about penetration by the infection peg of such pathogens (e.g. *Bremia* spp.). The main wall components which are degraded by biotrophs are the matrix polymers. However, the role of degrading enzymes is questionable for these fungi. Certainly polysaccharidases have been extracted from appressoria and conidia of the rust *Uromyces viciae-fabae* and the powdery mildew *Erysiphe graminis*, but the extent to which these contribute to wall perforation is probably limited.

Necrotrophic pathogens do cause rapid and extensive wall swelling and breakdown prior to and during wall penetration (e.g. *Botrytis* spp. and

Rhizoctonia solani) and have been shown to kill cells in advance of hyphal invasion. *Botrytis alli* produces multiple forms of endopolygalacturonases and endopectin lyase which are both detected in lesions. Hemibiotrophic fungi which initially invade plant tissues in biotrophic mode and later switch to necrotrophy (e.g. *Colletotrichum lindemuthianum*, Fig. 3.8c) may use this strategy as a means of delaying the activation of plant defense responses until they are better established in the plant. These fungi do produce endopolygalacturonases or pectin lyases which offers flexibility to these pathogens. It is not clear what triggers the change to necrotrophy and the production of larger amounts of cell wall degrading enzymes. Such a switch may be an environmental trigger, occurring in response to substrate depletion or changes in physiological conditions within host cells as a result of biotrophic invasion.

The sensitivity of modern techniques used to detect enzymes has greatly increased awareness of the multiplicity of enzyme forms liberated by pathogenic species. Ultrathin isoelectric focusing (IEF) gels and substrate overlays allow detection of very small (pg) quantities. Additionally, more precise techniques for biochemical measurements have led to a recent increase in the understanding of the regulation of enzyme synthesis. Most cell wall degrading enzymes are subject to induction by monomers of the substrate, or to catabolite repression. Increases in information relating to the rate, level and regulation of syntheses, together with the possibility of precise mutagenesis programmes, currently provide the opportunity to screen for deficient and regulatory mutants. The need for analysis of a range of strains, isogenic but for defined mutations affecting enzyme regulation, synthesis or secretion, has been pointed out (Cooper, 1987). Such strains would allow further examination of the contribution and the role of cell wall degrading enzymes in pathogenesis.

Toxin production

Toxins most certainly have a significant role to play in plant diseases, particularly with regard to the establishment of a pathogen within a host and in symptom development. In this context a toxin can be defined as a compound produced by a microbial pathogen which causes damage to the host plant and which is known to be involved in the development of plant disease (Scheffer, 1983). It is most likely that an invading pathogen may produce a range of toxic molecules, which may be chemically related, and which act on the plant in tandem.

In general, the compounds usually included in this category are low molecular weight compounds which disrupt the highly integrated physiological processes of the host and give rise to symptoms such as wilting, chlorosis and necrosis. Fungi produce a wide range of toxic compounds with varied biochemical structures and modes of action, including polypeptides, glycoproteins, amino acids derivatives, polyketides, terpenoids, sterols and quinones. Toxins are usually classified (Scheffer, 1983) as host-selective (host-specific) or non-selective (non-specific). Host-selective toxins are toxic to plant species which are susceptible to the toxin-producing pathogen but are not toxic to non-host plants. The production of host-selective toxins is important for the pathogen to induce disease, and often increases the virulence of that pathogen. Non-selective toxins are those for which the toxicity is not related to the host range of the fungus and these may be differentially toxic to plants which are not susceptible to the pathogen. Although a large number of toxins have been well characterised, in many cases the chemical nature of compounds which are known to be involved in disease development has yet to be determined.

It is not always easy to establish the role of a toxin in plant disease. It is obviously important that the toxin should be detected in naturally produced diseased tissue and ultimately the same compound extracted from lesion tissues. So far, however, this has been achieved for very few putative toxins. The concentrations present in the diseased plant may be low and lesion tissue may also contain other toxic compounds of plant origin. Much work, to date, has concentrated on the identification and purification of compounds from the culture fluid of pathogenic isolates. The use of increasingly sensitive biochemical techniques may now allow the collection of further information, for example immunochemistry will provide quantitative and spatial evidence for involvement in disease and the potential of molecular analyses has not yet been realised (Gilchrist and Yoder, 1984). Crude culture filtrates allow for easy bioassay and screening for toxic effects, usually making use of plants, cuttings, leaf discs or even cell suspension cultures of the host species. It is very important to use physiological concentrations in such experiments (Yoder, 1983), but these are not always easily determined, especially as toxins often exert profound effects in minute quantities. Some toxins may only be produced by a pathogen growing on host tissues and will not be produced in culture. Conversely, toxins isolated from pure culture may be artifacts of that culture and may not be produced in host tissues (Turner, 1984). In the characterisation of a toxin it is important to know the biochemical effects on host metabolism, in addition to the symptoms produced, and also to establish the stage of disease at which a toxin is liberated and has effect (Knoche and Duvick, 1987). Indeed some diseases may occur without the development of any symptoms. In other instances, toxins liberated by germinating spores (e.g. *Alternaria* and *Botrytis*) may influence the penetration and early establishment of the fungal invader. Alternatively, toxins may be liberated late in disease development, enhancing senescence in invaded tissues and adding to the overall severity of the disease. Fungi also produce a wide range of secondary metabolites, many of which exhibit phytotoxic effects, but which do not have specific roles in the establishment or course of plant disease.

The mechanisms of toxicity for those compounds which have been satisfactorily associated with disease development are still largely unknown. Some important toxins have, however, been structurally characterised and shown to be extremely diverse. Each has more than one chemically related toxic component, and some more than ten. This suggests a variety of molecular mechanisms for toxicity and specificity. The mechanisms underlying the physiological effects on the host plants also remain to be fully understood and characterised.

(a) Host-selective toxins

(i) *Helminthosporoside (HS-toxin). Helminthosporium sacchari,* which causes eyespot disease of sugarcane (*Saccharum officinarum*), produces a number of sesquiterpene glycosides which contain galactose. Three isomers with differing toxic activity and a number of non-toxic homologues are formed. HS-toxin gives rise to red/brown streaks on the leaves of susceptible sugarcane clones. Subcellularly the toxin causes changes in the outer membranes of chloroplasts, which disintegrate, giving rise to a corresponding decrease in fixation of CO_2. The susceptibility of plants to HS-toxin has been shown to correlate with the ability to bind the toxin; insensitive clones do not exhibit binding activity. It has been suggested that this binding is regulated by membrane lipids and that the binding protein is located on the plasma membrane. Toxin binding indirectly disturbs membrane ATPase activity and also causes an increase in permeability and gives rise to necrosis in host tissue.

(ii) *Hm T-toxin. Helminthosporium maydis* causes the very serious disease southern corn leaf blight. The toxin produced exerts a multiple effect on host plants which possess Texas cytoplasmic male sterility (T-maize), affecting leaves and inhibiting root growth on this widely planted host. Hm T-toxin consists of about ten linear polyketols. Treatment with toxin affects host plant phytosynthesis, respiration and stomatal closure. It causes profound effects on membranes, resulting in permeability changes and electrolyte leakage. In cells of susceptible root tissue active transport through the plasma membrane has been shown to cease and intracellular K^+ is lost. Recent evidence (reviewed by Knoche and Duvick, 1987) suggests that the toxin is not bound to specific sites but acts as a Ca^{2+} ionophore in mitochondria from susceptible plants. It is also suggested that in mitochondria from non-sensitive plants polypeptide interaction prevents toxic action.

(iii) *PM-toxin. Phyllosticta maydis* causes yellow leaf blight of corn also showing the same specificity for T-maize as Hm T-toxin. PM-toxin is produced in culture and has very similar chemical properties to Hm T-toxin. The chemical structures and primary sites of PM-toxin and Hm-toxin have not yet been identified, but much evidence suggests that the mechanisms of toxicity and sites of action are very probably the same. PM-toxin also inhibits root growth, induces chlorosis and causes electrolyte loss from T-maize leaves.

(iv) *Victorin (HV-toxin). Helminthosporium victoriae* is specific for the oat cultivar 'Victoria', causing foot and root rot and leaf blight. *H. victoriae* is a weak parasite and the production of HV-toxin is essential for the growth of the fungus in host tissues and for symptom development. The toxin probably acts prior to, or during, host penetration. The oat cultivar 'Victoria' carries a crown rust-resistance gene (Rpc) and is 10 000 times more sensitive to HV-toxin than cultivars which lack the gene. It is now thought that rust resistance and toxin sensitivity are conferred by one locus. The site of toxic action is, as yet, unknown and although it is thought to be a polypeptide linked to sesquiterpene the chemical structure is also largely unknown. HV-toxin causes general physiological changes in host plants, inducing rapid increase in the electrolyte permeability of the plasma membrane, rapid changes in membrane potential and respiration together with decreased growth and protein synthesis.

(v) *AM-toxin. Alternaria mali,* a pathotype of *Alternaria alternata* which causes blotch in susceptible apple cultivars, produces a mixture of three related toxins, AM-toxins I, II and III. Toxin I is a cyclic peptide with a 12-membered ring system and exhibits the most toxic activity of the three. Small chemical changes greatly reduce the toxicity and toxins II and III are much less potent. The action of these toxins is very specific and any possible cell receptor would also need to be highly specific. It has been suggested that AM-toxins may have more than one site of action, causing disruption to the plasma membrane and chloroplasts. These toxins have now been chemically synthesised which will undoubtedly lead to greater understanding of the relationship between chemical structure and toxic activity in the future.

(vi) *AK-toxin.* A pathotype of *Alternaria alternata* which was known as *A. kikuchiana* produces necrotic lesions (black spot disease) on leaves and fruits of Japanese pear (*Pyrus serotina*). Two related toxins (AK-I and II) are produced by the fungus. AK-toxin I is derived from phenylalanine and toxin II has a similar structure but with an extra methyl group on the phenylalanine residue which reduces the toxicity. AK-toxin is produced by spores at germination and prior to invasion of plant tissues. This toxin is therefore considered as an initial agent in disease, since it has a role very early in the infection process. Toxic activity causes the plasma membrane of susceptible host cells to leak electrolytes, particularly K^+ and phosphate, very rapidly. This results in a reduction in membrane potential and leads to necrosis.

(b) Host non-selective toxins

(i) *Fusicoccin. Fusicoccum amygdali* produces the toxin fusicoccin which is involved in wilt disease of almond and peach. The toxin is a diterpenoid glycoside and affects cellular transport processes, in particular the H^+/K^+ ion exchange pump across the plasma membrane, by activation of membrane-bound ATPase. Fusicoccin also causes increases in cell wall plasticity, respiration rate and cellular enlargement.

Transpiration rates are also increased by the toxin due to the disturbance of the solute balance in stomatal guard cells which results in the permanent opening of stomata. Since water balance is then upset the plant suffers water stress. Foliar desiccation leads to the death of the plant.

(ii) *Tentoxin. Alternatia tenuis* causes chlorosis in many seedlings, notably in cotton, by inhibiting chlorophyll accumulation. Tentoxin is a cyclic tetrapeptide which acts by inactivating the chloroplast-coupling factor which is involved in energy transfer into chloroplasts, and inhibits cyclic photophosphorylation. It also causes loss of polyphenol oxidase from sensitive tissues. The actual sites of action and molecular mechanism of the toxic activity are unknown.

(iii) *Ophiobolin (cochliobolin).* This toxin is produced by a number of fungi, in particular by *Cochliobolus miyabeanus* (*Helminthosporium oryzae*) which causes rice leaf spot disease. Ophiobolin is a phytotoxic sesquiterpene (Fig. 3.9) which alters plasma membrane potential and phenol metabolism in host cells. Polymerization of phenolics gives rise to the brown pigments which appear in necrotic areas of host tissues. The mode of action and sites of action are unknown.

(iv) *Fusarial wilt toxins.* Much controversy surrounds the role of non-selective toxins in the development of vascular wilt diseases caused by *Fusarium* species. A number of symptoms make up the vascular wilt syndrome, e.g. plugging of xylem, necrosis and wilting. The primary symptom, wilting, occurs due to lack of water movement through the infected plant. The xylem pathway becomes blocked by the presence of the pathogen itself, as mycelium or spores, by breakdown products (cell wall fragments) released by enzyme activity of the pathogen and by the presence of tyloses formed as a response of the host plant. Vascular wilt disease is therefore the result of wilting due to restricted water movement, but also to complex interactions between toxins, enzymes and hormones (Chapter 4).

A number of toxins are produced by *Fusarium* species. Fusaric acid (5-*n*-butylpicolinic acid), which is produced by a number of *Fusarium* species (Fig. 3.9), acts by binding metal ions and increasing respiration rates in infected plants. The primary effect is to increase cell permeability which gives rise to electrolyte loss. Lycomarasmin is produced by *F. oxysporum* f. sp. *lycopersici* and other species (Fig. 3.9). This toxin is a tripeptide which acts as a powerful chelating agent binding Cu^{2+} and Fe^{2+}. Naphthazarin is produced by *F. solani* f. sp. *pisi* , and acts by inhibiting the anaerobic decarboxylation of pyruvate. It is not clear just how important a role these toxins play in the development of vascular wilt disease but they certainly contribute to the severity of infections and give rise to some of the disease symptoms (Scheffer, 1983).

(c) Role of toxins in plant disease

Toxins have been seen to play very important roles in the development of disease epidemics. *Helminthosporium victoriae* produced a widespread epidemic on the Victoria oat variety in Europe in 1946–48. *H. maydis* spread very rapidly through the corn belt of the southeastern USA in 1970/71 causing vast crop destruction. However, the severity of the effect

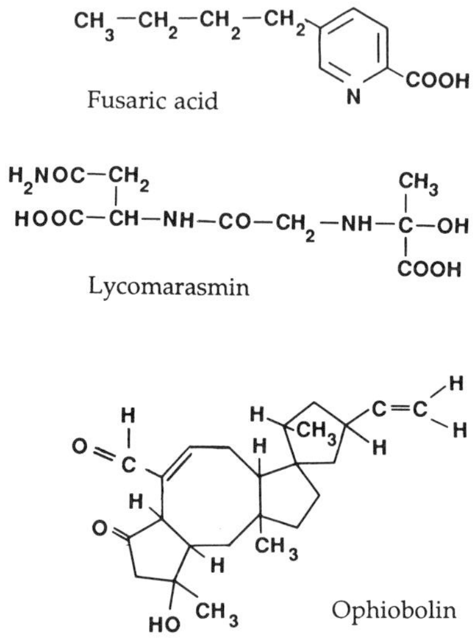

Fig. 3.9 Chemical structures of some host non-selective toxins.

in terms of crop losses was compounded by the planting of genetically uniform cultivars, produced by plant breeders for superior cropping. Repeated planting of such uniform stands increased the risk of sudden and massive infection by highly adaptable plant pathogens.

Toxins are very diverse molecules and so the categorisation as host-selective and non-selective is not always clear. There may be low correlation between toxin sensitivity and susceptibility of host plants, particularly where multiple forms of toxic molecules produced by a pathogen have different influences on the metabolism of that plant. Therefore, toxin production may not determine pathogenicity in all host varieties. As yet, the regulation of toxin production is not understood and very little information concerning the genetics of production have been accrued. However, it is likely that several genes would be involved in the synthesis of most toxins.

Host-selective toxins are very fast acting and often have their effect at very specific sites. These toxins also function at extremely low concentrations on susceptible cultivars but may not cause a response, even at much higher levels, on resistant cultivars.

It is not easy to identify exactly what benefits a plant pathogen derives from toxin production. Theroetically the killing of cells in advance of invasion and leakage of electolytes and nutrient materials from host plant cells may be seen as an advantage to the establishment of an invader. However, there is little experimental support for this view (Scheffer, 1983). Toxins may contribute to disease not by the direct killing of cells but by upsetting the normal metabolism of the host and by suppressing resistance mechanisms which would otherwise be induced. Further knowledge concerning the structure and specificity of toxins may help to identify modes of toxic activity (Daly, 1987; Kohomoto *et al.*, 1987) and to characterise the full effects within the highly integrated metabolic system of the host plant.

HOST PLANT COLONISATION

After host plant penetration a pathogen must spread within invaded tissues to colonise and establish a parasitic relationship with the host. Fungal invaders respond to compounds in the host plant and this regulates their subsequent development, switching on a new genetic programme and giving rise to the pathogenic growth habit. Specific infection structures (appressoria, infection pegs, haustoria) are formed after germination. Developments such as these are accompanied by an increase in the metabolic activity of the pathogen. There is a rapid rise in the synthesis of macromolecules, particularly RNA and protein, subsequently followed by DNA. Indeed nuclear division is an early event following germination and the outgrowth of germ tubes. Once inside host tissue a pathogen may exert effects in a variety of ways. General ultrastructural changes occur, often

often quite early after infection, in host cells. Cell contents become more granular in appearance with an increase in the amounts of rough endoplasmic reticulum and concentration of ribosomes. Host nuclei migrate near to penetrating hyphae and may become swollen. Nucleoli often become more pronounced. Physiological effects may also be profound, with alterations in the regulation of normal host metabolism, increases in cell permeability, alteration of gene expression and ultimately, cell death.

However, the outcome of an invasion is the result of interactions which are dependent on the properties of both organisms involved and mediated by a range of other factors. The resistance of the host plant will influence the spread of the pathogen and the severity of the disease which occurs in the host. Additionally, the characteristics of the growth habit and life cycle of individual pathogens will influence the invasion of a host. For example, in powdery mildew infections (*Erysiphe pisi* and *E. graminis*) almost all the fungal growth is superficial, occurring on the outer surface of the plant. Cell penetration occurs into epidermal cells alone, although in a heavy infection there may be many individual penetration points. In contrast, cereal smuts (*Ustilago tritici* and *U. coronata*) spread throughout the plant forming systemic infections. In other diseases, such as *Pythium* or *Phytophthora* root rots, infection sites may be localised but may sufficiently weaken the plant to kill it rapidly. Vascular wilt fungi are also restricted in their spread through a host, being confined to vascular tissues, but have devastating effects. However, if a plant is already weakened by drought or other environmental stresses, then the effect of a pathogen may be more rapid and/or more pronounced. The timing of an infection may also limit the spread and the effect of pathogenic invasion. Infections late in the season will not have such profound effects on yields and will not be as devastating to the plant as similar infections occurring earlier in the growing season.

It is important for biotrophic pathogens, particularly those which are obligate, to establish physiological contact with the host. Most species are able to invade parenchymatous tissues and grow between host cells. In addition to these intercellular penetrating hyphae, a variety of biotrophs form specialised intracellular hyphae, known as haustoria, which are thought to be involved with nutrient transfer from host to pathogen. Haustorial structures and their developmental sequence differ in some detail between biotrophic species, although, in general terms the organisation is very similar.

Formation of haustoria

The Chitridiomycete fungus *Olpidium brassicae*, which infects cabbage roots, is one of very few parasitic fungi to break through the host cell wall and membrane, penetrating directly into the host cytoplasm. In general, however, the terminal cell of an intercellular hypha penetrates the plant cell wall, probably by a combination of enzyme dissolution and mechanical pressure, to establish very close contact with that cell. However, the host cell membranes are not breached but remain intact. There is nothing particularly unusual about the ultrastructure of the invading hypha. A diagrammatic representation of an *Erysiphe* (powdery mildew) haustorium in an epidermal cell is given in Fig. 3.10. At the point of wall penetration the haustorium is constricted, and remains that way, causing minimal disruption to the integrity of that cell. This region is known as the haustorial neck and the position is often marked in the fungus by the formation of a septum. Sometimes the neck region becomes surrounded by a collar, probably of callose, laid down gradually by the host. After penetration the haustorium expands inside the host cell and the host plasma membrane becomes folded and invaginated around it. Haustoria are formed in many shapes, all of which increase the surface area of contact. The forms may be very variable even within a single fungal species. The region of the plasma membrane surrounding the haustorium is known as the extrahaustorial membrane and is quite distinct from the host plasmalemma which surrounds the invaded cell wall. It is thicker, relatively stronger, more convoluted and contains more polysaccharide. At the junction between the extrahaustorial membrane and the host plasma membrane are neck bands which act as seals around the neck of the haustorium. The region between the extrahaustorial membrane and the haustorial wall, delimited by the neck bands, is rich in polysaccharide and is known as the extrahaustorial matrix. The haustorium, matrix and extrahaustorial membrane together are known as the haustorial complex (Fig. 3.13). In some cases the host cell nucleus is closely associated with the haustorial structure. In cereal rusts (*Puccinia coronata*) the extrahaustorial membrane

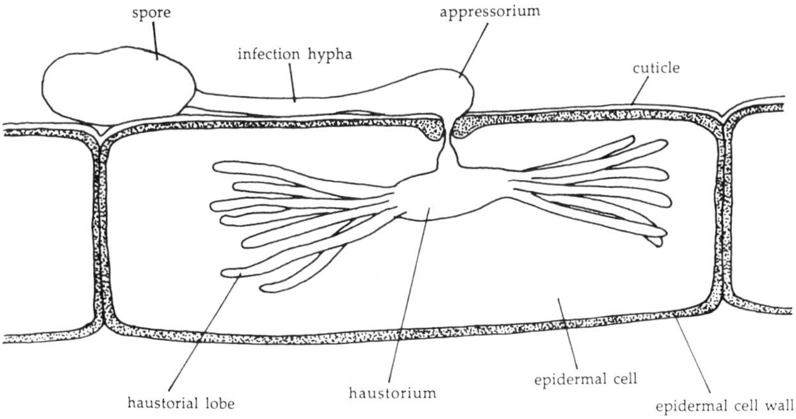

Fig. 3.10 Diagram to show *Erysiphe* haustorium in an epidermal cell.

and the matrix are connected to host endoplasmic reticulum, which has been suggested to facilitate the efficient passage of nutrients to the fungus (Harder and Chong, 1984).

The function and role of haustoria

Haustoria have long been thought to be involved in nutrient uptake, a view based only on circumstantial evidence. However, in the powdery mildew fungi haustoria are the only route by which nutrients can be assimilated from the host and in fact these fungi have provided excellent experimental material for the investigation of haustorial function. The bulk of *Erysiphe pisi* hyphal growth is on the outside of the host plant and it has therefore been possible to separate the fungus and the host for chemical analysis. In this way, contributions from vegetative hyphae have been analysed separately. It has also been possible to isolate haustorial complexes from infected plant cells, providing a means for the examination of the host pathogen interface *in vitro*. The protocol used is outlined in Fig. 3.11. Evidence has been collated from electron microscope studies (Spencer-Phillips and Gay, 1981) and the use of radiolabelling techniques, leading to an increase in our understanding of the biochemistry (Gay and Manners, 1981; Gay, 1984) and the role of haustoria (Gay and Woods, 1987) in biotrophic invasions.

Information concerning enzyme activities and evidence of solute exchange associated with isolated haustorial complexes has led to the development of hypotheses concerning solute transport mechanisms across fungal haustoria. The extrahaustorial membrane has been shown to be semi-permeable using isolated haustorial complexes from powdery mildew (*Erysiphe pisi*) infections of *Pisum sativum*. By the use of ^{14}C radiolabelling techniques Gay and Manners (1981) showed that sucrose, and possibly glycerol are the most likely translocates to be absorbed directly into powdery mildew haustoria. Spencer-Phillips and Gay (1981) showed that the distribution of ATPase activity was polarised in cells infected by *E. pisi* (Fig. 3.12a, b). ATPases are membrane-bound enzymes involved in the functioning of ion pumps at membranes which are responsible for pumping H^+ ions out and K^+ ions in to cells (H^+/K^+ electrogenic exchange). The region of host plasma membrane which invaginated to accommodate the haustorium (extrahaustorial membrane) lacked ATPase activity (absent or inactivated), while the region of host plasma membrane which continued to line the invaded cell showed ATPase activity. Therefore, two functionally distinct regions (or domains) of the host plasma membrane were recognised in infected cells. The haustorial plasma membrane (fungal) also showed ATPase activity. The extrahaustorial membrane is tightly bound to the haustorium at the neck region and the neck bands appear to act as seals preventing the escape of solutes from the extrahaustorial

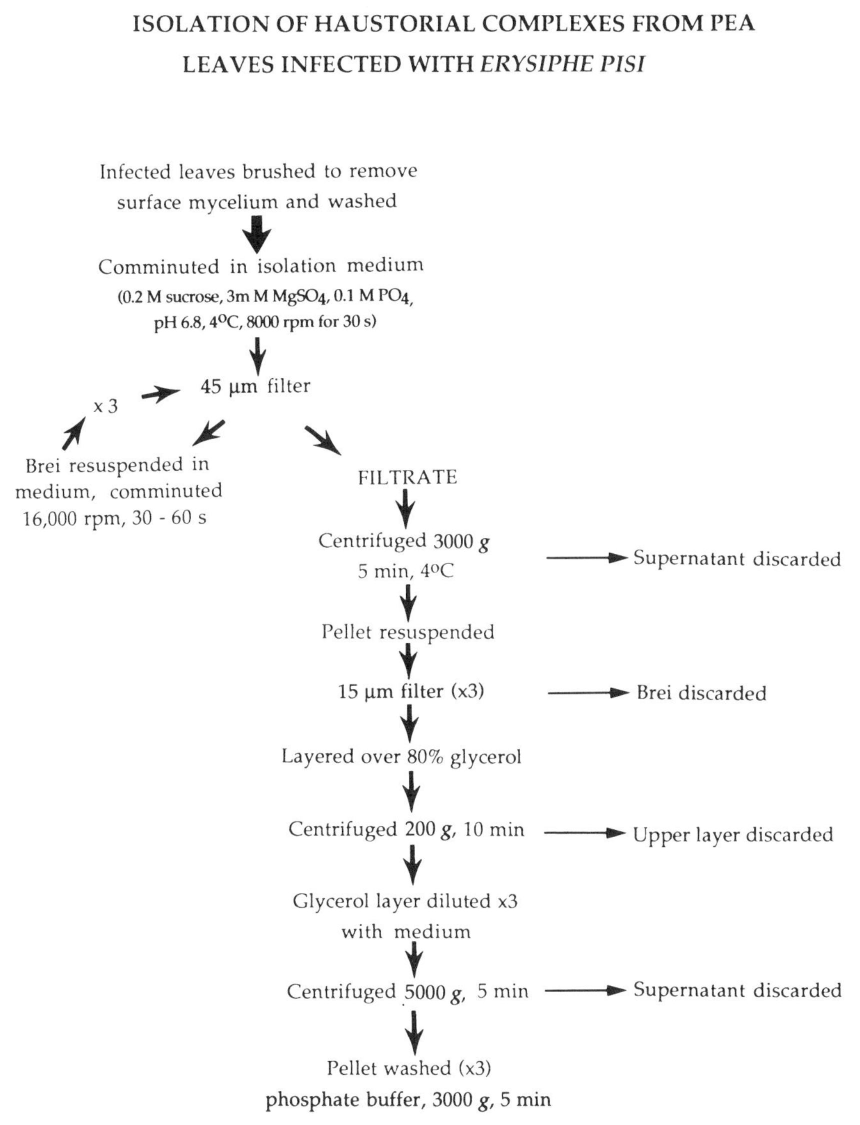

Fig. 3.11 Protocol for the isolation of haustorial complexes from pea leaves infected with *Erysiphe pisi.* (Redrawn from Spencer-Phillips, 1984, with permission.)

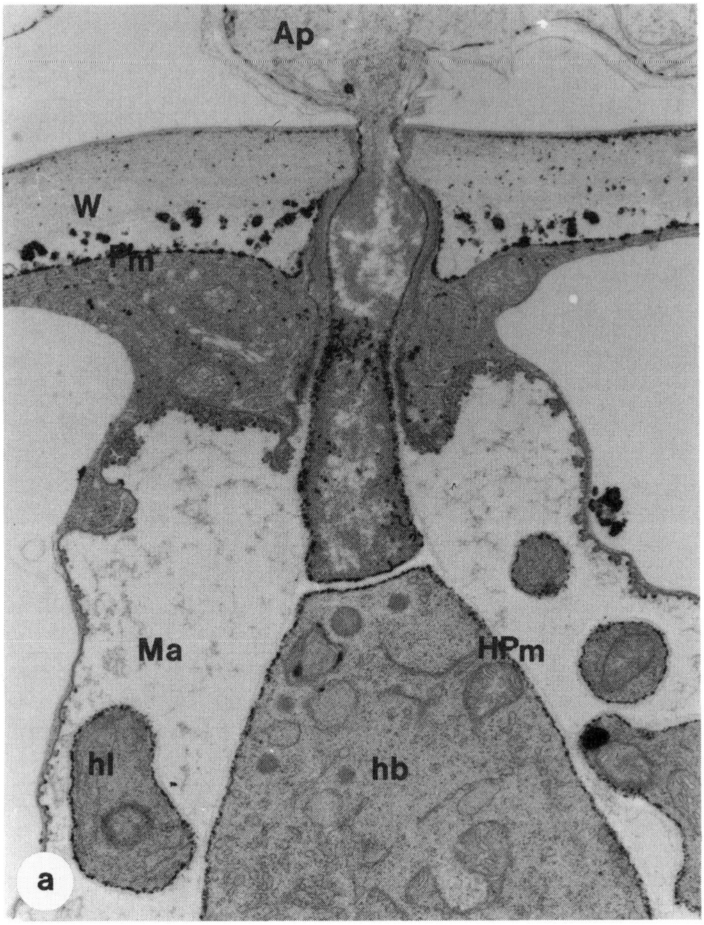

Fig. 3.12 Haustorium of *Erysiphe pisi* infecting an epidermal cell of *Pisum sativum*. (a) The epidermal cell wall (W) has been penetrated by the haustorial neck arising from the appressorium (Ap). The septum is perforated (not shown) and the haustorial body has lobes seen as separate in the extrahaustorial matrix (Ma). Electron-dense deposits, indicating ATPase activity, occur on the plant cell wall-lining region of the host cell plasma membrane (Pm) and on the haustorial plasma membrane (HPm) in the haustorial body (hb) and haustorial lobes (hl). (x16 900). (b) (p.179) Higher magnification (x43 800) to show the lack of ATPase activity at the extrahaustorial membrane (EM), the points of transition (arrowed) and the presence of two neck bands (N). Host cytoplasm (Cy), and host vacuole (v). Reproduced from Spencer-Phillips and Gay (1981), courtesy of the Trustees of the New Phytologist.

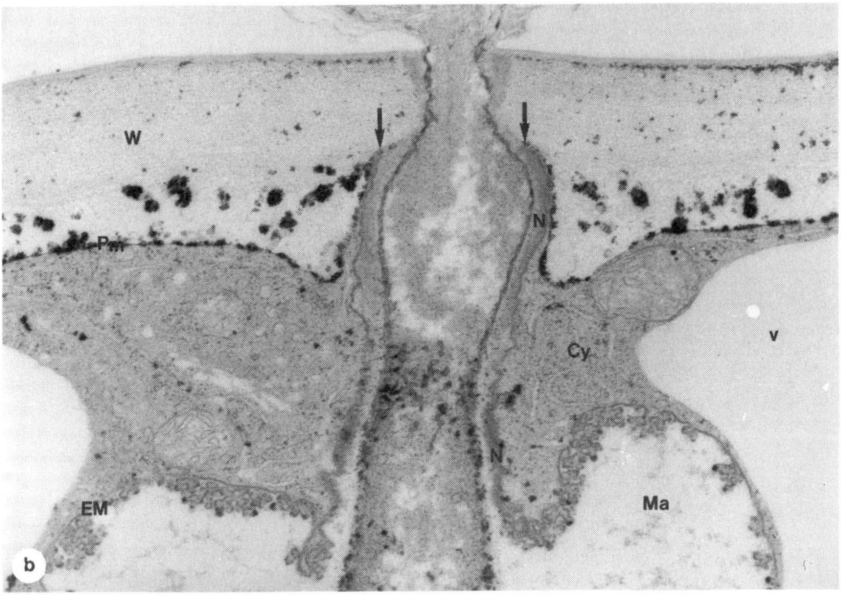

Fig. 3.12(b)

matrix. The matrix is not in direct contact with the host apoplast (intercellular spaces and cell walls). A diagrammatic representation of solute uptake into fungal hyphae is given in Fig. 3.13. More recent work (Beale *et al.*, 1990; Clark and Spencer-Phillips, 1990) has demonstrated the presence of ATPase activity on fungal plasma membranes within leaf tissues (Fig. 3.14) and the development of relatively gentle enzyme maceration techniques (Fig. 3.15) for isolating intact haustoria provides a useful means for further investigations.

The structural and functional modifications induced in the region of the host plasma membrane which surrounds the haustorium allows the fungus to control the solute efflux from the host. As the epidermal cells move solutes from the host apoplast into the invaded cell cytoplasm the ATPase activity at the haustorial plasma membrane allows solutes in the extrahaustorial matrix to become depleted and therefore maintains a high concentration gradient. Host cells lack the control to oppose this solute efflux from the extrahaustorial membrane. Spencer-Phillips and Gay (1981) suggested that the ionic products of the ATPase pumps may be complementary in host and fungus leading to the activation of transport of photosynthates across the membrane.

In other biotrophic infections the distribution of ATPase in both domains of the host plant plasma membranes has been shown to be similar. However, in *Uromyces appendiculatus* (rust disease of beans), *Albugo candida*

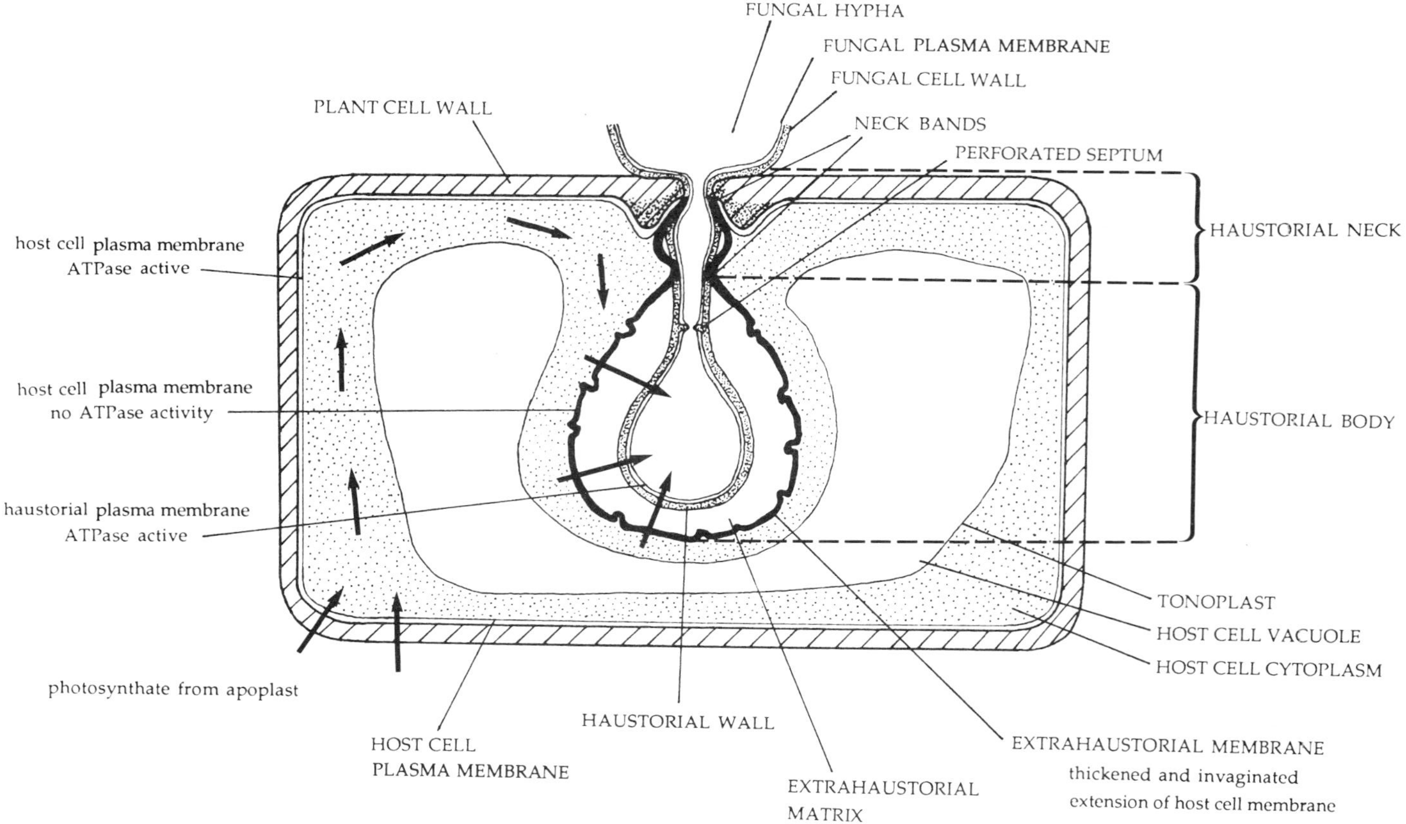

Fig. 3.13 Diagram of *Erysiphe* haustorium in an epidermal cell to show the major structural features and to indicate the probable route of solute transport into the fungal hypha (adapted from an unpublished drawing by Dr G. N. Greenhalgh, with permission).

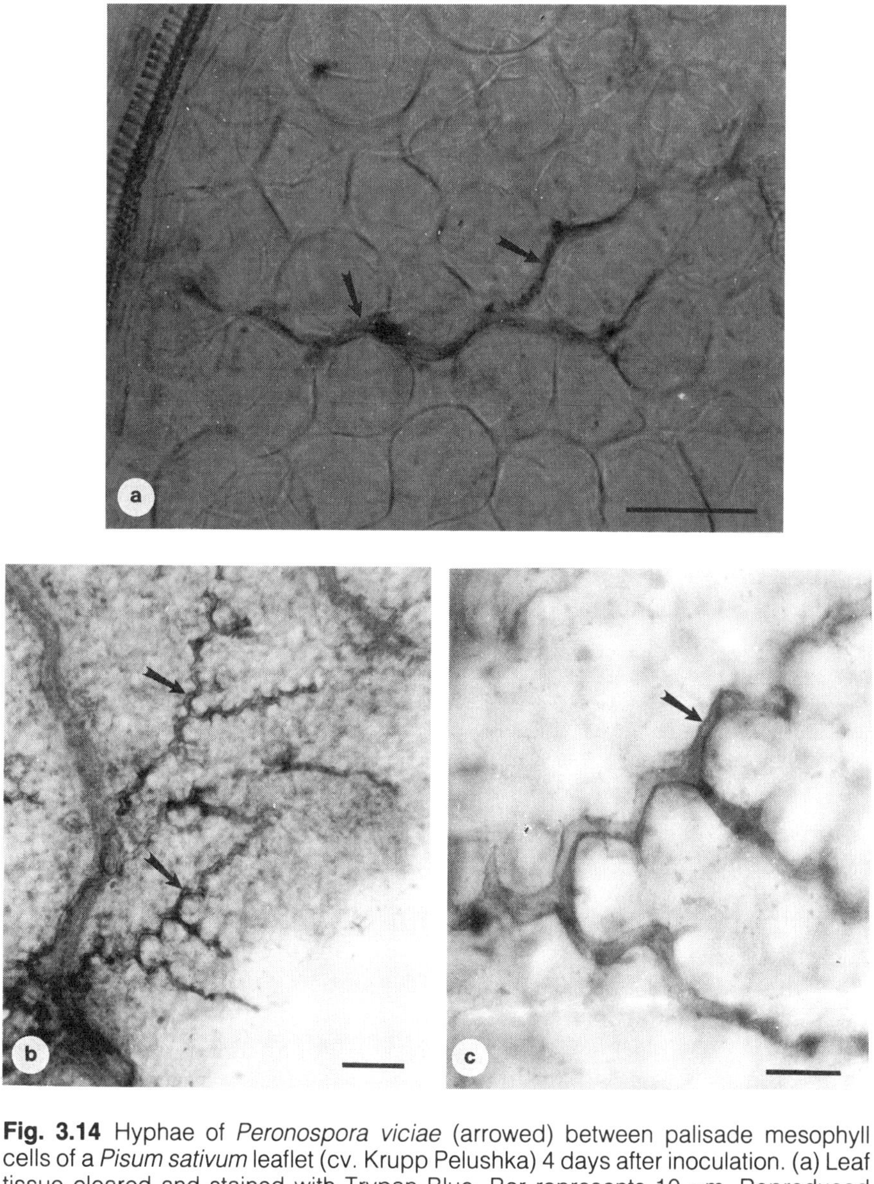

Fig. 3.14 Hyphae of *Peronospora viciae* (arrowed) between palisade mesophyll cells of a *Pisum sativum* leaflet (cv. Krupp Pelushka) 4 days after inoculation. (a) Leaf tissue cleared and stained with Trypan Blue. Bar represents 10 μm. Reproduced from Clark and Spencer-Phillips (1990), courtesy of P. T. N. Spencer-Phillips and Cambridge University Press. (b, c) Hyphae stained using ATPase cytochemistry. Bars represent (b) 50 μm and (c) 20 μm. (b) Reproduced from Beale *et al.* (1990), courtesy of IOP Publishing Ltd. (c) Beale, A. J. and Spencer-Phillips, P. T. N. (unpublished).

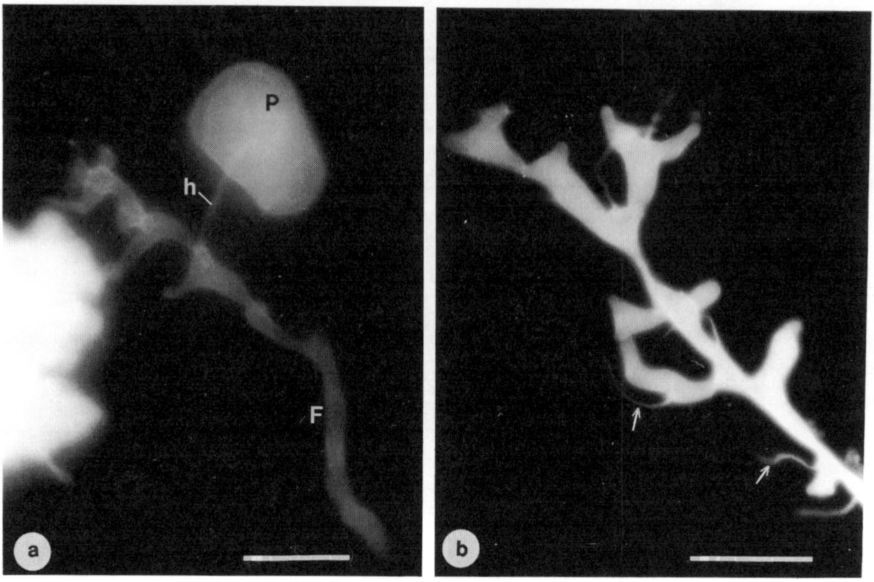

Fig. 3.15 Hyphae of *Peronospora viciae* and cells of *Pisum sativum* from enzymically macerated tissue, 4 days after inoculation. Cell walls were stained with fluorescent brightener and photographed using epifluorescence optics. Bars represent 20 μm. (a) Hyphal fragment (F) with palisade mesophyll cell (P) attached to distal portion of haustorium (h). (b) Hyphal fragment with haustoria (arrowed) completely detached from plant cells. Reproduced from Clark and Spencer-Phillips (1990), courtesy of P. T. N. Spencer-Phillips and Cambridge University Press.

(white rust of crucifers) and *E. graminis* (barley powdery mildew) the haustorial plasma membrane did not show ATPase activity. Gay and Woods (1987) have suggested that in these infections transport into the haustorium depends on the generation of a concentration gradient. This may be brought about by the combination of solute conversion in the fungus, solute synthesis in the host and ATPase activity at the wall lining domain of the host plasma membrane, together with the fact that the haustorium is linked to the transport system of the host by the seals at the haustorial neck region. The key to this hypothesis lies in the formation and efficient functioning of the neckband seals which maintain the distinction between the host plasma membrane domains and the coupling of the host and fungal cells. Not all invading haustoria have such effective seals and in these cases haustorial function may not be as efficient, in fact greater infection frequencies may be necessary to compensate. Alternatively, in some instances the extrahaustorial membrane and haustorial wall may be closely bound together, coupling the host and fungal transport systems and maintaining two domains.

It can be seen, therefore, that haustoria are very specialised structures by means of which the fungus is able to tap host resources very efficiently and causing only minimal disruption to that host.

GENETICS OF PATHOGENICITY

Amongst the main goals of plant breeders are two objectives. The production of crop plant varieties or cultivars which are high yielding and/or exhibit disease resistance. It certainly has been possible to introduce quite high levels of disease resistance and field grown cultivars may remain disease-free for several planting seasons, but over the course of a few years this resistance is lost. This has previously been described as the breakdown of resistance but it is now well accepted that this is in fact the result of alterations to the genotype of the pathogen rather than to changes in the host. There are a number of sources for such genetic variation in fungi, generated by a range of mechanisms.

Sources of genetic variation

Mutations

As with other organisms mutations, spontaneous changes in the genetic material of fungi, will occur and can be inherited by progeny. Such changes may be the result of additions, deletions, substitutions, inversions and/or a range of other alterations to the base sequences within the DNA of the chromosomes. Mutations are an important source of genetic variation. However, from a plant pathology viewpoint mutations towards loss of pathogenicity are often observed, especially after prolonged periods in culture. Ellingböe (1987) has pointed out that for *Kabatiella zeae* this may be a physiological change and pathogenicity can often be restored after passage through the host plant.

Recombination

Fungi reproduce by means of spores. Those spores that are derived by sexual processes (oöspores, ascospores and basidiospores) are also subject to variation through the segregation and recombination of genes during meiosis. Sexual processes probably have a key role in the evolution of host–pathogen relationships (Courtice and Ingram, 1987). For asexually produced spores (zoospores, conidia and uredospores) however, the degree of variability is reduced but does occur. These spores are formed in very large numbers so that even a very low frequency of variation gives rise to noticeable changes within a population. The generation of haploid spores (one set of chromosomes) which give rise to haploid mycelium means that, since there are no competing alleles present, every gene has the potential to be expressed and that changes are not masked but are expressed immediately.

Some fungi are homothallic, one individual producing both sets of gametes, and are capable of self-fertilisation. This removes the need for cross-fertilisation and gives rise to a population with the same genetic structure as for asexual reproduction. For heterothallic species gametes from two individuals must interact before sexual stages can be formed. This, combined with mating types (strains assigned to different incompatibility groups), limits the amount of inbreeding which can occur.

Fungal hyphae are also able to fuse together (anastomosis) and exchange nuclei, usually with genetically similar individuals. There is also the potential for some degree of cytoplasmic and organelle exchange. Extranuclear genetic material may be exchanged and give rise to cytoplasmic inheritance of characters controlled in this way. Tolerance to toxic materials and virulence to host plant species (e.g. *Puccinia graminis* f. sp. *tritici*) have been shown to be cytoplasmically inherited.

Heterokaryosis

Nuclear exchange following fertilisation or anastomosis between hyphae of differing genotypes leads to the formation of heterokaryons, in which hyphal compartments contain two or more genetically different nuclei. Two different nuclei may then fuse to form a diploid zygote nucleus or give rise to dikaryotic mycelium. Dikaryons may exhibit quite different morphological and physiological characters from halpoid mycelium, indeed the general life-style may be quite distinct. The importance and role of heterokaryosis in pathogenicity is not known although differences between genetic states have been observed and may have significance in terms of disease development.

For instance, in the Basidiomycete *Puccinia graminis,* which causes wheat rust, haploid basidiospores infect the alternative host *Berberis vulgaris.* Haploid, monokaryotic mycelium grows only within that host and heterokaryon formation occurs on *B. vulgaris.* Dikaryotic mycelium does grow on either host but dikaryotic aeciospores and uredospores infect only wheat plants.

The hemibiotrophic Basidiomycete *Crinipellis perniciosa* causes Witches' broom disease of cocoa in South America. It is a major pathogen of cocoa causing 90% crop losses in some regions but the mechanisms which control the unusual life cycle are still poorly understood. Basidiospores germinate on and give rise to the infection of actively growing meristematic tissue. Invasive hyphae are wide (5–20μm), rather irregular and distorted in appearance and grow intercellularly. This biotrophic stage (primary) is monokaryotic and causes massive proliferation of cocoa meristems accompanied by cell expansion and tissue distortion, giving rise to the production of green brooms in the host. Eventually broom tissue dies and the fungus dikaryotises, growing both inter- and intracellularly as characteristic (secondary stage) narrow hyphae (1.5–3μm) which have clamp connections. After extensive saprophytic growth basidiocarps form on dead brooms. In this instance the dikaryotic phase mycelium has been readily grown in culture whereas the primary monokaryotic phase has only been grown briefly in culture and limited growth achieved. It is clear that the two genetic states have very different nutritional requirements and operate under very precise control mechanisms.

Parasexual hybridisation

Genetic recombinations can occur in fungal heterokaryons by the process known as parasexual hybridisation. In instances where the vegetative phase is normally haploid the fusion of two nuclei within a heterokaryon will give rise to a diploid nucleus. The occurrence of this diploid phase in the asexual cycle is stable and may give rise to more diploid nuclei. Mycelium may remain dikaryotic but alternatively crossing over may occur during mitosis. Subsequently the diploid nucleus may revert to the haploid state as chromosomes are lost by non-disjunction during divisions. This whole process is a very rare occurrence and each event in the sequence takes place only occasionally. However, there are many nuclei within a colony and the chance of occurrence within the whole colony is high. The process is sometimes referred to as the parasexual cycle. It is important to realise that it is the result of random changes and does not occur as a regular cycle in any way akin to the sexual cycle. However, this mechanism enables genetic recombination where the sexual cycle is absent and it has been identified in the Deuteromycotina, e.g. *Fusarium*, *Verticillium* and *Alternaria*.

Variation among fungal pathogens

Within a fungal species some individuals may be pathogenic on a particular host species. Such fungi are often grouped together as *formae specialis* (f. sp.) or varieties. Within such *formae specialis* those individuals which are specifically pathogenic on a particular host variety or cultivar are grouped in to races. Individual variants within a defined race are termed biotypes.

The term race is used to designate genetically distinct mating groups or to describe different genotypes within a species. Such genetic differences give rise to reproductive isolation, either biologically or geographically. Physiologic races have been separated on the basis of the reactions of biotrophic fungi to different host cultivars. Described in this way the number of races is determined by the number of resistance factors and is therefore potentially enormous. A number of other concepts and criteria (e.g. toxin production, aggressiveness and morphological differences) have also been used to define races and there has been increasing controversy concerning the use of the term (Caten, 1987).

MECHANISMS OF DISEASE RESISTANCE

During its life a plant will be affected by the activities of a wide range of microbes, will be threatened by numerous pathogenic species and may be attacked by vast numbers of potentially pathogenic individuals. In general, many plants survive such onslaughts, often with very little appreciable damage although a number of factors may influence the eventual outcome of the confrontations.

A host plant may exhibit a range of responses to challenge by a specific pathogen. If no visible symptoms are manifested the plant is said to show high resistance (low susceptibility) and the disease reaction is termed incompatible. Where disease symptoms develop and the plant is completely overcome then it is said to show low resistance (high susceptibility) and the disease reaction is compatible. Plants may exhibit great variation in the degree of resistance expressed between these extremes.

The age and physiological state of the host plant and the favourability of the prevailing environmental conditions will also influence the general susceptibility of the host. Additionally, factors arising directly from the genetic make up of the host and of the pathogen will interact. Host resistance/susceptibility, i.e. the inherent or acquired ability of the plant to suppress invasion and successful colonisation by the pathogen, must be considered together with pathogen virulence/avirulence, i.e the ability of the pathogen to penetrate the plant, colonise host tissue and reproduce within that environment. The specific genetic interaction in and between host and pathogen, modified by the environment, has in fact been regarded as an entirely new biological system and called the aegricorpus by Löegering (1966). This is a helpful concept which emphasises both the importance and the degree of interaction between the genetic expressions of host and pathogen in confrontation.

Interactions begin at a very early stage in any encounter between a fungus and a plant. Disease resistance, or susceptibility, probably begins at or during the initial contact between the cells of the host and the pathogen; a stage often referred to as recognition. Theoretically, at least, recognition between cells must be the start of subsequent physiological interactions. Sequeira (1979) defined recognition, in plant pathological terms, as the specific event eliciting host response which facilitates or impedes further growth of the pathogen. Taking into account current molecular and genetical theories (Daly, 1984; Gabriel and Rolfe, 1990; Dixon and Harrison,

1990) it is likely that cell membranes are the sites of such recognition, in particular the plasma membranes, since these are the initial point of contact between pathogen and host. Some form of chemical sensor (recognition factor) must be involved; probably surface carbohydrates and/or membrane proteins or glycoproteins. A signal, transmitted to the cytoplasm by that sensor, initiates those metabolic responses in the host cell which subsequently restrict pathogen development or differentiation prior to infection. In plants the sensing system may not be very specific and may be triggered by abiotic as well as biotic factors.

Plants do not have any immunological defense systems of the kind described for animals. However, it has been possible to immunise plants and to protect them, to some extent, from disease using various abiotic and biotic agents. Natural and synthetic chemicals have been used successfully to induce resistance, e.g. D-phenylalanine, polyacrylic acid, salicylic acid (aspirin) and 2,2-dichloro-,3,3-dimethylcyclopropane carboxylic acid (DDCC). The influence of such chemicals combined with the natural defense systems of the host serve to enhance resistance. Similarly biotic factors have been used to induce protection (Dean and Kuc, 1987). For example, cucumber plants have been successfully protected against *Colletotrichum lagenarium*, which causes anthracnose, by inoculation of cotyledon leaves or early true leaves with the same fungus. Such treatment does not confer complete immunity on the plant but it can reduce the severity of the disease and/or delay symptom development very considerably. Immunity may also be detected in areas of the plant which have not previously been inoculated (systemic induced resistance). Additionally such immunity has been shown to persist under field conditions.

Most plants are in fact resistant to most pathogens and a range of resistance mechanisms have been identified. Mechanical barriers which act as physical inhibitors to pathogen invasion, and biochemical resistance mechanisms, including toxic or inhibitory compounds, undoubtedly function in combination. In some instances such structural and biochemical characters are already present in the plant prior to pathogen invasion as preformed (passive resistance) resistance mechanisms. However, in other cases resistance reactions are induced (active resistance) by pathogen invasion.

MECHANICAL BARRIERS TO PATHOGEN INVASION

Preformed structural barriers

The initial contact between a fungus and a plant most usually occurs at the cuticle, which overlies the epidermal cell walls. The waxes which constitute this layer provide the plant with a highly water-repellent surface.

Additionally structural features, such as general leaf shape, vein pattern, folding and the presence of hairs will influence the surface environment and also other factors, such as run-off from leaves. In some instances this may create a very hostile environment where fungal spores are subject to physical removal from the surface, and to desiccation. The thickness and quality of cuticle layers will vary on leaf surfaces and, as leaves age, waxes may be lost by damage and erosion. It is surprising, however, that there is little correlation between the thickness of the cuticle and underlying epidermal cell walls and the ease with which fungal pathogens are able to penetrate.

Many fungal pathogens are able to enter leaves through stomata but the structural characteristics of these openings do influence penetration. In particular, for example, stomata with very broad or elevated guard cells and small appertures may restrict the entry of hyphae. Although some pathogens can force entry through closed stomata, others (e.g. *Puccinia graminis*) must wait for apertures to open and therefore long periods of stomatal closure would be very restrictive to entry. The timing of opening during the day also influences the ease with which pathogens may penetrate. Rust spores germinate in the early morning and are prone to desiccation on the leaf surface if stomata do not open until evening.

Other natural openings, such as lenticels, may become suberised. Suberin is very resistant to microbial attack and will form an efficient mechanical barrier to entry particularly if it is laid down as a continuous layer throughout the lenticel. Cells of the endodermis, which are also suberised, restrict the spread of vascular wilt and other root-invading fungi. In more deeply seated tissues subepidermal sclerenchyma can act as a structural barrier. Such thickened, lignified, secondary walls are highly resistant to penetration.

Induced structural barriers

Cork layers

Cork layers are laid down in response to mechanical injury to cells. Activity in the cork cambium results in the formation of cork cells in areas surrounding penetration points, which has the effect of cutting off the pathogen from the rest of the plant. Nutrient supplies to the invader are thereby removed and additionally the plant is protected from any toxins released by the pathogen. The cells in the infected area die, forming a necrotic spot, and the isolation by cork layers gives rise to scabs which may later slough off and be lost (e.g. apple scab caused by *Venturia inaequalis*).

In infected leaves, cork cambium may develop around an infection site extending between the upper and lower epidermis. A gap (abscission

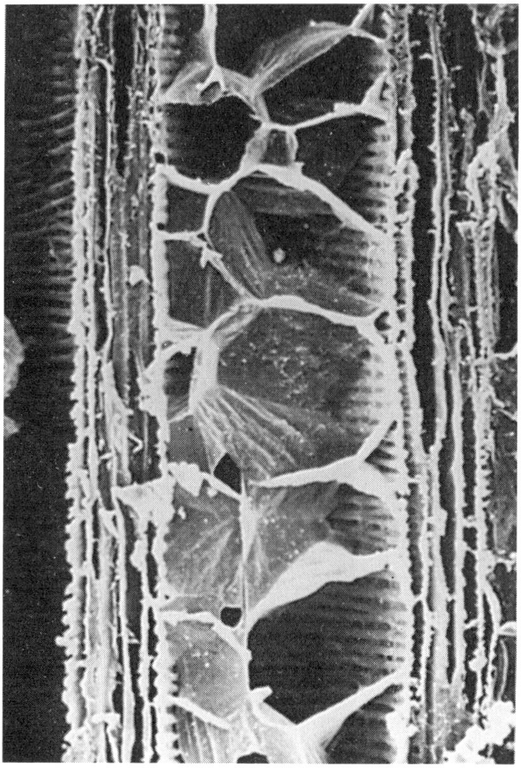

Fig. 3.16 Scanning electron micrograph of a longitudinal section through xylem vessels of resistant banana (cv. Valery) inoculated with *Fusarium oxysporum* f. sp. *lycopersici*. Lumen of xylem vessel totally congested by tyloses 8 days after inoculation. Reproduced from Pegg (1985), courtesy of Cambridge University Press.

layer) eventually forms between the cells surrounding the infection and the healthy cells, as the middle lamellae are dissolved. Eventually the central area is cut off and isolated and may ultimately fall out of the leaf to leave a 'shot-hole'. Healthy regions are therefore protected from toxic exudates and necrotic zones may eventually be lost from the plant. The speed of cork layer formation has also been related to resistance levels in plants.

Tyloses

Tyloses are formed in xylem vessels as a response to abiotic stress, invasion and/or aging. Protoplasts from the surrounding xylem parenchyma cells extend and protrude into the xylem through pits. The cellulosic walls of the tyloses expand to such a degree, and in such numbers, that the xylem

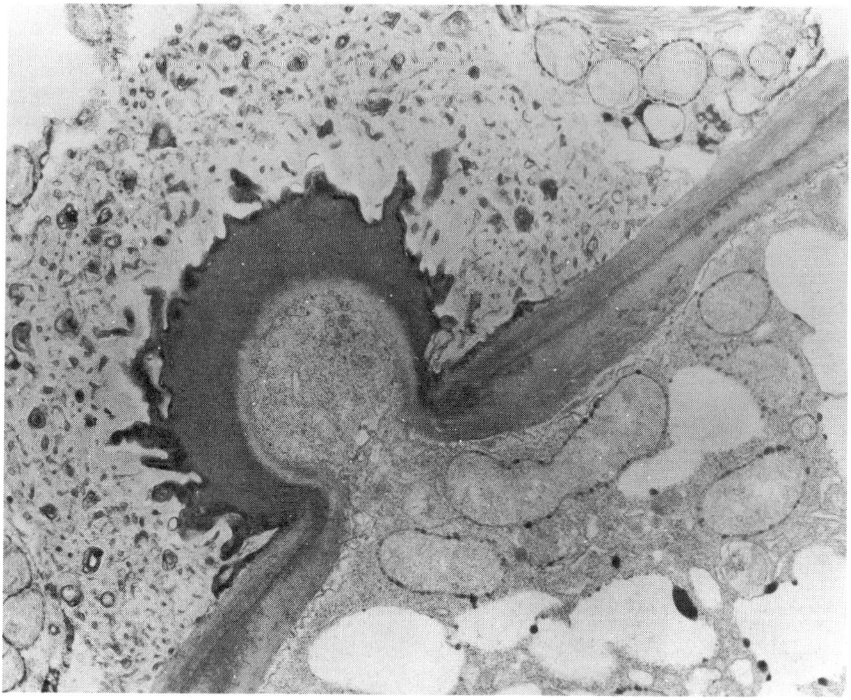

Fig. 3.17 Resistance response to *Peronospora parasitica* of Brassica. Transmission electron micrograph section through an attempted site of haustorial formation in a mesophyll cell. Penetration of the host wall has occurred and the cell has responded by producing a thick accretion of material sealing off the penetration site. The 'wall appositions' have been shown to fluoresce under ultraviolet illumination, suggesting the presence of β-(1, 3)-glucans, such as callose. x27 000. Reproduced courtesy of Dr J. A. Lucas and Dr M. Coffey.

vessels may become completely blocked (Fig. 3.16). Although this restricts water transport through these vessels the spread of vascular pathogens may also be curtailed. In fact plants which are resistant to vascular wilts have the capacity to form many tyloses, whereas more susceptible plants form relatively few. In a similar way hemicellulose gels and gums may be produced, in cells adjacent to infected tissue, blocking pathogen growth and spread through vascular tissues but also restricting solute movement through plants.

Lignitubers

As an invading hypha extends through the cells of plant tissues, the progress of further penetration may be progressively impeded by the deposition of extra deposits of cellulose, callose and lignin, often with additional insoluble phenolic compounds. A sheath, or lignituber, may form around the invading hypha increasing the mechanical strength of the cell walls involved and preventing water loss. Lignification and the laying down of suberin on lignified cell walls are thought to limit the proliferation of an invading fungus and are therefore important factors in disease resistance (Asada and Matsumoto, 1987). Lignification and suberinisation may be induced by both biotrophic and necrotrophic species. It is interesting that the exact composition of the materials laid down after damage has occurred differs from those deposited during natural plant development. It has also been suggested that greater resistance is afforded by these depositions as wounds age (Biggs and Miles, 1988; Bostock and Stermer, 1989). Lignin deposits may increase the mechanical strength of cell walls preventing further penetration and may also alter the susceptibility to lytic enzyme attack by invading fungal pathogens. Additionally, it is likely that phenolic precursors of lignin may act as antifungal compounds.

Additionally, callose (β-(1,3)-glucan) papillae may be deposited on the inside of cell walls as a response to fungal invasion, e.g. the response of wheat plants to *Gaeumannomyces graminis* (Hornby and Fitt, 1981) and barley to powdery mildew (Stoltzenburg *et al.*, 1984a, b). This may prevent further hyphal penetration (Fig. 3.17) or may serve as a repair mechanism following invasion, as a consequence of disease resistance. It has been suggested that the deposition of papillar material may begin before penetration (phase one) and continue after penetration (phase two) and that in some cases papilla formation occurs as a plant response to hyphal degradation in infected tissue (Aist, 1983).

BIOCHEMICAL RESISTANCE MECHANISMS

Preformed compounds

Resistance to invasion may be the direct result of the presence of substances in or on the plant before infection which are toxic to potential invaders. Such chemicals probably have a role in both non-host species and in resistant genotypes of susceptible cultivars. Some plants produce fungitoxic exudates on leaf surfaces which inhibit spore germination. The most cited example is that of resistant onion bulb scales. The water-soluble compounds catechol and protocatechuic acid (Fig. 3.18) interact in solution on the plant surface. Spores from *Colletotrichum circinans*, which causes onion smudge disease, lyse on contact. However, this mixture is not effective against spores of all onion pathogens.

Other phenolic compounds have also been implicated in resistance, either present in surface waxes or within host cells. Phenolic compounds (e.g. phenolic glucosides, glucose esters and wyerone) inhibit the action of fungal hydrolytic enzymes (Fig. 3.18). Other compounds which show

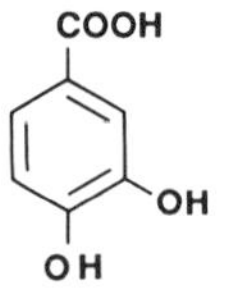

Protocatechuic acid

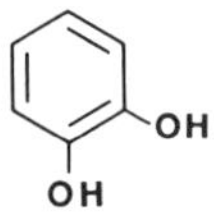

Catechol

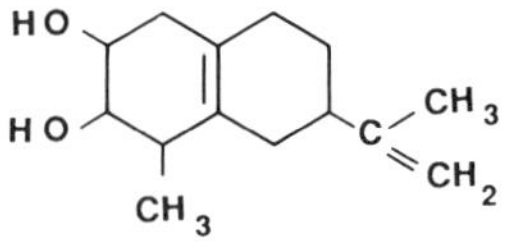

Rishitin

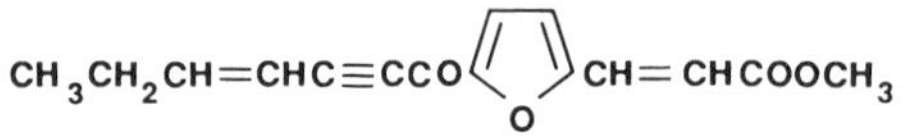

Wyerone

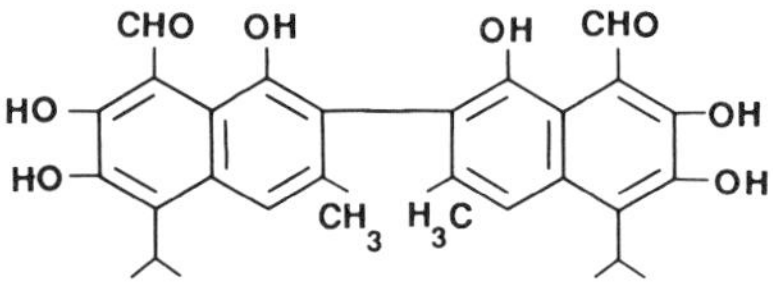

Gossypol

Fig. 3.18 Chemical structures of some preformed resistance factors that occur in plants. Levels of these compounds may increase after infection.

antifungal properties may also be important in resistance to infection, e.g. the saponins, tomatine in tomatoes and avenacin in oats. However, there is relatively little evidence that these substances are involved in the intact plant and it is difficult to prove the presence of such compounds prior to invasion and tissue extraction.

Hardening of plants to cold conditions has been shown to increase disease resistance, as the result of the stabilisation of membranes. Cold stress tends to increase the synthesis of phospholipid and the phospholipid/sterol content of membranes increases accordingly. It is also suggested that if the pathogen does not recognise the host plant (lack of recognition factors) then the normal infection process will not occur. Similarly if toxins are not recognised by receptor sites then no toxic action will result.

Induced biochemical responses

Biochemical inhibitors

The production and accumulation of phenolic compounds are often associated with injury responses and the wound healing of plants, but these substances also have fungitoxic activity. Rapid accumulation of such compounds (Fig. 3.19) may occur in resistant plants after infection, e.g.

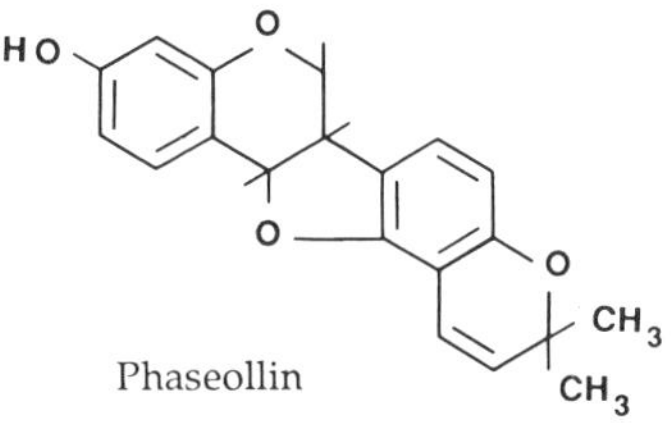

Phaseollin

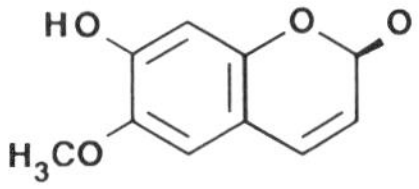

Scopoletin

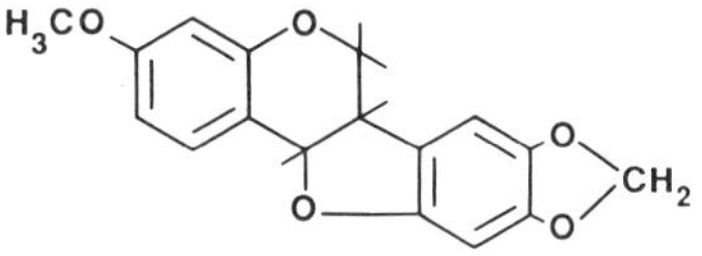

Pisatin

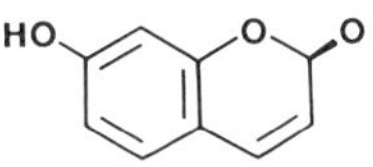

Umbelliferone

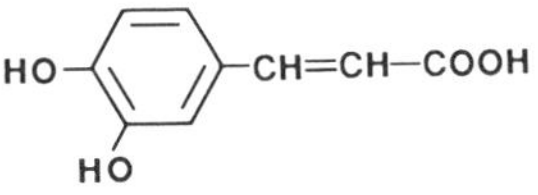

Caffeic acid

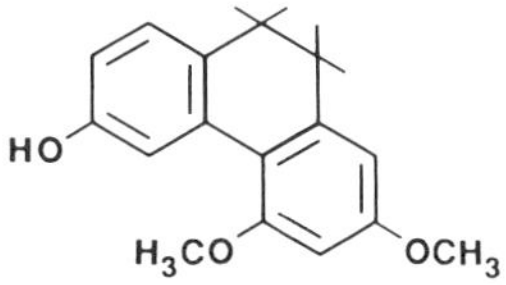

Orchinol

Fig. 3.19 Chemical structures of some phenolic compounds that are induced in plants as a response to fungal invasion. Low levels may be present prior to invasion.

caffeic acid, umbelliferone, scopoletin and orchinol. The concentration of any one such phenolic may not be sufficient to inhibit the development of infection but in combination the compounds present may well be effective antifungals.

In plant tissues some phenolics may be present in bound forms. On infection fungal invaders release glucosidases which hydrolyse these molecules. Phenolic compounds are released, often in toxic concentrations sufficient to inhibit further fungal development, although the role of these substances in disease resistance is not clear.

Hypersensitive response

The hypersensitive reaction observed in host plant cells was related to disease resistance many years ago (Ward, 1902; Stackman, 1915). It occurs as the result of the recognition of infection by the host plant and

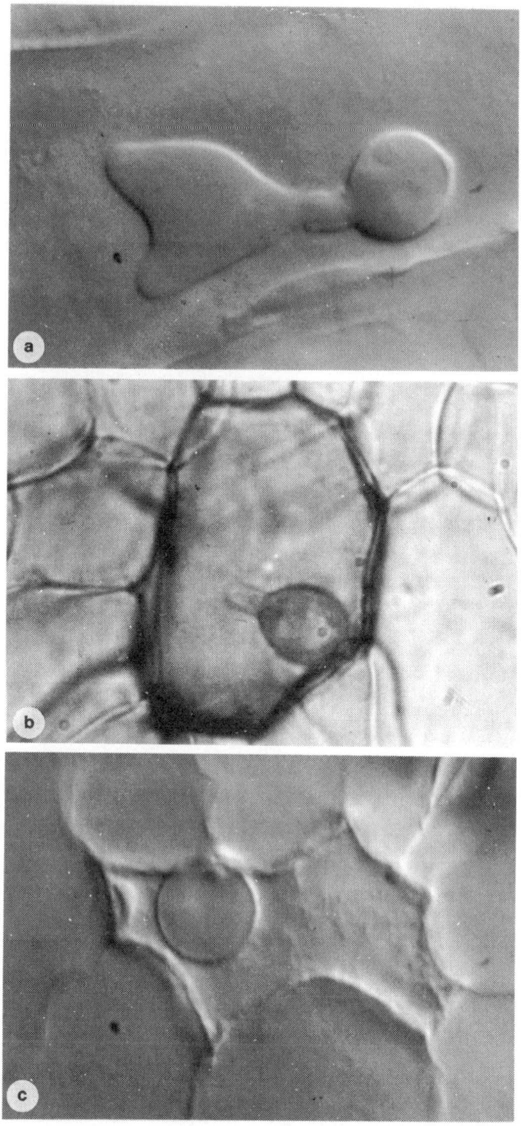

Fig. 3.20 Invasion of *Bremia lactucae* (lettuce downy mildew). (a) Primary and secondary vesicles in an epidermal cell of a compatible lettuce cultivar (Cobham Green) 24 h after inoculation. Note the lack of host response. (b) Single-cell hypersensitive reaction of *Brassica oleracea*, a non-host species following inoculation. The fungus has formed a spherical primary vesicle and growth has stopped before any further development (the early development of a secondary vesicle or hypha is present). (c) Primary vesicle associated with a single-cell hypersensitive response in the epidermal cell of an incompatible lettuce cultivar (Mildura carrying gene *Dm* 3) 24 h after inoculation. (a, c) Nomarsky interference contrast microscopy. Reproduced courtesy of Dr I. R. Crute, HRI, East Malling.

as a consequence of incompatibility between host and pathogen (Fig. 3.20). The term hypersensitive response relates to the visual symptoms of dramatic host cell necrosis associated with penetration in resistant host cultivars. Infected cells rapidly and suddenly loose membrane permeability and turgor, are subject to dramatic increases in respiratory activity, accumulate phenolic compounds and phytoalexins (Fig. 3.21). An increase in oxidation reactions and a decrease in reduction reactions (evidence of phenoloxidation) results in necrosis; a limited number of cells die and turn brown. The pathogen is therefore isolated in localised necrotic tissue and does not spread to healthy regions. It is interesting that the biochemical changes which are associated with hypersensitivity are very similar to those which occur after wounding, at senescence or as stress responses.

The sequence of events surrounding the manifestation of hypersensitivity has been, and is still, the subject of much controversy. Although it is difficult to measure accurately, particularly as the extent and timing of the response is very variable, there is evidence that the growth of the pathogen is inhibited before the macroscopic symptoms of necrosis become apparent. For example, the death of *Puccinia coronata* f. sp. *avenae* haustoria was shown to occur before the death of host cells (Prusky *et al.*, 1980). Additionally, the hypersensitive response has been shown to be induced in potato tuber tissues by high molecular weight materials leaked from cell free homogenates or dead *Phytophthora infestans* mycelium, whether from isolates normally showing compatibility or incompatibility with the host plant (Bostock *et al.*, 1982). However, chemical inhibitors and compatible pathogen isolates did not induce the hypersensitive response. This evidence therefore suggests that the hypersensitive response is probably a symptom of incompatibility rather than a primary determinant of disease resistance.

It has recently been proposed (Doke *et al.*, 1987) that host plant cells become activated for the hypersensitive response prior to pathogen penetration. Cells become sensitised during fungal spore germination or wounding. Subsequently, biochemical changes associated with hypersensitivity are triggered by adhesion between the outer surfaces of invading fungal hyphae and the host plasma membrane. In an incompatible reaction determinants on the surface of the invader, which are most probably hyphal wall components, initiate the early events of the hypersensitive response. Host cell membranes are depolarised and electrolyte leakage occurs, leading rapidly to cell death. Phytoalexin synthesis is then triggered in cells adjacent to those sustaining damage. It has been suggested that in compatible reactions specific inhibitor molecules act as suppressors, blocking recognition in the host plant and preventing the progress of the hypersensitive response. Such substances, probably water-soluble β-(1, 3)-linked glucans, have been found to be released from *Phytophthora*

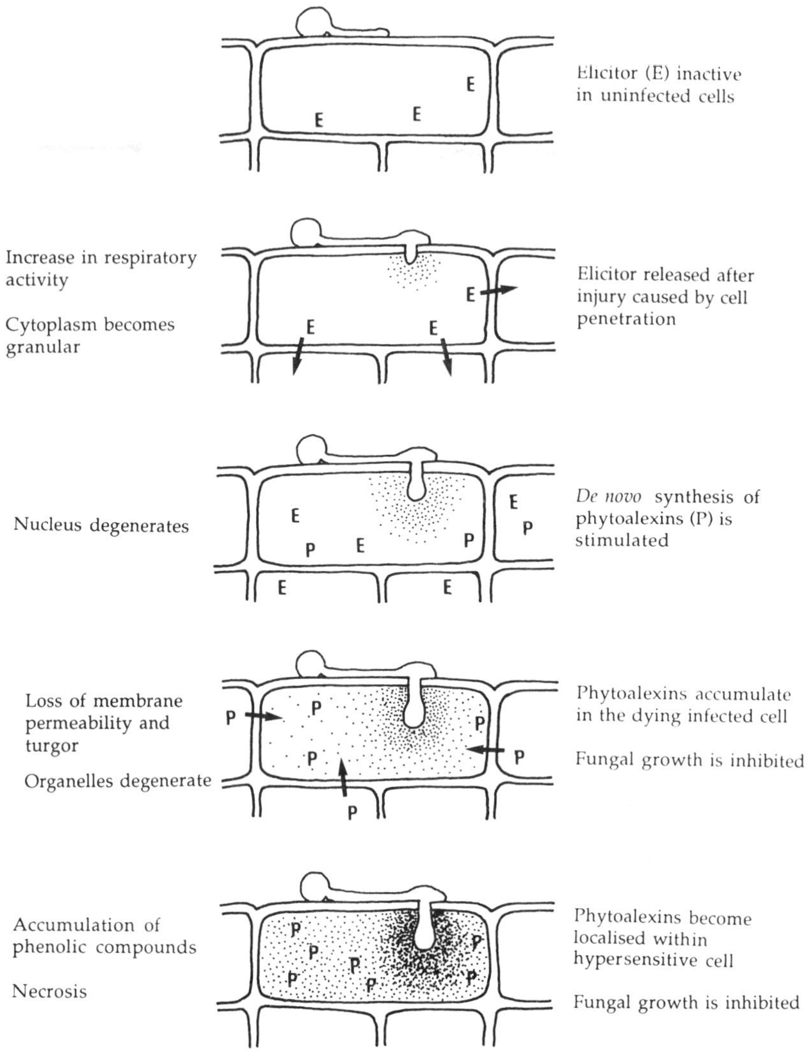

Fig. 3.21 The stages of morphological and physiological change occurring during development of the hypersensitive response of plant cells and the possible role of constitutive elicitors in phytoalexin accumulation. Redrawn after Bailey (1982).

infestans zoospores, and to specifically suppress hypersensitivity in host plants. Such a mechanism may contribute to compatibility, resulting in infection. There is still much controversy concerning the role of hypersensitivity in host plant resistance.

A great deal of evidence (Dixon, 1986; Dixon and Harrison, 1990) supports the view that phytoalexin accumulation accompanies the manifestation of the hypersensitive response. Phytoalexins are compounds, synthesised in host plant cells, after infection, which have antimicrobial properties. Accumulation of phytoalexins occurs in cells localised to regions showing the hypersensitive response (Fig. 3.21). High concentrations are often found in nearby necrotic cells, probably as a result of transport from live cells, which prevents further growth of the pathogen.

Phytoalexins

Phytoalexins are low molecular weight antimicrobial compounds which are synthesised by and accumulate in plant cells after microbial infection (Paxton, 1981). These secondary metabolites are produced as a response to the localised death of plant cells (hypersensitivity). Phytoalexin synthesis occurs *de novo* in cells adjacent to those which have been damaged, either by physical wounding or biochemical injury, as the result of penetration or toxin effects. Some may however, probably in low concentrations, be present prior to infection as a result of the plants response to abiotic stresses. Phytoalexins are produced by both resistant and susceptible necrotic tissues although resistance appears to be related to the concentrations that accumulate. The actual amounts of phytoalexin which are synthesised by plants which have been challenged are very variable and may depend on a number of factors including the physiological status of the plant and its genetic composition.

Many different phytoalexins have been characterised and they have been isolated from a range of plant families, e.g. Leguminosae, Solanaceae, Orchidaceae and Gramineae (Dixon, 1986). The compounds are diverse and belong to a number of chemical groups (Fig. 3.19). Phenolics, including flavonoids (pisatin and phaseollin) and coumarins, polyacetylenes (wyerone) and isoprenes, including terpenoids (rishitin and gossypol) and steroids. Most plants produce several different phytoalexins and although the compounds are structurally diverse (Stöessl, 1983), any one plant family tends to produce similar compounds. This correlation of structures with plant species is close and phytoalexins have been used as chemotaxonomic markers (Ingham, 1981). As Bailey (1987) has recently pointed out, the chemical structures produced are solely determined by the host plant genome although physiological and environmental factors may influence the amounts produced.

The presence of phytoalexins was first demonstrated by Müller and Börger (1940) in their experiments on the host parasite relationships between potato cultivars and races of *Phytophthora* . Slices of potato tuber were inoculated with a race of *Phytophthora infestans* to which the potato cultivar was resistant. This resulted in the protection of the tuber from subsequent infection by a compatible *Phytophthora* race, to which the potato cultivar was susceptible. They suggested that a compound(s) had been produced by the host cells which prevented successful development of the later infection. Since that time phytoalexins, their biochemical action and their role in disease resistance (Yoshikawa *et al.*, 1987; Dixon and Harrison, 1990) have been the subject of a great deal of investigation and controversy.

More recently Van der Plank (1975) suggested that the hypersensitive response in infected plants, and phytoalexin synthesis, occur after inhibition of an invading microbe (primary infection) and act as preformed, acquired resistance only to secondary pathogens. Neither therefore represents the primary mechanism of resistance. The occurrence of necrosis and phytoalexin accumulation are not necessarily related but are often localised together. In any case, it is difficult to precisely time events which occur. It is suggested that cell death stimulates phytoalexin accumulation. Although they are water-soluble, phytoalexins do not move within host plants. These compounds accumulate at sites of pathogen invasion and are not detected at a distance from infection sites. This may occur because phytoalexins become bound to plant cell walls or are metabolised in nearby cells (Kühn and Hargreaves, 1987).

Phytoalexins affect the growth of fungi, inhibiting germ tube elongation, colony growth (radial growth rate) and dry weight accumulation. The main effect of phytoalexins on fungi is via the membrane. The plasma membrane is rapidly disrupted and the structural integrity is affected, resulting in an often dramatic and massive loss of electrolytes. There is also evidence that phytoalexins disrupt respiratory pathways (Yoshikawa *et al.*, 1987). Cytoplasmic streaming is prevented and cell contents become granular. Membrane disruption and distortion has been observed in electron microscope sections of affected hyphae. Newly synthesised regions of hyphal tips are particularly vulnerable and sensitive and swell to bursting (Smith, 1982). There is also some evidence that phytoalexins inhibit the action of fungal cell wall synthesising enzymes but this may be primarily a disruption of the highly integrated lysis/synthesis (Chapter 1) system of wall growth at hyphal tips. However, the effects of phytoalexins certainly serve to restrict the growth of fungi within host plant tissues although the role this has in disease resistance is controversial.

The challenge of a plant by a compatible pathogenic fungal species may stimulate the accumulation of lower concentrations of phytoalexin than if the plant was challenged by non-pathogens. In many instances pathogenic species may be less sensitive to phytoalexins produced by the host plant than are non-pathogenic fungi, indeed such tolerance may have an important role in pathogenicity. Additionally some pathogenic fungi may be able to detoxify or inactivate a phytoalexin. These are mainly necrotrophic species. A correlation has been shown between the pathogenicity of *Fusarium solani* (perfect stage, *Nectria haematococca*) on peas (*Pisum sativum*) and

Table 3.2 Some examples of compounds with elicitor activity, derived from fungal pathogens

Pathogen	Host	Elicitor	Biological activity	Race specificity
Cladosporium fulvum	Tomato	Glycopeptide (extracellular)	Phytoalexin induction	–
Cladosporium fulvum	Tomato	Peptide	Necrosis	+
Colletotrichum lindemuthianum	Bean cotyledons	Polysaccharide	Phytoalexin induction	+
Fusarium solani	Pea	Chitosan	Phytoalexin induction	–
Phytophthora infestans	Potato tubers	Fatty acids	Phytoalexin induction	–
Phytophthora megasperma f. sp. *glycinea*	Soybean cotyledons	Glycoproteins	Phytoalexin induction	+
Phytophthora megasperma f. sp. *glycinea*	Soybean cotyledons	Glucomannan	Phytoalexin induction	–

the ability of the fungus to detoxify (demethylation) and be insensitive to the phytoalexin pisatin (Van Etten *et al.*, 1980, 1989). Other fungi are able to detoxify phytoalexins while the concentrations encountered are below those which cause direct inhibition. The mechanisms by which detoxification of phytoalexins is achieved by pathogenic fungi are varied and may involve a single metabolic reaction or several biochemical steps. Some fungi are able to tolerate phytoalexins without the need for degradation.

Compounds which induce phytoalexin synthesis in plants are termed elicitors and may be microbial in origin (exogenous elicitors) or plant-derived (endogenous elicitors). Such biotic elicitors are often complex molecules such as carbohydrates, glycoproteins, polypeptides, enzymes or lipids. Additionally, phytoalexins also accumulate in response to abiotic elicitors. These are usually factors which cause plant stress, such as cold, ultraviolet light or the presence of heavy metals in the immediate environment.

(a) Microbial elicitors

Microbial elicitors include intact fungal mycelium, cell-free homogenates, fungal cell wall carbohydrates such as glucans and glycoproteins, and some fungal enzymes (Table 3.2). Some fungal cell wall elicitors have been characterised from a range of pathogenic fungi (Ebel, 1986; Dixon and Harrison, 1990) but are not universal or species-specific and very few have been shown to be race-specific.

Many reports concerning the action of elicitors of microbial origin have been made. A polypeptide extracted from *Monilinia fructicola* has been shown to elicit the formation of phaseollin in French beans, and glycopeptides isolated from *Phytophthora megasperma* f. sp. *glycinia* caused the formation of glyceollin in soybean tissues. Chitosan (β-1,4-linked glucosamine) extracted from *Fusarium solani* was shown to be a very effective elicitor of pisatin in peas. Glycoproteins from *Cladosporium fulvum* act as elicitors of rishitin in tomato fruits. Some of these elicitors are also responsible for the induction and stimulation of other host defense responses in addition to phytoalexin biosynthesis.

(b) Plant-derived elicitors

It is also clear the fragments of plant cell wall material may act as endogenous elicitors for phytoalexin synthesis, after release by wall degrading enzymes originating from either the fungal pathogen or from damaged host plant cells. For example, the enzyme endo-α-1,4-polygalacturonic acid lyase, produced in culture by *Rhizopus stolonifer* (a pathogen of castor bean plants), has been shown to release oligogalacturonides from host plant cell walls, which then act as elicitors of phytoalexin synthesis. Molecules released from necrotic tissues may also act as elicitors in healthy tissues. Soybean plant cell walls contain β-glucosyl hydrolase which has been suggested to degrade β-glucan elicitors. It is possible that these enzymes act to localise the effect of elicitors in plant tissues (Darvill and Albersheim, 1984).

(c) The function of elicitors in host defence

It is clear, therefore, that a range of molecules trigger defense reactions in plants. These elicitors act on cells very rapidly (effects can be measured in minutes) and obviously operate very precise metabolic control. It is also clear that there is no universal mechanism by which such effects are achieved.

Elicitors appear to induce phytoalexin accumulation in plant tissues by stimulation of mRNA synthesis, through an increase in the transcription rate of the genes involved (Bailey, 1987; Dixon and Harrison, 1990). After plant infection or treatment of tissues with elicitors some mRNAs have been shown to increase in activity, while others are unaffected. It has also been shown that polymorphisms at gene, mRNA and protein levels, of enzymes involved in the biosynthesis of phenolics, do occur in infected plants which has given rise to the suggestion (Ryder *et al.*, 1986) that such gene amplification may enhance plant responses in terms of disease resistance. There is a great deal of evidence that many biochemical defense responses are controlled by *de novo* gene expression (Collinge and Slusarenko, 1987; Dixon and Harrison, 1990).

In compatible biotrophic infections, where the host does not apparently recognise the presence of the invader, no phytoalexin accumulation occurs. In some instances it seems likely that compatible fungi release suppressor molecules which inhibit elicitation but stimulate the plant to produce non-fungitoxic compounds and therefore render the host susceptible. Suppression, probably by glycoprotein molecules, may be induced in susceptible plants by the invading pathogen. A glucan elicitor from *Phytophthora megasperma* has been shown to exert biochemical effects on plants cells by causing changes in the metabolic fluxes across the plasma membrane, particularly changes in Ca^{2+} fluxes (Stäb and Ebel, 1987), although this effect may not be a universal mechanism (Kunoch, 1990). Metal ions (Cu^{2+}; Cd^{2+}, Hg^{2+}) are also thought to act as elicitors.

It has been suggested that endogenous elicitors may function synergistically with pathogen-derived elicitors (Davis *et al.*, 1986) to effect a resistance response in the plant. Relatively low concentrations of individual compounds may therefore induce a dramatic defense response when present in combination. Since a number of compounds obviously act as elicitors, various combinations of two or more different molecules may act synergistically. This theory can therefore explain the great variability in plant responses which are observed.

A number of models have been proposed to explain the function of elicitors. Early ideas suggested that (elicitor-receptor model: Albersheim and Anderson-Prouty, 1975) avirulence genes code for the synthesis of enzymes which modify molecules carried on the outer surface of a pathogen. These then act as elicitors of defense reactions in plants by binding to specific receptors on the host surface. Later on (dimer model: Ellingböe, 1982) elicitation was envisaged to occur as the result of the formation of protein dimers between host and pathogen. In this case primary protein products of resistance genes in the plant, and avirulence genes in the pathogen, would bind together in protein–protein interactions. This results in the stimulation of host defense mechanisms. Hypersensitivity then occurs as the result of this incompatibility between the partners (Gabriel and Rolfe, 1990). Slight changes in the ontogeny of such interactions can provide explanations for the modified responses which can be observed in some infected host plants.

More recently a further hypothesis has been proposed (ion-channel defense model; Gabriel *et al.*, 1988), which implicates that the plant cell plasma membrane has a key role in the communication of defense signals. A large number of transmembrane proteins, coded by resistance genes in the plant, provide specific receptor sites on the outer membrane surface, which bind to avirulence gene products from the potential invader. This results in membrane depolarisation, which has indeed been shown to occur very early in host–pathogen interactions, the opening of ion channels in the plasma membrane and the leakage of electrolytes. If a large amount of membrane leakage occurs then the collapse of membrane integrity would give rise to the communication of a wound response within the plant and to rapid host cell death. In developing this theory it was proposed that different functional classes of transmembrane receptors might be involved, each binding to different types of signal molecule from different origins (e.g. elicitors from microbes, elicitors of plant origin, etc.). Additionally, it has been suggested that the receptor-ion channels may be ion-specific (e.g. for K^+, Ca^{2+}, Cl^- and H^+) so that different fluxes would occur through the membrane. In this way alterations in cytoplasmic ion concentrations may affect exchanges with the vacuole and generate different signals within the plant. Calcium ions are implicated particularly in the functioning of this model, since these are so important in plant cells and very small increases in calcium ion concentrations cause rapid responses in cells. Different types of cells, at different locations in the plant, may respond to signals in different ways and cells surrounding one which has been invaded may respond to secondary signals from their damaged neighbour.

Enzyme activities

On mechanical injury or infection increased levels of the key enzymes and precursors of phenolic compounds are found in plants. These are found in higher levels in infected resistant cultivars and have therefore been associated with disease resistance. Phenylalanine ammonia-lyase (PAL) levels have been reported to increase on infection. Synthesis of PAL can be prevented by inhibitors of protein synthesis and it has therefore been suggested that the synthesis of this enzyme is normally repressed in undamaged plants and that it is synthesised *de novo* in damaged plants. PAL is involved in the biosynthesis of lignin and also in the production of compounds acting as phytoalexins, e.g. the synthesis of pisatin and phaseollin are dependent on PAL activity. Polyphenoloxidase enzymes (phenolase, polyphenoloxidase and laccase) have key roles in the oxidation of phenolic compounds to quinones. These quinones are much more toxic to fungi compared with the non-oxidised forms and probably contribute to disease resistance in plants. However, the source of synthesis is not always clear since these enzymes may be produced by either the host or the pathogenic invader. Peroxidases catalyse the oxidation of phenolic compounds and aromatic amines and also contribute to the polymerisation of phenyl propane substances to lignins. The role of peroxidases in resistance can probably be attributed to their contribution to increased rates of lignification.

Increased activities of chitinase and β-(1, 3)-glucanase are also found in wounded and infected plant tissues. It is thought that these may act to degrade fungal cell walls and may therefore release elicitor molecules in infected areas. Additionally, purified enzymes with cell wall degrading activity (e.g. pectate lyase, pectin lyase and xylanase) have been shown to kill plant cells in suspension cultures (Mauch *et al.*, 1988). It is now thought that these enzymes may act synergistically in killing cells (Bucheli *et al.*, 1990); certainly chitinase and β-(1, 3)-glucanase are most effective in combination.

Disease resistance may relate to increased levels of activity, the speed of synthesis of these enzymes and the production of fungitoxic compounds. However, the role of these enzymes may be as much involved in responses to injury, wound repair and the production of highly thickened and toughened tissues for mechanical protection.

GENETICS OF PLANT RESISTANCE

Types of resistance

Non-host resistance

Most plants are resistant to most pathogens. For a potential pathogen most plant species are outside its host range. These non-host plants are therefore completely immune to infection, even under optimum conditions for disease development and in the presence of a large inoculum. In the future, the study of non-host resistance may well reveal the mechanisms by which host range is determined, provide information concerning the evolutionary basis of parasitism and may hold the key to breeding for highly durable resistance in crop cultivars (Heath, 1986).

Apparent resistance

Many plants escape disease even though they are theoretically susceptible. In other words, not all healthy plants are immune, but may exhibit apparent resistance. Much of this may be due to the developmental sequence of the plant and the pathogen under particular conditions. If a plant is susceptible only when young, or at fruiting for example, and the pathogen is not active or not present at this stage, then the plant will simply miss becoming infected. Temperature fluctuations, lack of moisture leading to desiccation, etc., will affect the development and dispersal of the pathogen. Although some plants may escape, such climatic changes can give rise to the start of an epidemic.

Some plants may be tolerant to disease. These may be susceptible to a pathogen but are not overcome by it. This tolerance characteristic is heritable. Invasion by the pathogen may cause little apparent damage to a tolerant plant or alternatively such a plant may give a good yield in spite of any damage which is caused. Such plants are often vigorous and hardy and are usually regarded as good stock.

True resistance

True host resistance is genetically controlled and has been the subject of many studies, and much controversy.

(a) Vertical resistance

It has been shown that resistance may be controlled by a single, or very few (2–3) genes (oligogenic resistance). Van der Plank (1963) introduced the term vertical resistance to describe the situation in which plants showed very high levels of resistance to a particular physiological race of pathogen but very little resistance to other races. Such hosts often show the hypersensitive response, usually very early in infection, and the pathogen is not able to establish or multiply within the plant tissues. This is also known as race-specific resistance.

(b) Horizontal resistance

Plants have a level of general resistance to pathogens, which is race non-specific and controlled by a large number of genes (polygenic resistance). Equal resistance is shown to all pathogen races. This horizontal resistance (Van der Plank, 1963) may not totally protect plants from infection, and the actual level of resistance is mediated by the environment, but it does reduce susceptibility and slows down the establishment of the pathogen. A high horizontal resistance is useful in a crop cultivar, to slow down the spread of disease. It has been argued that horizontal resistance is an artifact and in fact corresponds to vertical resistance. Additionally, it is likely that the molecular mechanism of resistance is the same.

(c) Cytoplasmic resistance

In some instances true resistance is controlled by genetic material which is contained in cell cytoplasm and has been termed cytoplasmic resistance (Day, 1974). The most famous example is resistance to leaf blight in corn (caused by *Phyllosticta maydis*), which is present in the cytoplasm of

Table 3.3 Quadratic representation of the host-pathogen gene interaction (gene-for-gene hypothesis for one gene in operation)

Pathogen genes	Host genes	
	R (resistant, dominant)	**r** (susceptible, recessive)
A (avirulent, dominant)	RA	rA
a (virulant, recessive)	Ra	ra

normal corn varieties but absent from the cytoplasm of Texas male-sterile corn.

Gene-for-gene hypothesis

Interactions between a host and a pathogen are mediated by environmental conditions but are ultimately determined by genotype. Flor (1956) studied the genetics of flax cultivars and isolates of the rust pathogen *Melamspora lini* in field observations. He demonstrated a close genetical relationship between the host and the pathogen and formulated the gene-for-gene theory which is still regarded as an important hypothesis in plant pathology. He showed that for each gene determining disease resistance in host flax plants there existed a specific, complementary gene in the rust fungus which determined virulence. For a host cultivar with a single resistance gene there is a single complementary virulence gene in the pathogen and for host cultivars with 2, 3 or 4 resistance genes there are 2, 3 or 4 complementary virulence genes in the fungus. Many studies have now provided support for this hypothesis; as well as the rust fungi, *Phytophthora infestans, Erysiphe* spp., *Cladosporium fulvum, Colletotrichum lindemuthianum* and their hosts conform to such gene-for-gene interactions.

Usually, although not always, genes for resistance in the host are dominant (R) and those for susceptibility (lack of resistance) are recessive (r). In the pathogen genes for avirulence are dominant (A) and those for virulence (a) are recessive. Where the gene-for-gene hypothesis operates therefore, four gene combinations are possible in a host–pathogen interaction (Table 3.3). Three combinations R–a, r–A and r–a give rise to compatible reactions (susceptible) and infections are successful. One combination, R–A, results in an incompatible reaction and no infection occurs (resistant). Genes for virulence appear in pathogens in response to new genes for resistance in the host. In other words, new resistance in the host creates a situation which selects for new virulence in the pathogen. Hosts and pathogens are in a continuing state of co-evolution.

Exceptions to this hypothesis have been recorded, particularly where virulence or resistance is controlled or modified by more than just a few genes, and possibly not all host–pathogen interactions may be represented in this way. The gene-for-gene concept remains largely theoretical and provides information which is important in epidemiological terms but it does not consider the biochemical or molecular bases of resistance.

Gene functions in host–pathogen interactions

Although a number of theories have been put forward to explain the specificity of host–pathogen interactions in biochemical terms (Bailey, 1983) very little is known concerning the genes involved and their regulation. Few gene products have yet been identified but modern molecular biological techniques now provide means by which these factors may be investigated (Gilchrist and Yoder, 1984; Collinge and Slusarenko, 1987; Lamb *et al.*, 1989; Dixon and Harrison, 1990).

The expression of resistance to a particular pathogen is determined by resistance genes in a plant. The molecular nature of such genes and the regulation of their expression is not yet known, although information accrued to date suggests that these are complex units which probably function with a high degree of genetic variability, keeping pace with changes in pathogen populations. No plant resistance genes have yet been cloned.

In an incompatible reaction molecular recognition leads to the activation of a range of defense mechanisms. In a compatible reaction however, recognition does not occur, or is suppressed, and successful host penetration and colonisation may be achieved by the invading pathogen. Structural and biochemical host defense mechanisms are initiated by the activation of defense response genes (disease resistance response genes). The activation of these genes rapidly results in phytoalexin biosynthesis, production of lytic enzymes, hydroxyproline-rich glycoproteins, lignin biosynthesis and other defense activities in the plant. A number of such defense genes, encoding for enzymes involved in defense pathways, have now been cloned from plants and future investigations will provide information concerning the activities and control mechanisms which are likely to operate in intact plants.

In a potential pathogen genes must be expressed which code for the production of suitable molecules to enable the preparation of the host surface for physical penetration (e.g. cuticle degrading enzymes, differentiation of appressoria and infection structures); for the dissolution of host tissues (plant cell wall degrading enzymes); for phytoalexin detoxification and for the damage and killing of cells (toxins). These are termed pathogenicity genes. Avirulence genes may code for the production of race-specific elicitors which may be recognised by a potential host plant and elicit defense reactions. Little is known concerning the regulation or functioning of these genes. The cloning of some such genes has now been achieved although most success has been with bacterial systems. However, recent developments with transformation systems for fungi may lead to greater success in the near future.

Effects of pathogenic fungal invasion on host plant physiology 4

INTRODUCTION

Although interactions between a host and a pathogen begin as soon as physical contact is established, effects remain largely localised until host tissues have been penetrated. However, increasing colonisation of the host leads to increasing levels of physiological interaction as direct or indirect effects of pathogen activity. There is a great deal of variation in the nature of the contact set up between a host and a fungal pathogen. Some fungi are highly invasive whereas other species penetrate localised areas or colonise particular types of plant tissues, e.g. vascular wilt fungi which preferentially inhabit xylem vessels. Biotrophs may invade host tissues intracellularly, penetrating the membrane and growing into the cytoplasm directly (*Olpidium brassicae*), or may penetrate the cell wall without breaching the membrane as in haustorium formation (e.g. rusts and powdery mildews), or may remain intercellular (Fig. 4.1). On the other hand, host cells may be killed in advance of mycelial spread by necrotrophic species. In general, necrotrophs are highly destructive to plant tissues, causing maceration of plant cell walls and disruption to physiological processes as a consequence. Biotrophs exert more subtle effects on host functions and metabolism, often causing a minimum of physical disruption. These species may have efficient means for diverting host nutrient supplies for their own use, to support their growth and metabolism.

Visible symptoms provide indication that plant tissues are suffering effects of disease. In some cases such changes appear in cells in close proximity to the fungus and coincide with the extent of fungal invasion. In other infections, however, symptom development and therefore physiological changes, extend to cells beyond the mycelial front. In powdery mildew infections the bulk of fungal biomass remains on the leaf surface,

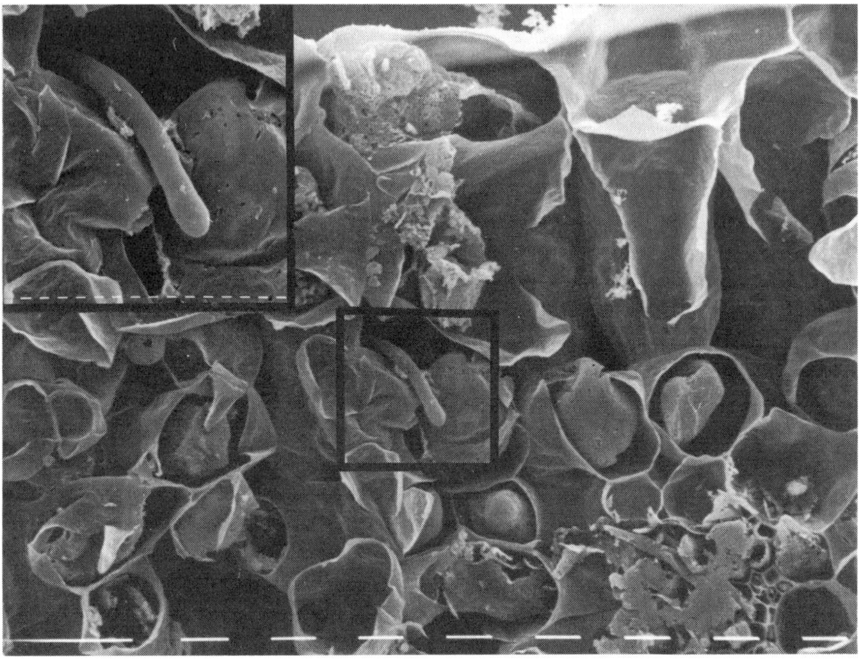

Fig. 4.1 Intercellular hyphal growth of *Septoria apiicola*, causal agent of leaf spot disease, between palisade cells of celery (*Apium graveolens*, L.) leaf tissue. Inset: higher magnification of hyphal tip as indicated. Bars represent 10 μm; inset 1 μm. Reproduced from Donovan *et al.* (1990), courtesy of Cambridge University Press.

with haustoria penetrating into epidermal cells. These haustoria must draw on nutrients from other leaf tissues. Since epidermal cells do not contain chloroplasts these compounds must be supplied by surrounding mesophyll cells. Physiological effects of the fungus therefore extend beyond the regions of hyphal invasion. Similarly in rust invasions, 'green islands' extend beyond the mycelial front in infected leaves. These are regions of metabolically and photosynthetically active cells which surround infection sites and occur in regions of otherwise necrotic cells.

Plant tissues are a complex and heterogeneous substrate for fungal invaders and are very rich sources of nutrients. Organic and inorganic materials and complex metabolites are readily available for extracellular digestion or for uptake directly into fungal mycelium. Plant cell walls are well recognised as leaky and many compounds are therefore present within intercellular spaces even in uninfected tissues. Invading fungi also have some influence on plant cell membranes and tend to increase leakage from surrounding plant cells. The mixtures of nutrients available to a pathogen will vary in different plant tissues too. Interactions between host and fungal pathogen, and the extent of influence which an invading fungus may exert on plant cells, will depend on the type of plant tissues invaded and the status of those tissues. Sieve tubes of the phloem are highly nutrient-rich tissues, with a constant supply of new materials from the circulation through the plant. Xylem vessels are dead cells, in their functional state, but contain good supplies of nutrients and water from the transpiration stream and can be readily colonised, especially by vascular wilt fungi. The nutrient status of other cells may be one factor determining the extent of the spread of a fungus within a plant.

Fungi which invade plant roots may maintain some contact with the soil and obtain nutrients directly from it or may be supplied from passive uptake by root tissues. Leaf-invading species however, must obtain all nutrients and water for growth and development from the host plant. Any diversion of nutrient supplies and/or stores must affect the plant at the tissue and also at the whole plant level. Similarly, the production of toxic compounds by a fungal invader or disruption of the structural integrity of tissues and cells will cause stress to that plant. These direct and indirect influences tend to compound effects within a host, often making a plant more vulnerable to disease and secondary invasion. Some of a plants responses to stresses (e.g. wounding and water stress) give rise to the production of compounds with growth regulator activity and hence to characteristic symptoms of the disease syndrome. Such stresses may aggravate disease development.

Early after infection the fungal invader has a low biomass in the host and plant tissues contain sufficient nutrients to support the growth of the pathogen. Plants often have a greater photosynthetic capacity than that normally used for growth and maintenance and are therefore easily able to support the extra drain on resources (Hancock and Huisman, 1981). In some instances plants are also capable of extra growth to replace damaged leaves or part of a root system and can compensate for losses due to pathogen activity, although the location of an infection may influence the impact of a disease. For example, some tissues are more dispensable, or can be more easily compensated for, than others. Later in infection, however, the fungal biomass becomes greater and the drain on plant resources is such that nutrients must be diverted from the rest of the plant to support further fungal growth. More massive disruption to the plant occurs when the fungus sporulates. The plants control of water loss is destroyed when the cuticle is ripped by the spore-bearing structures and spores are released from mycelium seated within leaf tissues.

Pathogenic invasion affects the highly organised and balanced metabolic pathways in plant tissues such that all physiological processes are influenced throughout the plant.

PHYSIOLOGICAL CHANGES IN HOST PLANT TISSUES

The biochemical pathways in a plant are very highly integrated and interdependent and any disruption represents a loss of control by the host. Even relatively small and localised changes may have drastic consequences to the plant as a whole. Although metabolic differences can be detected between healthy and infected plants and changes can be measured as disease progresses it is not always possible to identify the causes of such alterations in an infected plant. It may not be clear whether the fungus exerts a direct influence on biochemical pathways or to what extent measured changes are attributable to the plants response.

The techniques employed, methods of measurement, experimental scope and design and ways of expressing data obtained, for infected plants, have a great deal of influence on the interpretation of apparent differences. In isolation, measurements may show clear differences between infected and uninfected control plants and between compatible and incompatible disease reactions, but may give little indication of the factors responsible for such changes. In order to investigate the influence of an invading fungal pathogen on a host plant it is important to understand the natural, normal metabolic processes of the uninfected plant and to consider those changes caused by disease in comparison. Some metabolic pathways are better understood than others but in many instances details are being pieced together, and in many cases are still unknown. Recently, modern techniques, with increased levels of sensitivity, have improved our understanding of many pathways; in particular, the use of radioactively labelled tracer compounds. Although it is now possible to calculate metabolic budgets for the whole plant this has been attempted in only a few instances. However, many studies have considered individual aspects of plant metabolism and the changes occurring after fungal infection.

RESPIRATION

A very early effect of infection is the general stimulation of host metabolism and biosynthesis, which leads to the rapid use of energy. The response in the host is an increase in the rate of respiration, which rises, reaching a maximum (may be 100% increase) at the time of sporulation of the fungus, although subsequently the level may reduce. In a normal, healthy plant only a small proportion of the synthetic capacity is used for maintenance, corresponding to less than 20% of the total carbon assimilated by the plant (Kosuge and Kimpel, 1981). After infection however, respiration rates must be sufficient to support the growth of the invading fungal pathogen and the defense response of the plant together with the maintenance of both fungus and host. Such increased respiratory levels lead to the rapid depletion of the plants carbohydrate reserves. It has been suggested that for biotrophic infections this increase in respiration could be directly attributable to the presence of respiring fungal tissues. However, in experiments where powdery mildew mycelium was removed from leaf surfaces, leaving comparatively little fungal material, the rates of respiration measured for infected leaves were not significantly altered.

Disease disrupts the normal processes of carbon turnover and recycling. Kosuge and Kimpel (1981) have suggested that this disruption may be minimised in the host by the efficient use of energy for biosynthetic processes which are concerned with defense reactions. The pathogen acts as a metabolic sink draining host resources. However, increased rates of respiration and an increased flow of nutrients may enable the plant to repair injuries and synthesise components which confer resistance. The metabolism of nutrients also provides a good supply for the use of the invading fungus.

Respiration in healthy plants

Respiration is the enzymatic oxidation of organic materials to simple compounds providing energy for cellular metabolism. The aerobic respiration of glucose follows the equation:

$$C_6H_{12}O_6 + 6H_2O \longrightarrow 6CO_2 + 6H_2O$$

with the release of energy, mainly in the form of phosphate bond energy of adenosine triphosphate (ATP). In the first phase of this reaction, in healthy plant tissues, glucose is converted to pyruvate, in the presence or absence of oxygen, primarily via glycolysis (Embden–Meyerhof–Parnas or EMP system) or, to a lesser extent, via the pentose phosphate pathway (PP). In the second phase, pyruvate is oxidised to CO_2 and H_2O via the Krebs cycle (citric acid or TCA cycle), accompanied by terminal oxidation only in the presence of oxygen (Chapter 1). During glycolysis, the PP pathway and the citric acid cycle, electrons and H^+ ions are removed and used to reduce the pyrimidine nucleotides, nicotinamide adenine dinucleotide (NAD), nicotinamide adenine dinucleotide phosphate (NADP) and flavin adenine dinucleotide (FAD). Reoxidation of these coenzymes gives rise to the formation of the high energy phosphate bonds of ATP, by the attachment of a phosphate group to adenosine diphosphate (ADP) during oxidative phosphorylation. Normally ADP is maintained at a low level in cells owing to this conversion to ATP via the electron transport chain (phosphorylation). When higher levels of ADP accumulate the rate of respiration is stimulated.

Mechanisms of respiratory increases induced by fungi

In a diseased plant there are a number of changes which accompany increased respiration, including changes to the pathways used, increases in the concentrations of enzymes involved and changes to the coupling of respiration to other, related, cellular processes. Plant reactions are non-specific and probably involve several mechanisms which may be different for biotrophic and necrotrophic pathogens.

Uncoupling of oxidative phosphorylation

The uncoupling of electron transfer from phosphorylation can be brought about by chemical inhibitors, e.g. 2-4-dinitrophenol (DNP). As a result the synthesis of ATP via the electron transfer chain is prevented and ADP accumulates, stimulating respiration and increasing the rate of oxygen uptake. There is evidence to support the view (Allen, 1953) that uncoupling is, at least in part, an explanation for the changes measured in diseased plants. The effect has often been attributed to diffusible substances or toxins produced by invading fungi, e.g. *Helminthosporium maydis, H. victoriae* and *Gibberella saubinetii*. In some instances however, the timing of the respiratory increase suggests that uncoupling may be a secondary effect.

In the presence of oxygen, fermentation is suppressed and glucose is used more efficiently through respiration. This is known as the Pasteur effect. In diseased plants however, aerobic carbohydrate breakdown becomes less efficient, but at the same time energy requirements are high and ADP levels rise. This results in an increase in the respiration rate. The Pasteur effect is lost, in a similar way to treatment with DNP. This is considered to be further evidence for uncoupling.

Increased activity of the pentose phosphate pathway

In a healthy plant the normal respiratory pathway is via glycolysis, the TCA cycle and the cytochrome electron transport chain. In diseased plants however, there is a shift from glycolysis to the PP pathway, which is normally used much less. There is also a concomitant increase in the activity of the enzymes involved in that pathway. Evidence for such a shift has been derived from the C_6/C_1 ratio of $^{14}CO_2$ generated by plants from glucose specifically labelled at the C_6 (glucose-6-^{14}C) or C_1 (glucose-1-^{14}C) carbon atom. C_6 and C_1 are derived about equally if only glycolysis and the TCA cycle are operating ($C_6/C_1=1$) and in healthy plants a ratio of about 1 has been obtained. However, if glucose is oxidised via the PP pathway then C_1 contributes most to the CO_2 which is generated and the C_6/C_1 ratio would be less than 1. In rusted plants the measured ratio was around 0.25 (Shaw and Sambourski, 1956). Such a shift to the PP pathway may be quite significant (up to 90%), especially in some biotrophic infections (Shaw and Sambourski, 1957).

Further evidence of such a shift in pathways has been gained by the investigation of the effects of sodium fluoride (NaF). NaF inhibits glycolysis but sensitivity to it is removed in plants infected by rust fungi. This suggests that respiratory activity has moved away from the glycolytic pathway in infected tissues. However, in contrast, barley plants infected with the necrotroph *Pyrenophora teres*, or treated with toxin from that fungus, showed increased levels of respiration which were sensitive to NaF. Smedegaard-Petersen (1984) suggested that in these infected tissues little switch to the PP pathway had occurred.

Production of energy via the PP pathway is very much less efficient than via the formation of ATP. However, the PP pathway is also the main source of phenolic compounds, upon which activated defense mechanisms call, in plants. The increased synthesis which occurs via this pathway may therefore be very important to a wounded, stressed or infected plant.

Increased activity of the terminal oxidases

Cytochrome oxidase is the main terminal oxidase in healthy plants. However, the reduced coenzyme NADPH, produced via the PP pathway, may be oxidised by non-cytochrome oxidative systems in diseased plants. With an increase in oxygen uptake in diseased plants there may be an increase in the activity of polyphenol oxidase, e.g. in diseases caused by *Fusarium*, *Phytophthora* and *Ceratocystis* spp. Ascorbic acid oxidation has also been shown to increase as oxygen uptake increases, e.g. in rusted wheat plants, in oat plants treated with toxin from *Helminthosporium victoriae* and in cabbage infected by *Fusarium oxysporum*. Activity of these oxidases has often been connected with tissue disorganisation.

Respiratory changes in host plants

Resistant plants

Resistant and susceptible cultivars have been shown to react differently to infection by fungal pathogens and it has been suggested (Smedegaard-Petersen, 1984) that different mechanisms may operate. In infected resistant cultivars (incompatible reaction) oxygen uptake has been shown to rise sharply quite early in infection, although subsequently the level reduced to near normal. In infected susceptible cultivars (compatible reaction) however, there was much less change initially and a rise in oxygen uptake occurred a few days after infection (Fig. 4.2). In resistant plants the manifestation of the hypersensitive response may involve host reactions which are very energy expensive and may lead to marked

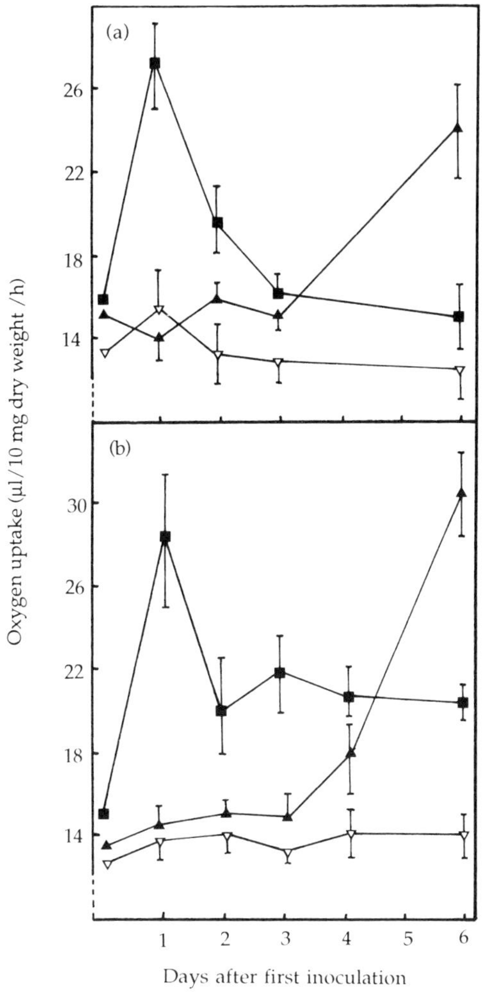

Fig. 4.2 Respiration in compatible and incompatible barley/powdery mildew infections: (■) incompatible, (▲) compatible, (▽) control. (a) Plants inoculated once, at zero time. (b) Plants inoculated three times, at 0, 2 and 4 days. Bar markers represent standard deviations. Reproduced from Smedegaard-Petersen (1984), courtesy of Blackwell Scientific Publications.

reductions in yields over uninfected plants. Certainly the activities of polyphenoloxidase and peroxidase enzymes are high in resistant plants and phenolic compounds are accumulated and oxidised. There is also evidence that the uncoupling of oxidative phosphorylation lowers plant resistance. However, the extent to which oxidative metabolism has a role in disease resistance is not clear.

Increases induced by biotrophic or necrotrophic fungi

Different mechanisms may also operate in plants infected by biotrophic or necrotrophic fungal species. In the infection of barley by the biotroph *Erysiphe graminis*, measurements of respiratory activity have shown a marked increase 3 days after infection. Oxygen uptake rates remained high until leaf senescence. In this instance the fungus probably acts as a metabolic sink, causing an increase in the flow of nutrient through the plant and raising the general level of biosynthetic activity.

In barley plants infected by the necrotroph *Pyrenophora teres* an increase in respiration rates became noticeable much earlier (24 h) after infection, and reached a maximum after 7 days. With increasing necrosis the respiratory level reduced with no measurable respiration after 10 days. In infections by such necrotrophs the fungus and/or the effect of the toxin(s) produced, tends to uncouple oxidative phosphorylation having both marked and rapid effects on host cell membranes.

Increases induced by saprophytic fungi

It is also recognised that leaf surface microflora, to which all plants are subject, may also have a marked effect on respiration. Tolstrup and Smedegaard-Petersen (1984) have shown that *Cladosporium* spp., normally regarded as saprophytes on barley, penetrated epidermal cells causing a significant increase in the respiration rates of plants. Although no visible symptoms of disease developed, field trials showed up to 16% reduction in grain yield as a result of such fungal activity (Smedegaard-Petersen and Tolstrup, 1986). The implications of this are far-reaching. All plant species are likely to be influenced in this way and all crop plants affected, whatever resistances are carried.

PHOTOSYNTHESIS

The loss of chlorophyll from leaf tissues, formation of chlorotic symptoms and reduction in yield, indicate disruption in the control of carbon balance in infected plants. However, the control of photosynthetic mechanisms by healthy plants and how these are modified by invading pathogens is not yet fully understood.

Photosynthesis in healthy plants

Photosynthesis takes place in the chloroplasts of higher plant cells and is the means by which sunlight energy is converted into chemical energy for biosynthesis and metabolism. Solar power is used to generate energy in the form of ATP and reducing power as NADPH, which are then used for the assimilation of CO_2, forming carbohydrate and releasing O_2 according to the equation:

$$6CO_2 + 6H_2O \xrightarrow[\text{chlorophyll}]{\text{light}} C_6H_{12}O_6 + 6O_2$$

The total process of photosynthesis is a complex series of interrelated and highly integrated reactions, occurring as two distinct phases. Firstly, light energy is captured and converted to chemical energy in light reactions. Carbohydrate is then formed from CO_2 by the use of chemical energy which can take place without the need for the use of light, in the dark or carbon-reduction reactions.

Light reactions

The light reactions depend on photosynthetic pigments to absorb light energy and to transfer high-energy electrons to the electron carriers. Photosynthetic plant cells contain chlorophylls *a* and *b*, together with other pigments (e.g. carotenoids) which are located in the thylakoid membranes of the chloroplasts. They are grouped into two photosystems. Chlorophyll *a* is the characteristic pigment of photosystem I (PS I) which is activated by light of 680–700 nm wavelength, and photosystem II (PS II) is activated by 650 nm wavelength light. The way in which the two photosystems interact is shown in Fig. 4.3. The photosystems absorb light energy which is transferred to chlorophyll *a* (reaction centre). Some of the energy is used to form ATP by photophosphorylation with inorganic orthophosphate (Pi). Radiant energy is converted into phosphate bond energy by a process known as cyclic photophosphorylation (supported by PS I only):

$$\text{ADP} + \text{Pi} \xrightarrow[\text{chlorophyll}]{\text{light}} \text{ATP} + H_2O$$

catalysed by the iron-sulphur containing protein ferredoxin. Photophosphorylation of ADP also occurs coupled to oxidation/reduction reactions in the presence of NADP. This process is known as non-cyclic photophosphorylation and involves both PS I and PS II. Electrons are transferred from water to ferredoxin photochemically with the evolution of oxygen.

$$\text{NADP} + \text{ADP} + \text{Pi} + H_2O \xrightarrow{\text{light}} \text{NADPH} + \text{ATP} + H^+ + O_2$$

Carbon-reduction reactions

The ATP and NADPH which are generated by the light reactions are used in another series of reactions in which CO_2 is reduced to glucose, and subsequently other carbohydates, principally via the Calvin

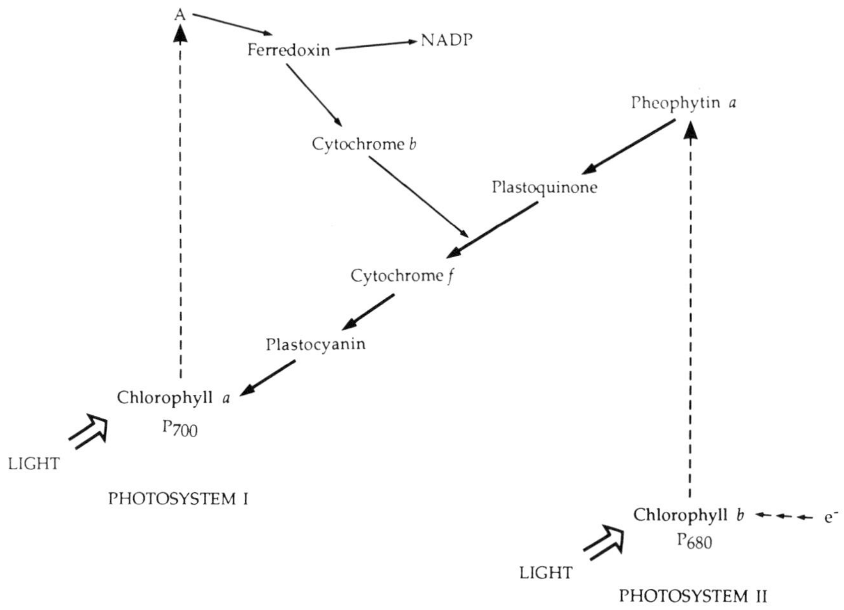

Fig. 4.3 Pathways of electron transport in plant photosynthesis, to show the relationship between photosystems I and II.

cycle. The main primary product of the pathway is triose phosphate. 3-Phosphoglycerate (a three-carbon compound) is formed initially from CO_2 and ribulose-1,5-bisphosphate. Plants which fix CO_2 in this way are known as C_3 plants. The CO_2 concentration may become low in leaves in certain conditions, making efficient photosynthesis impossible, e.g. if it is hot and stomata close to conserve water. Some plants have an additional pathway for CO_2 fixation which allows for the concentration of CO_2. In these plants, mainly tropical species (e.g. maize and sugar-cane), the Hatch–Slack pathway operates, preceding the C_3 pathway. Bundle sheath cells, which occur adjacent to phloem and within mesophyll tissue carry out this pathway. CO_2 is obtained from mesophyll cells and is fixed to produce oxaloacetate (a four-carbon compound). Plants which do this are known as C_4 plants.

Effect of pathogens on photosynthesis

Much of the influence which pathogens have on photosynthesis arises as a result of damage to chloroplasts, together with increased ageing and premature senescence of leaf tissues. Facultative parasites, particularly necrotrophs, bring about early effects on the structure and integrity of chloroplasts. The loss of chlorophyll leads to reduced rates of photosynthesis. Obligate parasites do not necessarily cause the overt expression of disease symptoms or necrosis but may alter the rate of CO_2 uptake and eventually cause chloroplast degeneration. In general there is, as yet, only limited information concerning effects which fungi have on specific reactions. If there is any general effect of fungal invasion it is the destruction of photosynthetic tissue rather than the presence of the pathogen or any metabolites which it may produce.

A number of approaches have been made to the measurement of photosynthesis in infected and uninfected plants. In relatively long-term experiments, measurements of dry weight increase per unit time and per unit leaf area, or carbon dioxide uptake or oxygen evolution have been made to estimate rates of photosynthesis. These methods do not allow more short-term determinations. More recently it has become possible to feed plants with $^{14}CO_2$ in the light or ^{14}C-labelled sugars, enabling the determination of photosynthetic rates during the course of disease development. In order to establish the fate of carbon assimilated after infection, host storage carbohydrate accumulation has been measured (e.g. hexoses, starch and sucrose). Additionally, the accumulation of carbohydrates which are specifically fungal metabolites (e.g. polyols, trehalose and glycogen) has also been measured in order to establish the relative proportions of turnover.

Rate of photosynthesis

It is not easy to determine the effects of pathogen invasion on photosynthesis. For any plant the rates of photosynthesis which can be achieved will depend on cultivar type, species, etc. The photosynthetic rate for any leaf will be influenced by the age of that leaf, the position on the plant and within the canopy and factors relating to the capture of light energy. Additionally, changes in the environment, both general climatic and local fluctuations, will have important influences.

Habeshaw (1984) has pointed out that in infected leaves the rate of photosynthesis, for the whole leaf, declines as the area of infected tissue increases, expressed relative to an uninfected control leaf. However, while there is loss of photosynthetic activity from infected regions there is a compensatory increase in infected tissues. Expressed in this way, there is a rise in photosynthetic rate in uninfected tissues of infected leaves (Fig. 4.4) as infection progresses, until senescence predominates. Initially the increase in rate is related to alterations in photorespiration (Ayres, 1979), whereas later the increase relates to changes in transport of photosynthates from that leaf to the plant (Daly, 1976) since photosynthetic

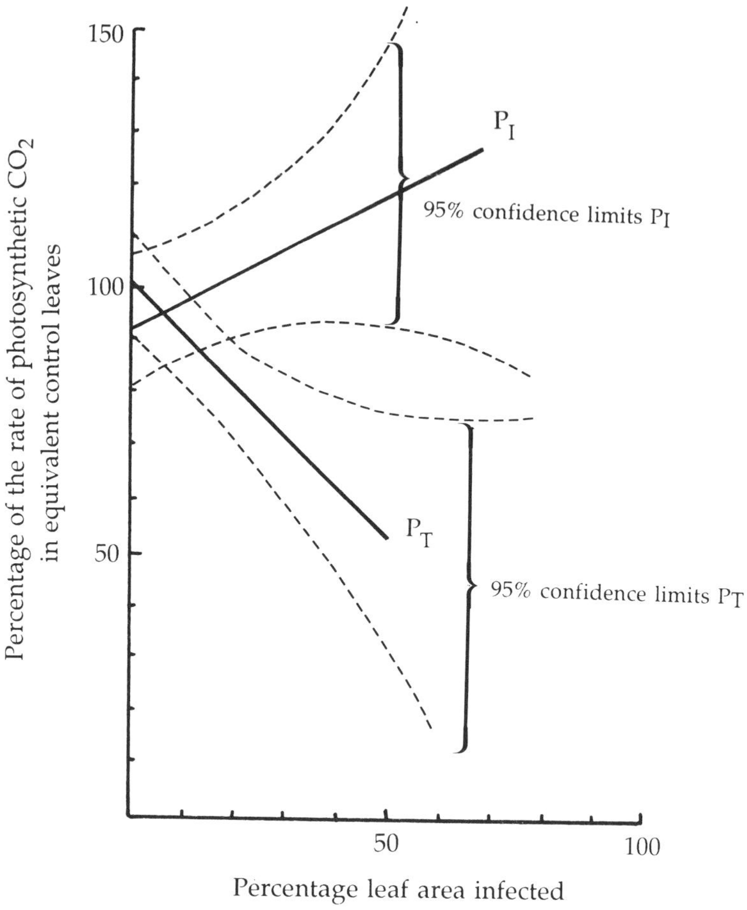

Fig. 4.4 Changes in photosynthetic rates with increases in leaf area infected (regression lines using individual leaf values; green island leaves omitted). Rate of photosynthesis in infected leaves under standard conditions expressed as a percentage of the rate in the equivalent control leaf (P_T). Rate of photosynthesis in the uninfected area of infected leaves under standard conditions expressed as a percentage of the rate in the equivalent control leaf (P_I). Reproduced from Habeshaw (1984), courtesy of Blackwell Scientific Publications.

activity depends partly on the rate at which photosynthates are removed (Habeshaw, 1973).

Evidence concerning more specific effects of fungal invasion has been gained from a series of experiments using chloroplasts isolated from sugar beet plants infected by powdery mildew, and also from rusted broad beans (Buchanan *et al.*, 1981). In both infections chloroplasts showed marked changes in ultrastructure. Infection also decreased the flow of electrons from water to NADP, and in powdery mildew infections photosynthetic CO_2 assimilation was inhibited and the capacity for sucrose production was reduced. The invading fungi caused a block in the non-cyclic electron transport chain by alterations to the components of that chain. The photosystems (PS l and PS ll) remained intact but the amounts of electron carrier molecules, the cytochromes, were reduced.

In brown rust infections of barley the use of radioactive tracer ($^{14}CO_2$) showed that CO_2 fixation occurred within pustules (infection sites) on infected leaves (Scholes and Farrar, 1985) but declined between pustules. An increase in gross photosynthesis occurred in those host cells invaded by haustoria. In pustules formed on bluebell leaves by *Uromyces muscari,* chlorophylls were lost and other changes also occurred (Fig. 4.5), suggesting that non-cyclic electron transport was reduced in those infected tissues (Scholes and Farrar, 1985; Farrar and Lewis, 1987). Levels of photosynthesis were apparently unaltered between pustules.

In terms of carbon fixation fungal invaders exert very different effects on host plant carbon-reduction reactions. Powdery mildew infections of barley cause a two-phase inhibition of CO_2 fixation. The initial inhibition, early after infection, occurs before any disease symptoms can be seen and is not apparently related to any loss of chlorophyll. This effect may possibly be due to alterations in glycolic acid metabolism. The second phase of inhibition in CO_2 fixation occurs later and is definitely related to the degradation and loss of chlorophyll. In rust diseases however, photosynthetic rates are almost unaltered immediately following infection, although dark fixation of CO_2 is enhanced at infection sites. Later, usually after sporulation by the fungus, CO_2 fixation is greatly reduced, probably as a result of the destruction of chloroplasts (Fig. 4. 6).

Much of the reduced rate of photosynthesis has been attributed to reductions in the photosynthetic machinery of leaf cells. Chloroplasts show ultrastructural changes after infection, sometimes loosing structural integrity. Water and solute relations change in infected cells and the observed reduction in the volume of chloroplasts has been attributed to increases in solute concentrations in host cells (Farrar and Lewis, 1987). The chlorophyll concentration has been shown to be reduced in mildewed barley plants, and in barley with brown rust infection the fluorescence kinetics of the chlorophyll is altered in infected cells. It has also been shown for barley plants with powdery mildew infections and for rusted bean plants, that uninfected leaves exhibit stimulated CO_2 fixation. This suggests that uninfected leaves are able to compensate for the reduced capacity of infected leaves by increasing photosynthesis.

The mechanisms by which fungal invaders affect photosynthesis are uncertain. It is also very probable that different pathogens have very different effects on host plant photosynthesis and photosynthetic pathways.

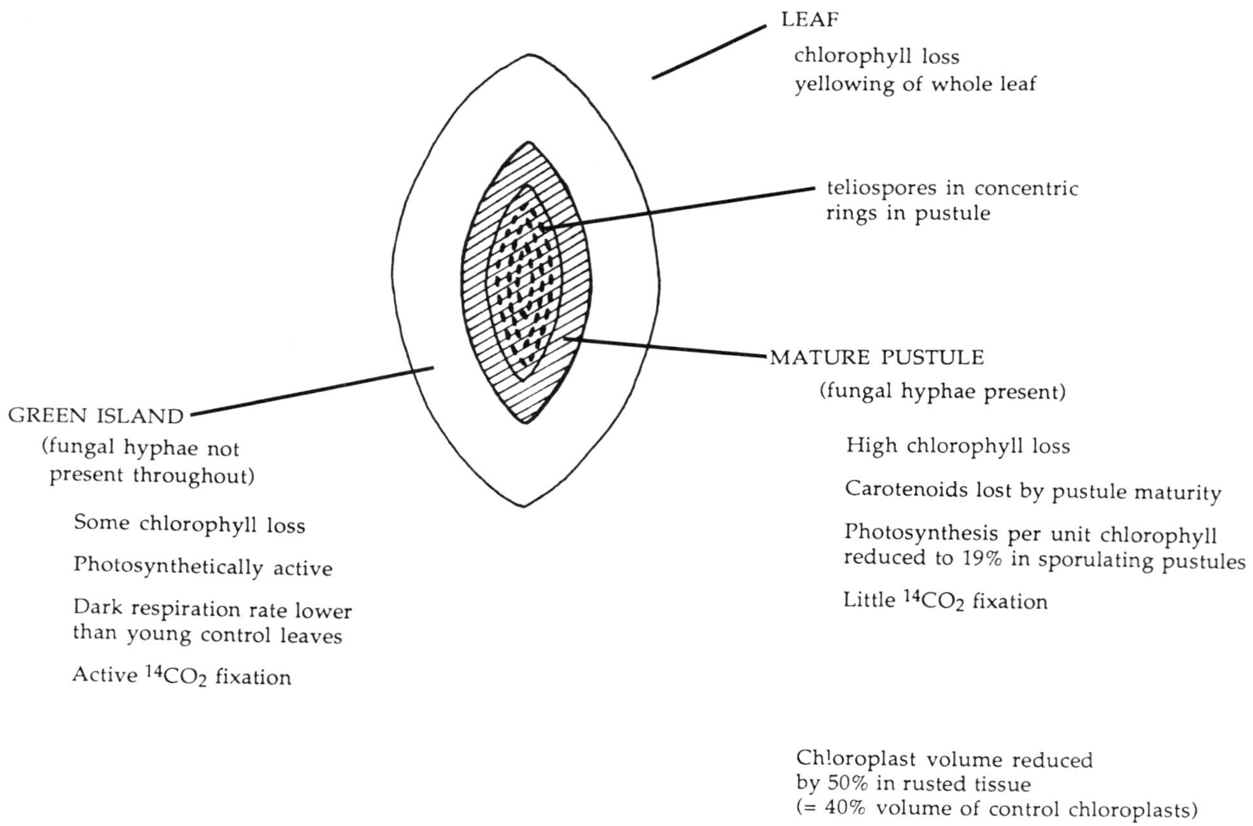

Fig. 4.5 Diagrammatic representation of the physiological effects of rust infection of bluebell leaves by *Uromyces muscari* (information from Scholes and Farrar, 1985).

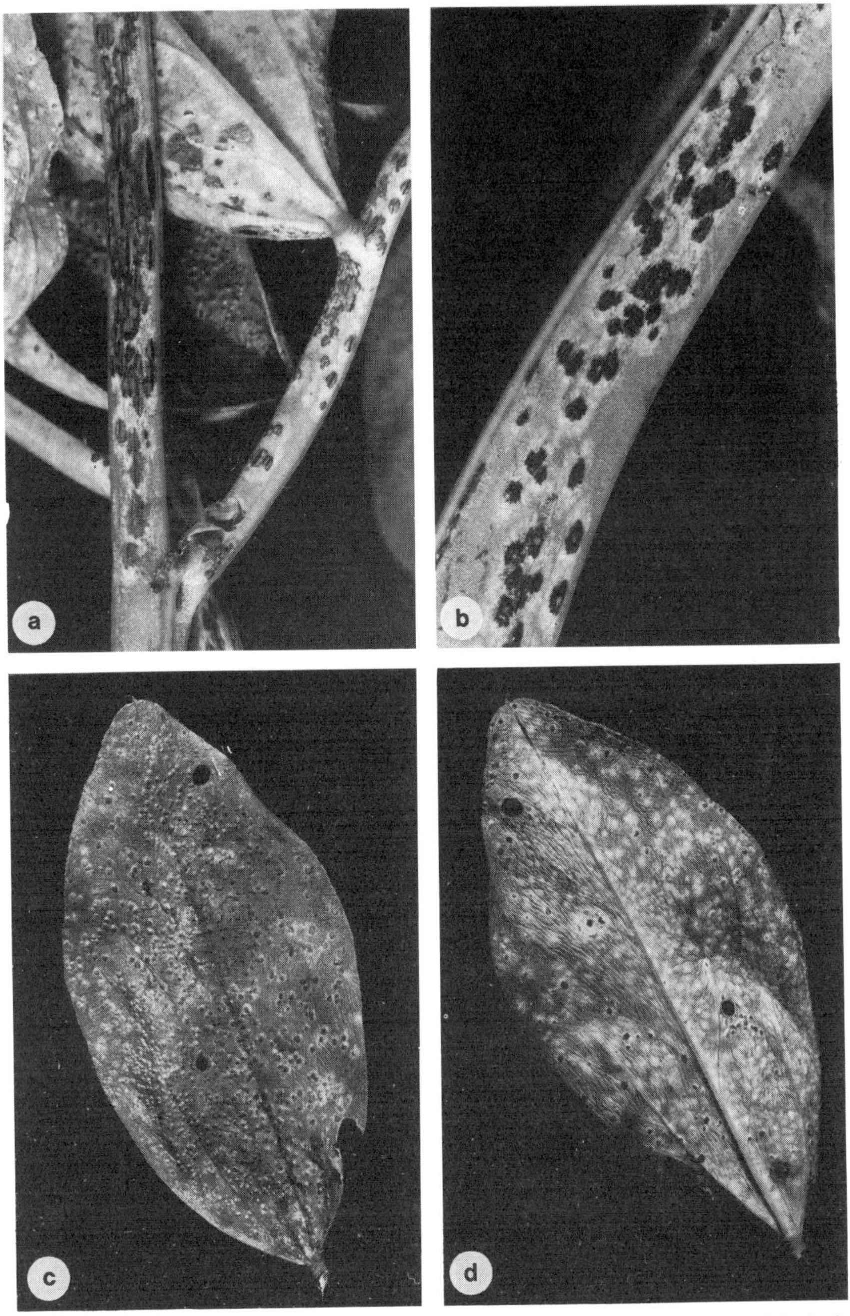

Fig. 4.6 Rust infection of bean (*Phaseolus vulgaris*). (a, b) Stem and petiole lesions well developed. (c, d) Adaxial and abaxial surfaces, respectively, of the same leaf showing lesions.

Translocation of photosynthates

The presence of a pathogenic invader upsets the carbohydrate balance of an infected plant by competition for and diversion of nutrient supplies. In general, for an infected plant, the translocation of photosynthates from uninfected tissues and leaves is enhanced whereas movements of materials from infected leaves are reduced.

In rust infections colonised leaves become chlorotic (loose chlorophyll) but immediately around the infection sites chlorophyll is retained, appearing as a green halo, known as a 'green island'. In powdery mildew infections, however, chlorophyll is initially lost from the leaf but later resynthesised around infection sites (regreening). Photosynthesis in the green island areas occurs at a rate comparable with that in uninfected tissues, or even slightly higher, and probably serves to support the invading fungus through sporulation. This characteristic green island symptom may be associated with the activity of growth regulator compounds, e.g. cytokinins; when these are applied to the surfaces of leaves similar symptoms develop. It is not clear if such activity is involved, and whether the invading fungus is the source or whether host plant cells are stimulated to produce increased levels.

Habeshaw (1984) has suggested that the effect may depend on the physiological state of the leaf and be the result of differential competition for nutrients between infected cells and uninfected tissues. A growing leaf acts as a nutrient sink within a plant, becoming less of a drain on resources as it reaches full development. Cells in a leaf compete for nutrients with each other and with the rest of the plant. The presence of a pathogen must add to the competitive pressure on an infected cell. Under these conditions, the synthetic capacity of the cell will be reduced and some degeneration of the photosynthetic machinery may occur. Chloroplasts may degenerate in infected cells, showing characteristics similar to senescent chloroplasts. The competitive ability of these cells is lost and since they no longer function as nutrient sinks, components pass to other cells. However, Habeshaw (1984) has argued that where the pathogen creates extra sink capacity, above that of the rest of the plant, then this will be met by redistribution of nutrients and senescence will be delayed in those cells.

Effects of fungal toxins on photosynthesis

High concentrations of tentoxin, a non-specific toxin produced by *Alternaria tenuis*, causes chlorosis in sensitive host plants and has been shown to specifically inhibit photophosphorylation in isolated lettuce chloroplasts, as an inhibitor of energy transfer. Rates of photosynthesis and translocation were inhibited in soybean plants infected by powdery mildew (*Microsphaera diffusa*) due to toxic effects acting directly on the chloroplasts. Increases in dark CO_2 fixation have been attributed to the toxin victorin (*Helminthosporium victoriae*). However, the extent to which the effects of fungal toxins on photosynthesis are responsible for symptom expression in naturally infected plants is not clear.

WATER TRANSPORT

Water is a vital component of living cells in terms of hydration of cytoplasm and cell walls and also as a solvent for essential components and metabolites. Very large quantities of water are continously absorbed from the soil through roots, translocated through the plant and lost to the atmosphere by transpiration. Water flow and transport therefore also bring about the movement and redistribution of nutrients and metabolites within the plant. Transport of water occurs over short distances by cyclosis within a cell and active transport across membranes. Over longer distances solutes pass through the xylem and phloem. The rate of water transport is related to the rate of transpiration and the rate of water uptake from the soil.

Water transport probably takes place via two pathways in the plant. The symplast pathway is through the cell cytoplasm, including the plasmodesmata by which cells are interconnected. The mechanism of transport across plasmodesmata is not known, but cell to cell transport probably requires a concentration gradient for solute flux. The apoplast pathway consists of the cell walls, dead cells and conductive tissues (xylem vessels and tracheids) and can be considered as corresponding to free space. Variable amounts of water flow through the symplast and apoplast, inversely proportional to their resistances. Symplastic water is bound within cell walls and cavities of dead cells.

Water potential

The free energy of water is important in terms of cellular function and in determining water availability. Water potential is defined as the free energy of water, per unit volume, related to the free energy of a reference pool of free water under the same conditions. Free water is considered to have zero water potential while that associated with living cells has a different potential energy in comparison. Water is at a negative potential when salts or sugars are dissolved in it or when it is adsorbed to cell walls or soil particles. The more negative the potential energy, the less physiologically available is the water. Evaporation from leaf surfaces, by transpiration, generates negative water potential in those leaves towards which water flows. The concept of water potential is now used in quantifying the energy of water in living cells (Papendick and Mulla, 1986) and it is the gradients in energy which are important in water movement.

Inside an intact cell, water potential (ψ_w) can be expressed as:

$$\psi_w = \psi_s + \psi_p$$

where ψ_s refers to the solute potential and ψ_p represents turgor pressure. In some instances this equation may need to be corrected for gravitational potential (ψ_g). This is a useful concept since it allows the component potentials, in any situation, to be itemised and the total water potential calculated, i.e. water potential is the sum of the component potentials. ψ_w and ψ_s are always negative, since energy would be required to move water from the cell to the free reference state, but ψ_p is always positive or zero. Metabolic processes in the cells are sensitive to changes in water content, which also alters ψ_w and other component potentials. Plant water potentials range from −0.5 to −4 MPa (Papendick and Mulla, 1986).

Water moves through the plant from the soil to the atmosphere in response to differences in water potential.

$$\text{Flux of water} = \frac{\psi_{soil} - \psi_{leaf}}{r_{plant}} = \frac{\psi_{leaf} - \psi_{air}}{r_g}$$

where ψ_{soil} is the water potential at the soil–root interface, ψ_{leaf} is the leaf water potential, y_{air} is the water potential of ambient air, r_{plant} is the total (combined) resistance of the root, stem and leaves (i.e. total for the plant) and r_g is resistance to water movement in the gas phase (Ayres, 1981, 1984a). It can be recognised therefore, that an invading pathogen, affecting any of these components, will alter the transpirational flux of the plant and hence the water relations. There will be a resistance to the flow of liquid water through the plant at all levels (soil, roots, stem, petioles and leaves). However, most resistance occurs in the root (50%), petioles or leaves (up to 35%) of the plant, whereas relatively little resistance is attributable to the stem (10–20%) (Jones, 1986). It is interesting to note that the effects of changing any component of resistance, even a very large increase, will have a small effect on the total resistance for the whole plant, particularly where other parallel pathways are present, e.g. in the stem where xylem vessels provide many routes for water movement.

Route of water movement

Transpiration from leaves generates a water potential gradient through cells of leaves and xylem elements causing water to be absorbed from the soil through root surfaces. Most water is absorbed through root hairs and young root tips which have a thin epidermis and are less suberised than older regions. Sites of lateral root emergence are also points at which water is easily taken up. Water passes from the soil–root interface into the epidermis, crosses (by mass flow) the root cortex and passes through the

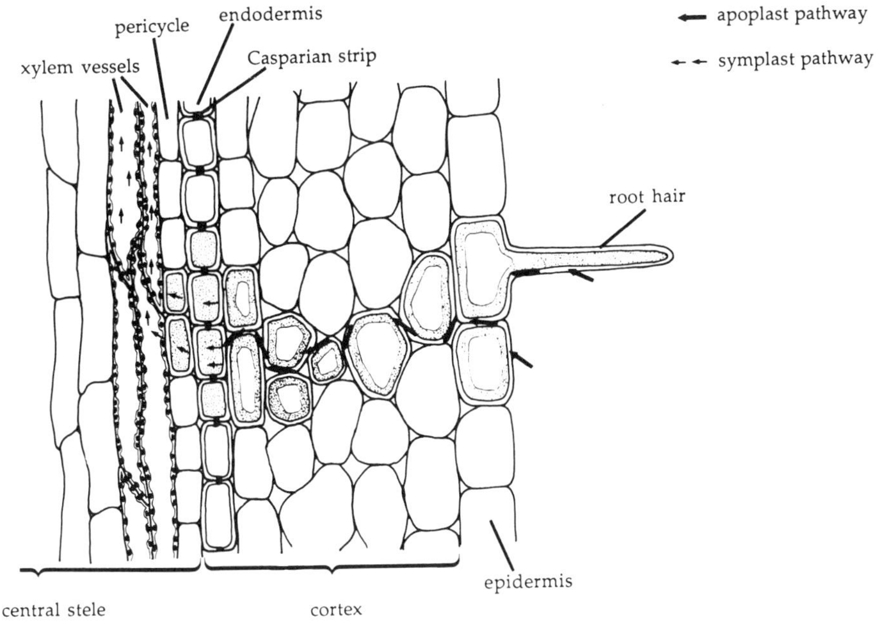

Fig. 4.7 Diagram of a root in longitudinal section to show the pathway of water transport from the soil to the vascular system. Water and solutes pass from the soil through the cortex by the apoplastic route (heavy arrows) and via the symplastic route (small arrows) from the endodermis to the xylem vessels.

endodermis and the pericycle to the xylem cells of the root. In mature regions of the root the cells of the endodermis develop suberised deposits which form the Casparian strip which limits water movement through the apoplastic route (Fig. 4.7). Water is therefore forced to pass through the symplast, by active uptake across membranes, to reach the cells of the pericycle and eventually to pass into the xylem cells. Mass flow of water will continue through the xylem elements, into leaf veins, passing through the apoplast of leaf mesophyll cells from where it evaporates into intercellular spaces of the leaf and is lost through stomata to the atmosphere. This transpiration stream is very important to the plant in maintaining metabolic processes, in fact less than 1% of the water absorbed by a plant is actually retained within the plant tissues.

Xylem vessels and tracheids provide a route by which large amounts of water can pass upwards through the plant. The cells are long, with lignified, thickened secondary walls and are dead when functional. Vessel elements have perforated end walls which allow water to pass from one cell to another, essentially froming a network of pipes which supply all plant tissues with water. Water passes up through the plant via xylem vessels to the xylem symplast and to the phloem from which, by virtue of the multidirectional translocation in the phloem there is a good circulatory system for the whole plant.

Xylem fluid is a perfect growth medium for invading fungi, with a pH at about 5.0. The solute concentration is generally fairly low although fluctuations do occur and with the increased transpirational pull which occurs in light conditions, xylem sap is usually more dilute during the day. Sap contains mainly mineral salts and ions with very low levels of amino acids and organic acids. Very little carbohydrate is carried by this route and usually only traces are found in xylem fluid.

Phloem tissue is composed of sieve tube elements and phloem parenchyma. It is via the phloem parenchyma pathway that nutrients move in the plant and these cells also have an important role in the synthesis, translocation and storage of sugars in the plant. Sieve tube elements are cells which are lined up vertically, the separating cross walls are perforated (sieve plates) and cytoplasm is continuous through these end walls.

In comparison with xylem fluid the phloem sap is more alkaline (pH 7.5–8.5) and solute concentrations are higher. The main components are sugars (90%), but absolute concentrations are variable, particularly during the day. Phloem sap also contains amino acids, hormones and minerals. Organic materials are translocated both up and down through phloem tissue, at the same time, but minerals are only transported downwards. Translocation through the phloem depends on a source-sink relationship in the plant. Solutes, under positive hydrostatic pressure, are loaded at the source and removed at the sink, thereby establishing a flow. Flow rates are mediated by temperature, water potential, and the presence of light, which alter the rate of photosynthesis. Water stress will also affect rates of photosynthesis and nutrient loading into the phloem. Nutrients pass from the phloem apoplast to the phloem symplast, which raises turgor pressure and results in mass flow through plasmodesmata. Movements through phloem tissues are multidirectional and together with xylem elements form a circulatory system.

Effect of pathogen invasion on water transport

Any effects on water movements and water balance in a plant are likely to have very great physiological significance to that plant. Some pathogens can bring about changes in water relations which have both rapid and profound effects, often resulting in the death of the host plant. Indeed devastating crop losses occur annually, world-wide, owing to the effects of pathogen disruption of water transport. Necrotrophic pathogens tend to produce drastic effects by the blockage or destruction of vascular tissues leading to plant cell death and enhancement of the growth and establishment of that fungus. Biotrophic pathogens however, affect gradients of water potential, altering the patterns and rates of water flow in the host plant. More major effects occur at sporulation after the reproductive potential of that invader has already been realised.

The major visible symptom of water stress is wilting. If loss of water exceeds uptake then wilting of leaves and shoots may result. For a while this effect is reversible and cells and tissues may recover. Plants may even become 'hardened' or adapted to the effects of prolonged water stress (Schöenwiss, 1986). Mild water stress, caused by a pathogen, may have an effect on the plant akin to environmentally induced water stress and the plant may be able to compensate. If, for example, an infection occurs during active growth the plant may be able to alter the root/shoot ratio accordingly (Ayres, 1984a). However, later in the season, when such a change is not possible, infections are more likely to cause water stress. Ayres and Paul (1986) have pointed out that water stress, in an infected plant, is also likely to influence the growth of the invading fungus itself. Fungal growth may be inhibited leading to reduced reproductive capacity, both in terms of the numbers of spores produced and the infectivity of those spores.

Changes in water transport are the result of a range of effects on normal host plant function and may be caused by a combination of blockage in the vascular system, alterations in the transpiration rates, changes in gradients of water potential and resistances to water flow through the plant.

Resistance to water flow

Most water flow occurs through wide xylem vessels, especially in the plant stem. Considering the whole plant, the smallest resistance to water flow is the stem component, where many parallel routes for water movement are present. Even if some vessels become obstructed alternative pathways are easily used. In fact, a 500-fold increase in the resistance of stem tissues in tomato plants suffering from *Fusarium* wilt increased the total plant resistance by only two-fold. The main resistance to liquid water flow is in roots (Davies, 1981), probably due to the presence of the endodermis which acts as a barrier. Water must cross living membranes at the Casparian strip in the endodermal cells. Roots become more suberised as they age, particularly after infection, therefore forming more effective barriers to water movement from the soil and increasing resistance to water flow. Endodermal cells are also a barrier to pathogen invasion. These are the most resistant cells in the plant to the macerative effect of fungal cell wall degrading enzymes, by virtue of the Casparian strip which is particularly resistant. Endodermal cells also have a high phenol and quinone content, substances which tend to inactivate cell wall degrading enzymes. Spaces through the endodermal barrier, resulting from root branching, are vulnerable points at which pathogens may invade more easily.

After invasion, increased resistance to water flow may occur due to physical blockages which obstruct the vascular system. These blockages may originate directly as the result of the presence of the invading pathogen, by the effects of the invader on host resistance mechanisms, from toxic or chemical effects of the invader or as a result of physical damage. The destruction of xylem vessels causes water shortages in leaves leading to water stress and wilting.

(a) Obstruction by fungal biomass

Water stress may occur as the result of physical blockage in xylem vessels by the growth of fungal mycelium and/or spores produced in those vessels. Fungal vascular wilt diseases are caused by soil-borne species (e.g. *Verticillium* spp., *Fusarium oxysporum* and *F. solani*), which invade roots and preferentially colonise vascular tissues. These fungi also sporulate quickly after invasion, as a means of continued plant colonisation, producing large numbers of spores which tend to clog perforation plates in the end walls of xylem vessel elements. In vascular wilt disease this is certainly a major cause of wilting although other factors do contribute to the syndrome. However, vascular invasion may not necessarily be accompanied by the development of wilting syndrome, e.g. *Rhizoctonia solani* may invade vascular tissues but does not usually cause wilt as a primary symptom.

(b) Action of fungal toxins

Invading fungal pathogens produce a range of chemical agents to attack plant functions and to break down cellular structures. Such toxic compounds have a range of physiological effects although in some instances this activity leads directly to the obstruction of xylem vessels and the disruption of water flow. Wilt-inducing toxins were thought to be the main cause of disease symptoms in vascular wilt disease, but although toxins cause chlorosis, stunting and other symptoms, the primary cause of vascular wilt is now thought to be mechanical blockage by mycelial biomass (Scheffer, 1983). Some toxic compounds are water-soluble and are therefore immobile in the plant, e.g. β-(1,3) glucans released by *Phytophthora* spp., which cause leaf wilt due to cellular toxicity. The production and presence of other polysaccharide materials may also lead to obstruction of xylem vessels by increasing the viscosity of xylem sap (Hall, 1986). Additionally, fungal cell wall degrading enzymes may release fragments of plant cell wall material into the transpiration stream which then cause blockages in the xylem pathway.

(c) Host plant responses

The invasion of pathogenic fungi may lead to the induction of host plant reponses which result in xylem vessel blockage. The formation of tyloses is often reported, particularly as host response to vascular wilt infections. Xylem parenchyma cells balloon through pit membranes and very effectively close xylem vessels. Gels and polysaccharide gums are also produced by the host plant, probably in response to the release of cell wall degrading enzymes by the invading fungus. These may also obstruct xylem vessels, often accumulating at perforation pits. Additionally, the distension of primary walls and middle lamellae, as a consequence of degradative enzyme activity, may lead to total obstruction.

(d) Blockage of xylem vessel pits

Vessel to vessel continuity is provided by the occurrence of pits in xylem cell walls, allowing efficient water transport. The resistance of pit membranes is probably very low (Jones, 1986), but since the size of pores is small the risk of mechanical blockage by materials derived from host resistance mechanisms or fungal invasion is very high.

(e) Xylem vessel cavitation

Breakages in the continuity of the xylem pathway will reduce conductivity. Small air bubbles may enter the system, possibly as a result of degradation and damage by fungal enzymes, and act as blockages in the system. The plant may respond to the presence of such embolisms or cavitations by producing tyloses, gums, etc., to isolate affected regions. In fact, such cavitations are difficult to detect (Jones, 1986) but are probably very important in the obstruction of water flow.

Reduction in the size of the root system

Plant root systems absorb water and nutrients from the soil. A substantial root system provides a large surface for absorbtion and therefore good access to soil moisture. Roots also provide important anchorage for the plant which may collapse or lodge if a large proportion is destroyed or weakened. Invasion by root-infecting fungi can severely affect the normal functioning of the root system. For example, penetration and destruction of vascular tissue may cause the death of the root distal to lesions, effectively reducing the size of the root system and the available surface area, resulting in a drastic reduction in water transport. This may result in increased water stress to the plant, particularly under conditions of reduced soil moisture. A reduction in root growth caused by infection may not necessarily result in water stress, however, providing that the main roots are not lost and that not too large a proportion of the system has been destroyed. In some cases further growth of other roots may be stimulated and compensate for the loss. On the other hand, the reduced growth rate of roots may have a pronounced effect on the water uptake of the plant. Root tips lack the suberised barrier of more mature roots and water uptake via the apoplast, in this region, is subject to less resistance. In slower growing roots however, the endodermal barrier becomes suberised in regions nearer to the growing tip and therefore the region of freer water uptake is less in such roots.

Reduced root growth may result from the effects of foliar pathogens. A reduction in the supply of photoassimilates may quickly lead to water stress, as has been shown for plants affected by rust diseases (Ayres, 1984a). The proportionally reduced root system must maintain the supply of water to leaves and shoots and under some conditions the system easily becomes strained and the total water balance for the plant may be upset.

Effects on transpiration

It has been shown that diseases which cause wilt symptoms lead to a reduction in transpiration rates in infected plants. Sugar-beet plants infected by *Aphanomyces cochlioides* showed root symptoms within a few days, followed by wilting. After 9 days uninfected plants showed normal transpiration rates, but rates for resistant cultivars were reduced by 60% and by 84% for susceptible plants (Safir and Schneider, 1976). Similar results were also found for transpiration rates of *Isopogon ceratophyllus* infected by *Phytophthora cinnamomi* and although, in this case, symptoms occurred much later after inoculation (3 months) the reduction in transpiration rate coincided with visible wilting. Where leaf integrity is disrupted by the invading fungus (e.g. at sporulation) a dramatic rise in respiration rate would occur locally, however, the rate per plant would probably be reduced. Such an effect might be attributable to differences in growth of new leaves and leaf expansion rates since water stress delays leaf expansion. Water is not used efficiently in such infected plants; comparatively more water is used to achieve a lower yield than uninfected plants.

(a) Leaf water status

Infections, derived from both foliar and root-invading pathogens, markedly affect the water status of host plant leaves. Root infections have been shown to lower leaf water potential and this has been considered a symptom which indicates the onset of water stress and which occurs prior to wilting. In healthy plants leaf water potential falls as transpiration rates rise, unless the plant is water stressed. Under conditions of water stress the patterns of plant response are different and often rather erratic. In avocado trees infected by *Phytophthora cinnamomi* the measured leaf water potentials were lower than those of healthy trees for given rates of transpiration (Sterne *et al.*, 1978).

(b) Cuticle rupture

Water balance is upset by factors which alter water uptake and transpiration rates. The disruption of this balance may depend on the amount of leaf surface area which is affected by pathogen invasion and, to some extent, which regions of the leaf are affected (Ayres and Paul, 1986). The major component of resistance to water flow from leaf surfaces is attributable to the cuticle. The normal contribution to water loss by cuticular surfaces is estimated to be about 1% of the total loss from a plant. The hydrophobic cuticle layer covers leaves and stems of plants and offers a very large resistance value to water flow. In a normal leaf the only gaps in this layer are stomata, which provide pores for the escape of water vapour and for the diffusion of CO_2 to sites of photosynthesis within the leaf. Sporulation of a foliar pathogen (e.g. *Uromyces phasoeli,* which causes rust disease of beans) ruptures the leaf cuticle and damages this protective layer, with drastic results for the plant (Fig. 4.8). Dramatically enhanced rates of transpiration ensue, often with high rates of water loss, leading to wilting. Sporulation of the invading fungus therefore lowers leaf resistance to water flow and causes loss of control of transpiration in the host plant.

(c) Interference with stomata

Healthy plants can protect against development of water stress by closure of stomatal pores, preventing any further water loss. The movements of guard cells therefore provide a very important means of varying the resistance (Ayres, 1981) to water flow through the leaf. In infected leaves guard cells sometimes loose their mobility and in situations where guard cells become stuck open this is a major cause of water stress and wilting. Ayres (1981) has also suggested that subcuticular fungal growth causes leakage of solutes from cells of the epidermis, which upsets turgor, preventing guard cell closure and disrupting leaf water balance due to increased transpiration. *Erysiphe graminis,* which causes powdery mildew of barley, inhibits the opening of stomata late in disease, thereby reducing transpiration rates as the result of a combination of effects. The increased permeability of cells leads to loss of solutes and thereby to reduced turgor. In powdery mildew and rust infections stomatal opening is inhibited in the light and severe wilting results after sporulation of the fungus has ripped the cuticle leading to the loss of control of transpiration.

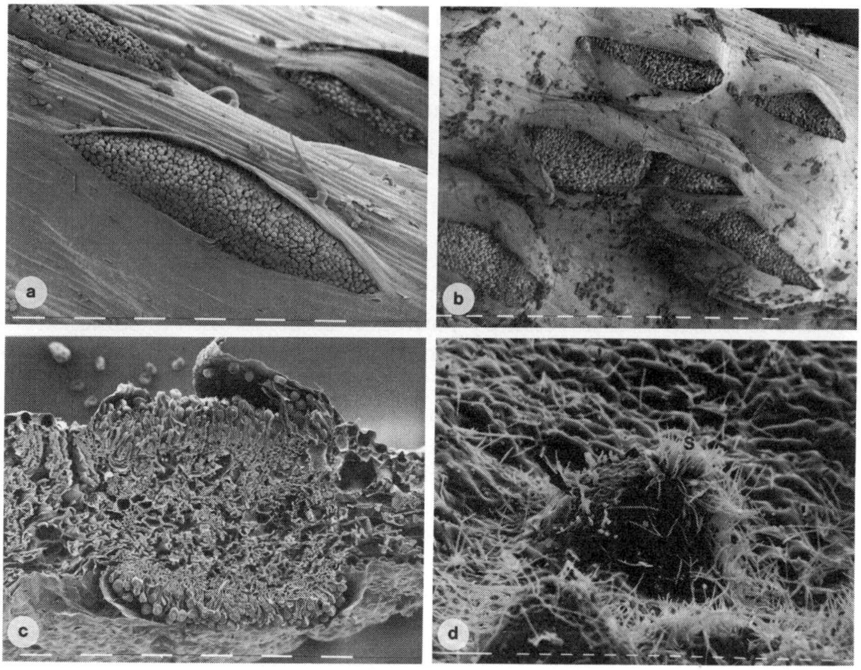

Fig. 4.8 Rupture of leaf cuticles by the sporulation of foliar pathogens. (a, b) Scanning electron micrographs showing rust infection on leaves of broad bean. Subcuticular spore formation has resulted in the rupture of the protective cuticle. Surface view showing rupture of upper cuticle. (c) Transverse section through leaf to show pustule rupture of both upper and lower cuticle by pustules. A great deal of intercellular mycelium can be seen within the leaf tissue. (d) Development of pycnidia in celery leaf tissue infected by *Septoria apiicola* (leaf spot disease). Mature pycnidium ruptured through epidermis (arrowed) of celery. Pycnidiospores (S) are visible in the ostiole of the pycnidium. Bars represent (a) 100 μm; (b) 62.5 μm; (c) 55.6 μm; (d) 10 μm. (d) Reproduced from Donovan *et al.* (1990), courtesy of Cambridge University Press.

The production of toxins by invading pathogens has also been reported to affect stomatal closure. *Fusicoccum amygdali*, which causes canker of peach and almond trees, produces a host non-specific toxin, fusicoccin. Transpiration leads to the distribution of the toxin through the plant. Fusicoccin causes stomatal opening by stimulation of solute accumulation in guard cells, which leads to loss of water and therefore to wilting.

Phytoalexins, which are produced by host plant cells in reponse to infection, have also been shown to have pronounced effects on stomatal movements. The effects of some phytoalexins have been assessed using epidermal strips from leaf surfaces (Willmer and Plumbe, 1986). Phaseollin and wyerone acid were the most effective of those tested and prevented stomatal opening. Pisatin and rishitin exerted much less effect. It is interesting that the effects of each of the phytoalexins was greater on closed stomata than on open stomata, including the extent of physical damage caused.

Ion transport

Ions are absorbed by the plant in solution from the soil and pass through the root to xylem vessels for more long distance transport. Uptake into the root may be active or passive processes but passage across the plasma membrane into cortical cells can only occur by active transport. Ions accumulate inside cell vacuoles at concentrations very much higher than those found in the soil. Ions diffuse differently through solutions than through pure water and their diffusion coefficients are greatly reduced in soil. In fact some ions, notably PO_4^{3-}, diffuse through soil very slowly indeed and a depletion zone may occur around active roots. New root growth, to exploit fresh soil, is therefore very important to a growing plant. Ca^{2+} moves through plant tissues with water, entering the symplast at the Casparian strip. Suberisation blocks the uptake of Ca^{2+}, Mg^{2+} and Fe^{2+}. Other ions, K^+ and PO_4^{3-}, are taken up into the symplast in the root cortex, so that increased suberisation of maturing roots does not affect the uptake of these ions (Ayres, 1984).

Increases in transpiration cause increases in ion uptake and it has been shown that concentrations of K^+ increase and accumulate at infection sites in leaves. Ca^{2+} and Mg^{2+} are not as mobile in the phloem and are not retranslocated. Take-all disease of wheat, caused by *Gaeumannomyces graminis* var. *tritici*, has been shown to lead to the reduction of K^+, P and Ca^{2+} contents in infected plants. In rust disease P has been reported to accumulate more than K^+, Ca^{2+} or Mg^{2+}, as the result of reduced export from infected leaves.

Ahmad *et al.* (1984) calculated phosphorus budgets for brown rusted barley plants. Phosphorus was found to accumulate in the fungus. However, there was little change in the concentration of phosphorus in the xylem sap of infected plants. Phosphorus accumulation by the fungus was supported by the input from the xylem, i.e. was sustained by transpiration and not derived as the result of transfer from host cells.

Vascular wilt diseases

Vascular wilt diseases (Fig. 4.9) require a special mention in the context of water transport, particularly since they are such devastating crop diseases and have been the subject of much research. The causal agents, e.g. *Fusarium oxysporum* f. sp. *lycopersici*, *Ceratocystis ulmi* (Dutch Elm disease) and *Verticillium albo-atrum*, are soil-borne and invade plant roots,

Fig. 4.9 Potted tomato plant showing symptoms of infection by the vascular wilt fungus, *Verticillium albo-atrum*. Reproduced courtesy of Dr I. R. Crute, HRI, East Malling.

preferentially colonising xylem vessels, and do not emerge from the host until after the infected plant has died. The mycelium grows and proliferates through the vascular elements of the plant, exploiting the xylem environment, living in the nutrient solution of the xylem sap, which Pegg (1985) has aptly likened to a continuously supplied culture vessel. As the result of mycelial growth (Fig. 4.10) and rapid sporulation, and also owing to the production of tyloses and gels, as a response of the host plant, xylem elements become physically blocked which gives rise to wilting, the primary symptom of vascular disease. Chlorosis, epinasty and stunting also occur. Toxic compounds are produced by vascular wilt fungi and although the action of these has an important role in disease development toxins are not now considered to be the main cause of symptom development.

It is, however, the complexity of the host–pathogen interaction which is of interest here. Physical blockages occur in the vascular systems of infected plants as the result of integrated metabolic reactions between fungus and plant (Fig. 4.11). In addition to the production of toxins and some cell wall degrading enzymes, mentioned previously, the invading fungus produces, and causes the plant to produce, other additional compounds during infection. Ethylene (ethene) and IAA (indole-3-yl-acetic acid) are produced by both the pathogen and susceptible host plants. Both of these factors stimulate the production of tyloses in host plant xylem vessels, by plasticising cell walls. Ethylene induces enzyme activity leading to increased phenoloxidising and pectolytic enzyme levels. These are associated with vascular wilt disease and give rise to the production of phenolic compounds which are deposited in vascular elements. IAA and ethylene also induce the formation of gels which accumulate at perforation plates and contribute to vascular blockage.

CARBOHYDRATE METABOLISM AND TRANSLOCATION

Carbohydrate metabolism and translocation in plants are part of the complex of highly integrated physiological processes which lead to the distribution of photosynthates and compartmentalisation of nutrient materials and storage compounds throughout the plant. Pathogenic fungal invaders certainly bring about alterations in the patterns by which these processes occur, both directly and by effects which cause disruptions in water transport and photosynthetic activities. Invasions by biotrophic fungi tend to increase the amount of carbohydrate imported to infected tissues and decrease the export of photosynthates from infected leaves. As a result, therefore, the amount of transport to the rest of the plant is diminished. Necrotrophic fungi do not affect levels of nutrients imported but reduce the amount of carbohydrate leaving infected leaves (Whipps and Lewis, 1981). Any such changes must affect the carbohydrate budget of the plant and

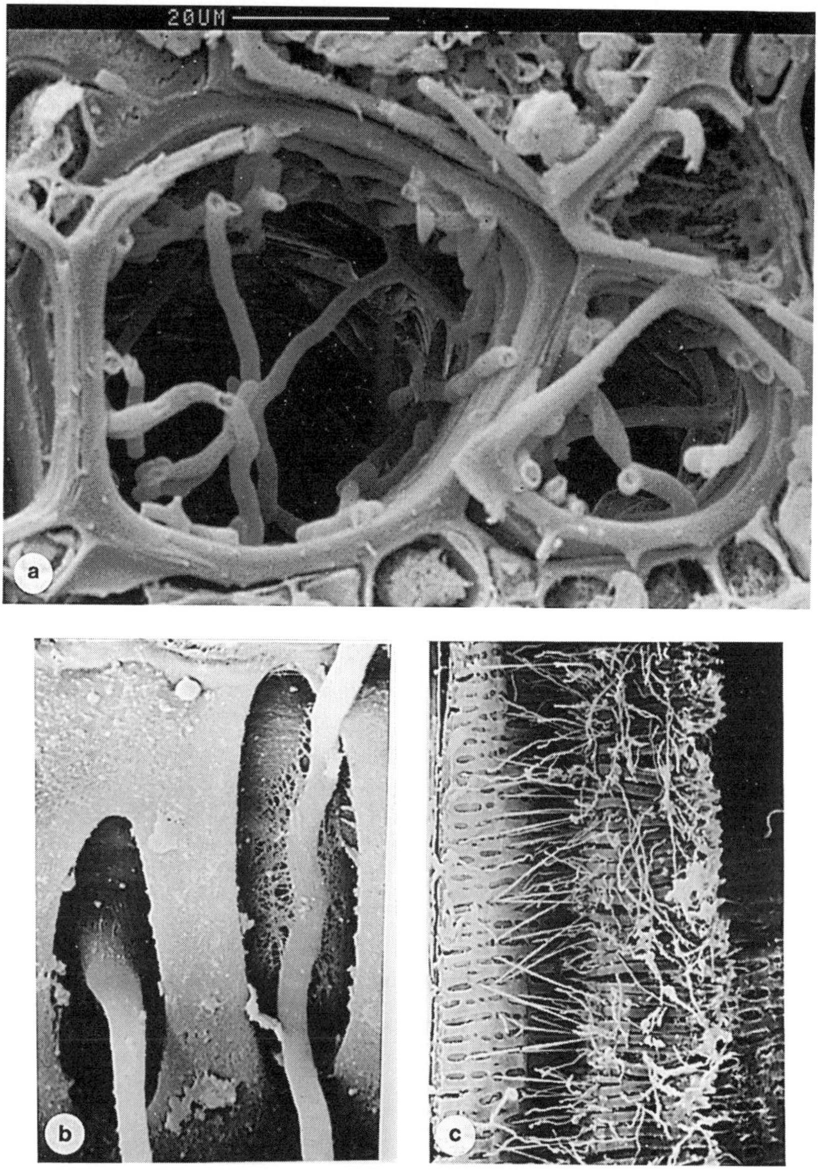

Fig. 4.10 Scanning electron micrographs of plants infected by vascular wilt pathogenic fungi. (a) Hyphae of *Verticillium albo-atrum* within vessels (transverse section) of hop cv. Northdown (x1060). Reproduced courtesy of J. H. Carder, IHR East Malling. (b) *Fusarium oxysporum* f. sp. *cubense* breaching pit membranes in banana root xylem vessels showing enzymic degradation. (c) Longitudinal section of banana root xylem vessels colonised by *Fusarium oxysporum* f. sp. *cubense*. (b, c) Reproduced from Pegg (1985), courtesy of Prof. Pegg and Cambridge University Press.

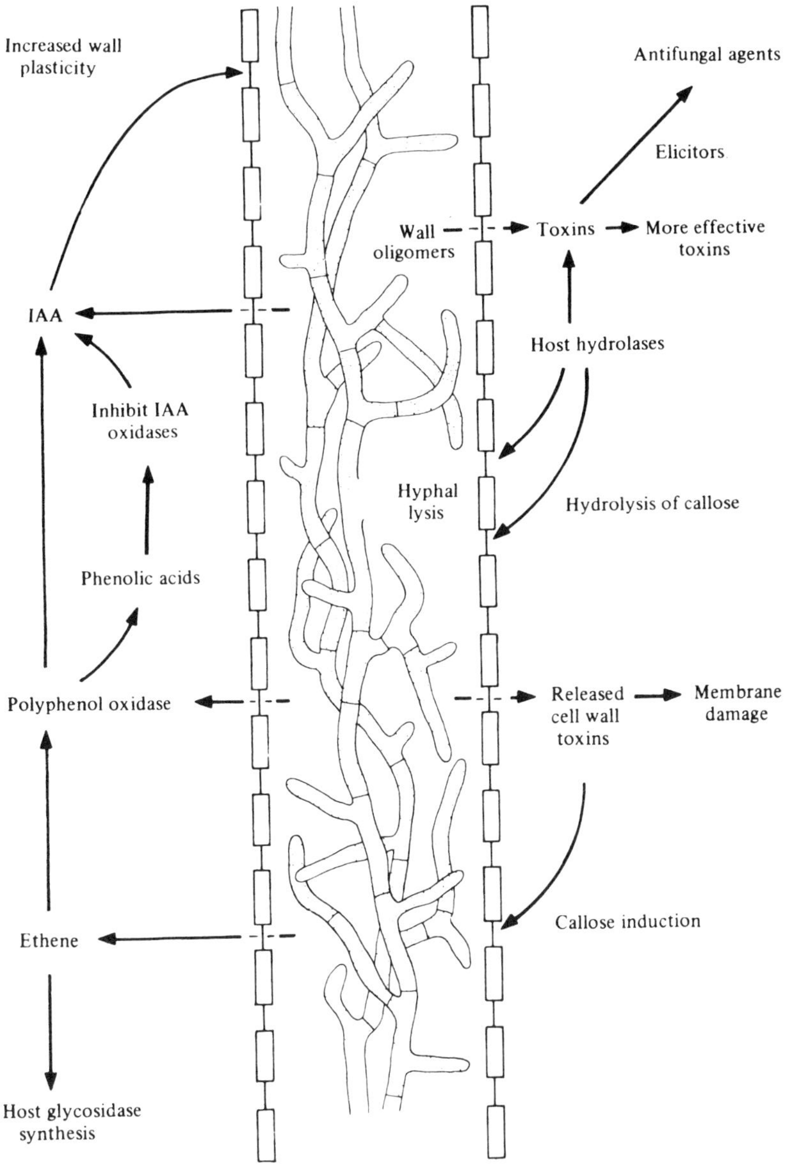

Fig. 4.11 Diagrammatic summary to show the interaction between host and fungal metabolic reactions in vascular pathogenesis. Reproduced from Pegg (1985), courtesy of Prof. G. F. Pegg and Cambridge University Press.

will therefore have consequences for tissues far removed from infection sites. The degree of effect for the whole plant will relate to the intensity or severity of the infection and the virulence of the pathogen. In recent years the use of $^{14}CO_2$ and other radiolabelled compounds have allowed the detection of sites of accumulated ^{14}C and the measurement of the fates of labelled sugars. Great advances have been made in the understanding of carbohydrate metabolism in healthy as well as diseased plants, and the information which has been collected has led to the formation of theories and suggestions regarding the influence of pathogenic invasion on carbohydrate metabolism. However, there is still relatively little definitive experimental evidence in this area.

Translocation in healthy plants

In vascular plants most carbohydate materials are transported through the phloem and movements via this route are multidirectional. Movements through the phloem require a source of photosynthates and a consuming sink (source/sink relationship), thereby setting up gradients through the plant. It is this relationship which regulates the rate and direction of the transport. The source may be considered to be leaf tissues, where photosynthates are generated, or more strictly, the chloroplasts since these are the sites of photosynthate production. The sink is the rest of the plant, which requires and uses those products, or alternatively on a cellular level, it is the consumer cytoplasm of living cells and tissues.

Membrane transport is required for the loading of materials into the phloem for translocation. It is important to understand that not all photosynthates are free to move via the phloem pathway and the forms in which materials are transported and stored will vary. Most carbohydrates are transported via the phloem symplast. Photosynthates are produced in chloroplasts and of those triose phosphates, which are mobile, pass to the cytoplasm. Some starch may be stored in the chloroplast temporarily. The non-reducing disaccharide sucrose which is synthesised in the cell cytoplasm, is found throughout the plant and it is in this form that most carbohydrate is translocated through the plant. Most of this movement is via the symplast although some may occur through the apoplast. Non-reducing sugars are mobile in the phloem and are transported via the symplast. However, the reducing sugars glucose and fructose are located in the cytoplasm and apoplast but are never translocated. Although these movements are regulated by the feedback control of production and consumption of materials, it has also been proposed (Whipps and Lewis, 1981) that plant growth regulators and inorganic phosphate have a regulatory role to play in the intact plant.

Carbohydrates are stored in plants during the day to be utilised at night. The main forms for this storage are as starch (Lewis, 1984), which is laid down in amyloplasts (particularly in tomatoes, soybeans and peppers), or as sucrose, which is stored in vacuoles (particularly in cereals, e.g. barley and wheat). An understanding of the carbohydrate budget of plants and how this is affected by pathogens, requires an understanding of the localisation and compartmentalisation of metabolites. It is the fluxes of carbohydrates which are most important (Farrar and Lewis, 1987) in the maintenance of the carbon balance in a plant.

Translocation in diseased plants

After fungal infection the export of nutrient materials from leaves is reduced and this alters the transport patterns throughout the whole plant. Some of the effects on transport will be derived from physical damage to the phloem which impares the transport pathway. In instances where infection and disease development have effects on the machinery and rates of photosynthesis then the supply of photosynthates will be reduced and as a result phloem transport will be decreased. It is in this context that the importance of carbon flux measurements can be seen (Farrar, 1984; Farrar and Lewis, 1987). In barley plants infected with brown rust disease, Owera *et al.* (1983) found that the pool of available carbon was reduced by 50% in infected leaves. However, the rates of translocation were not significantly altered. This suggests that infection reduced the carbon supply but did not affect the loading of the phloem. From calculations of the carbon fluxes for healthy and diseased plants (Fig. 4.12) it was concluded that the normal fluctuations which occur in carbon status of the plant were greater than alterations which were induced by the disease.

It has been suggested (Farrar and Lewis, 1987) that, for biotrophic infections, where sufficient detail is available, e.g. mildew of oats, that the supply of carbohydrate to the invader is sufficient to maintain the growth of the fungus. In other words, the production capacity of a photosynthetically active plant cell is greater than the carbohydrate consumption of a haustorium, and the growth of the fungus is unlikely to be limited due to lack of carbohydrate.

Carbohydrate accumulation

It has been known for many years that soluble carbohydrates accumulate within malformations which form in host plants after fungal infection. Recently, radiolabelling techniques have shown that, in general, carbon compounds (photosynthetic products) accumulate at sites of biotrophic fungal infection. Invading colonies act as a sink for plant resources.

Green islands, regions of chlorophyll retention or resynthesis which form around infection sites in otherwise yellowing leaf tissue, are a visible symptom of infection. These areas are often regarded as sites of increased

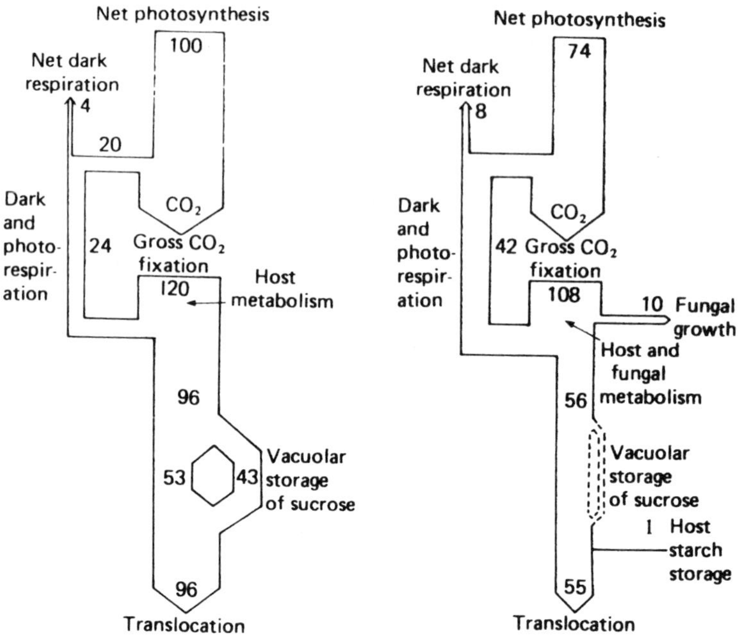

Fig. 4.12 Diagram summarising the main carbon fluxes in barley leaves infected by brown rust; uninfected material (left), infected tissues (right). Arrow widths proportional to flux, figures are percentages of net photosynthesis in control leaves. Reproduced from Farrar and Lewis (1987), courtesy of Cambridge University Press.

vigour in plant cells, since photosynthetic activity is retained. The invading fungus acts as a sink for the compounds produced and the infected plant tissues must compete heavily with the invader for available nutrients at these sites. There are also accumulations of some carbohydrates which are not available to the host plant. The fungus acts as a sink for host sugars and metabolises these to support its own growth. Some of these host resources are also used to form fungal storage compounds which the host cannot use and are therefore rendered unavailable to it, e.g. the non-reducing disaccharide trehalose and the sugar alcohols mannitol and arabitol.

Starch levels may undergo quite dramatic fluctuations in the host after infection. Some of these changes may be associated with sporulation of the fungus and indicate the degree of disruption occurring in the host plant.

Mechanism of effects on carbohydrate metabolism and translocation

There are various means by which fungi disrupt the carbon metabolism of the normal host plant. Some fungi liberate toxic compounds in the host plant which are mobile in the normal transpiration system and are therefore transported away from infection sites. Fusicoccin (*Fusicoccum amygdali*) impaires the function of membrane ATPases, which may affect the loading of phloem and will therefore disrupt translocation in an infected plant. Any factors which affect cell permeability, causing membrane leakage, will also alter phloem loading.

Altered translocation patterns which occur in infected plants can be mimicked by the application of plant growth regulators to leaf surfaces. This implies that such hormones (e.g. gibberellins, cytokinins and auxins) have a role in controlling rates of photosynthesis and translocation, and the increased levels and changes in the distribution of these occur in infected plants. In particular, cytokinins are implicated in the retention of chlorophyll in green islands.

Changes in enzyme activities which affect carbohydrate metabolism also occur in infected tissues, some as a result of changes in distribution owing to cellular damage and altered permeability. Enzymes will be lost from cells as disruption occurs. In cases where increased levels of enzymes (e.g. invertases and amylases) have been measured the source of that increase is not clear. Enhanced levels may be produced by the host plant in response to physical wounding, caused by the pathogen. Alternatively, the pathogen may provide enzyme activators which stimulate the host to increased production or the invading fungus may liberate enzymes directly.

PATHOGEN-INDUCED CHANGES IN MEMBRANE PERMEABILITY

Changes in membrane permeability are often quoted as the first effects of wounding and pathogenic invasion. While this is certainly true in some cases, it is not always clear if membranes are indeed the primary target of a pathogen or whether changes in permeability occur as a result of infection, after organelle disruption for example, and therefore are a secondary effect. Extensive membrane damage will affect the protoplast and cause cell death, but even small amounts of damage will cause alterations in localised solute and water relations which may have important consequences for the rest of the plant.

Membranes are composed primarily of lipids and proteins, with small amounts of carbohydrate. These are arranged as a bilayer of phospholipid molecules with the ionic groups facing outwards and the non-polar ends of the phospholipids facing inwards with embedded protein molecules protruding on either side. These proteins may move within the fluid bilayer. The membrane acts as a highly selective permeability barrier lining the plant cell wall. Similar membranes also occur in cell cytoplasm as endoplasmic reticulum and delimiting organelles. All these membranes may be affected by fungal activity. Membranes are responsible for ion transport.

Ions may pass through membranes by passive diffusion in response to concentration gradients or may be actively transported against such gradients and against electrochemical gradients. Most ions are actively pumped across membranes using metabolic energy, in the form of ATP, mediated by ATPases. The importance of the H^+ extrusion pump is well recognised for the co-transport of amino acids and sugars across membranes and for the maintenance of cellular pH. Alterations in ion fluxes may result in marked biochemical changes within cells and may have an important role in the recognition of invasion and intercellular signalling for defense response activation.

Biotrophic fungi which form haustoria establish a very specific relationship with host cell membranes. Invasion of host cells causes minimal structural disruption and the invaded cells remain fully functional and active. The fungus penetrates into the host cell without breaking the host membrane. As the fungal haustorium grows the host membrane becomes invaginated (extrahaustorial membrane) to accommodate the fungus (Chapter 3). The plasma membranes of host and fungus remain in very close contact near the point at which the cell was penetrated (haustorial neck) and seals form at this point. These neckband seals prevent solutes escaping and allow the fungus to control solute efflux from the host. The extrahaustorial membrane lacks ATPases, membrane-bound enzymes which control pumping of H^+ ions out and K^+ ions in in exchange across the membrane. The fungal haustorium can therefore deplete solutes from the matrix which surrounds it and maintains a high concentration gradient which the extrahaustorial membrane cannot oppose. The fungus can therefore maintain tight control over membrane permeability and transport processes.

It is obviously not an advantage to biotrophic fungi to cause massive membrane damage since this would kill host plant cells. Many plant pathogens however, produce toxins which have pronounced effects on the permeability of cell membranes. Examples of host-selective and host non-selective toxins cause increases in permeability and losses of electrolytes (see Chapter 3 for details relating to toxins). For example, fusicoccin, produced by *Fusicoccum amygdali*, specifically activates membrane ATPase activity and therefore disrupts solute fluxes. This toxin also causes stomatal opening. The effects are well known because the toxin is mobile in the transpiration stream. The toxin cercosporin alters the ratio of saturated to unsaturated fatty acids in membranes which causes a decrease in membrane fluidity. This results in an increase in permeability, loss of electrolytes and subsequently cell death.

Some membrane damage may result from enzymic attack by pathogens. *Sclerotium rolfsii* and *Botrytis cinerea* produce extracellular enzymes which degrade lipids and *Rhizoctonia solani* has been shown to degrade protein. However, although there is some evidence to support direct enzyme attack on membranes it is likely that much damage occurs as a result of wall degradation by the invading pathogens. Pectolytic enzyme activity causes gelation in middle lamellae and plasmodesmata are then exposed to progressively more liquid substances and may therefore become ruptured by osmotic forces. It is not clear if the rupture of plasmodesmata is fatal to cells but it seems unlikely. It is possible to release plant protoplasts by enzymic lysis of plant cell walls, into an osmotically stabilised solution which prevents bursting. A proportion of these isolated protoplasts have the capacity to regenerate new cell walls and are then able to grow and divide, given a suitable growth medium and carbon source (Cocking, 1983).

To retain functionality membranes must be maintained by cells and therefore require inputs of energy and reserves. Where pathogenic invasion has caused cellular disruption to such an extent that resources are not available then membrane damage will result and is likely to compound the effects of the pathogen on the plant. In general, alterations to membrane permeability are not reversible. Plants also respond to pathogen invasion by the production of compounds with antimicrobial properties, the phytoalexins. The presence of these compounds in the plant may have consequences for membranes. Some phytoalexins cause membrane damage as well as other effects on the host plant, e.g. pisatin influences stomatal opening by affecting membrane function.

NUCLEIC ACID AND PROTEIN METABOLISM

Changes in nucleic acid and protein metabolism are often some of the earliest physiological effects induced in host plants after invasion. The implications of the observed changes are not well understood but have often been associated with the expression of resistance genes in the host, and the changing patterns of photosynthesis induced by the pathogen. However, the sources of these variations are not clear. For example, some alterations will be due to the response of the host plant to mechanical injuries caused by infection and therefore related to repair mechanisms but not directly to any disease interaction. It is also extremely difficult to determine, experimentally, the relative contributions of the host and the pathogen to the measurements which are made and the changes which are seen to occur. There is certainly a great deal of evidence (Dixon and Harrison, 1990) suggesting that *de novo* gene expression gives rise to biochemical defense responses in plants (Fig. 4.13).

In terms of the research which has been carried out in this area, much interest has been associated with the nitrogen metabolism of infected host plants, particularly in relation to the differences between resistant and susceptible cultivars. After infection, resistant cultivars often show decreased levels of total nitrogen and protein nitrogen, although in established biotrophic infections on susceptible cultivars it is more likely

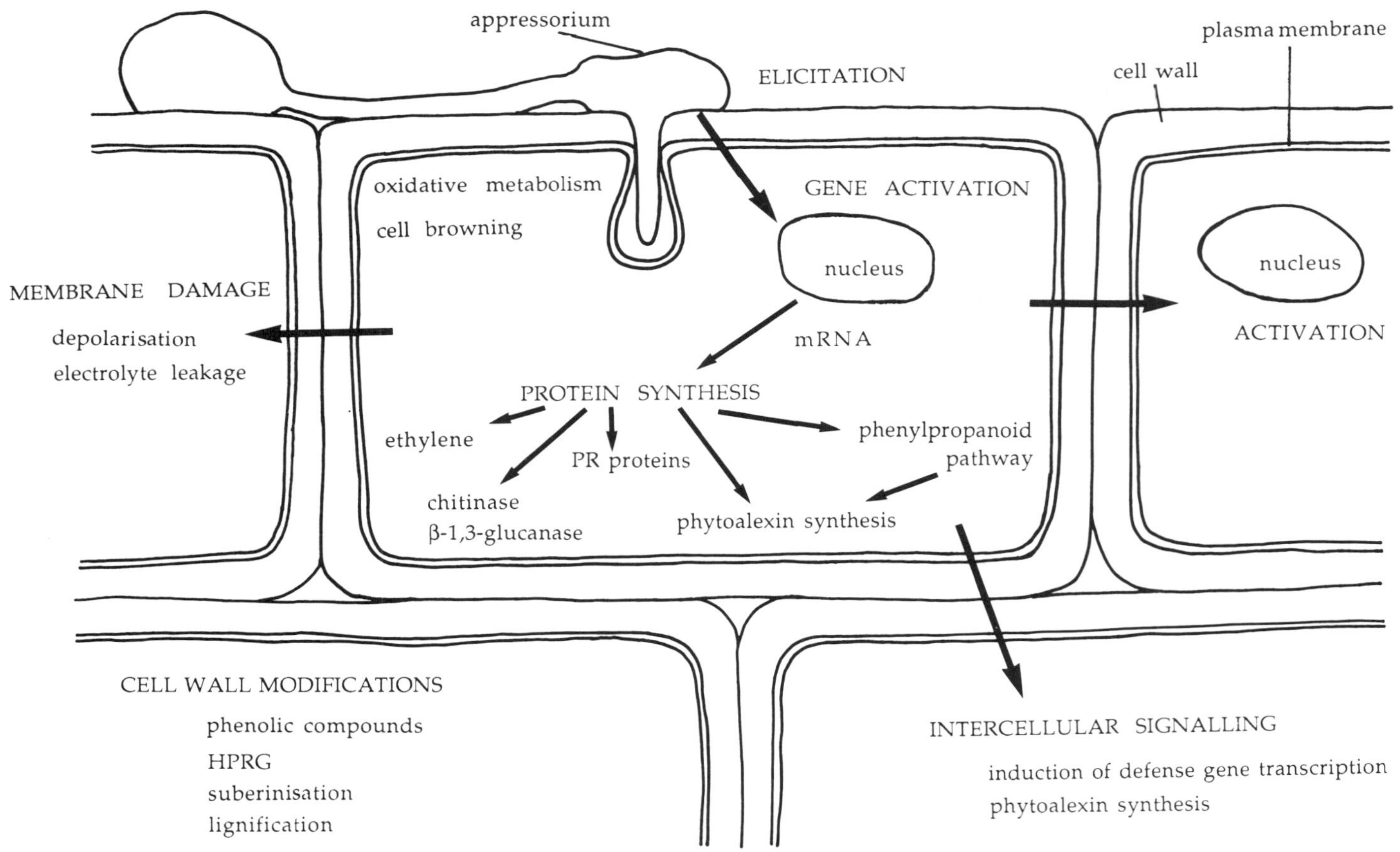

Fig. 4.13 Generalised diagram to show major elicitor-induced changes in plant cell metabolism.

that levels of total nitrogen are increased. However, host plants exhibit a range of responses in terms of the uptake of nutrients. Rates of uptake and concentrations of nitrogen measured in infected tissues (Paul, 1989) vary considerably between disease interactions and differ throughout the various stages of disease development.

Nucleic acids

In a wide range of biotrophic infections (e.g. *Plasmodiophora brassicae*, cabbage; *Fusarium solani*, pea; powdery mildew, barley; rust disease, wheat) host nuclei and nucleoli are seen to increase in size very early after infection, particularly in those cells nearest to infection sites. This is accompanied by an increase in the amount of nuclear RNA and the ratio of RNA to DNA measured in susceptible infected leaf tissue. However, later in infection, after disease establishment nuclei are seen to collapse.

There are conflicting reports for changes in ribosomal RNA (rRNA) levels but most records indicate increases in the amounts of rRNA synthesised in infected tissues. A loss of chloroplast rRNA at the time of visible symptom expression has been shown to accompany disease establishment. However, the changes which occur directly after infection and during disease development are not clear. In rusted oat plants chloroplast rRNA levels fall, perhaps as a result of chloroplast disintegration, but cytoplasmic rRNA levels increase, possibly due to the presence of fungal mycelium. It has been suggested that in plants infected by powdery mildew fungi, toxic materials released from the pathogen must be the source of affect on chloroplasts and contents of chloroplast rRNA. Haustoria of powdery mildew fungi only penetrate epidermal cells, which do not contain chloroplasts. Loss of nucleic acids from chloroplasts must therefore be caused by diffusable metabolites. Increases in total RNA contents have not been measured in experiments using resistant plants. Additionally, no changes in cytoplasmic or chloroplast rRNA have been recorded. However, it has been suggested that in incompatible reactions of soybean plants infected by *Phytophthora megasperma* f. sp. *glycinea*, increases in mRNA levels may precede the formation of phytoalexins.

Proteins

Levels of protein synthesis measured at infection sites in susceptible plants are usually increased over levels in healthy or resistant plants. Much of this increase can be attributed to the synthesising activities of the invading pathogen. Radiolabelling techniques have allowed the sites of protein synthesis to be traced, and shown them to be localised in the fungal mycelium and spores formed in infected leaf tissues. Synthesising activity is stimulated in areas around infection sites in some plant species.

The most general effects reported are changes in normal protein turnover in infected leaves. The overall balance of protein synthesis is disturbed, with changes to the levels of substrate supply, activators and coenzymes, etc., so that disruption to normal synthetic activities would be expected. It is likely that the amounts of synthesis of some proteins would be reduced while others would be produced at higher levels than normal. Some proteins may be degraded more readily after infection. Many of the changes measured in infected plants are likely to be attributable to the plants response to the repair of mechanical injuries.

Increases in protein synthesising activity of green island tissue are dramatic whereas in the chlorotic tissues between these sites protein synthesis is depressed to very low levels. As disease progresses and infected tissues become more physically disrupted, increases in proteolytic activity can be measured and levels of total nitrogen are often reduced in the later stages of disease.

Some new proteins, novel to the host, may be synthesised in resistant plants and are usually associated with the hypersensitive response (Chapter 3) and resistance mechanisms. Additionally, the application of protein synthesis inhibitors has been shown to decrease the level of resistance shown in some host–pathogen combinations (Heath, 1979). These novel proteins are termed pathogenesis-related proteins (PR proteins). Such polypeptides are usually of low molecular weight and have been detected in microbially infected plants and also in tissues which have been treated with elicitor molecules from microbes or with abiotic factors such as ultraviolet light, heavy metal ions, or other stress factors. Characteristic PR proteins are often formed by particular plant species. In some instances the appearance of PR proteins has also been correlated with developmental changes such as flowering, or senescence. Some information has now been accrued concerning the timing of synthesis in treated plants and it has been shown that these proteins occur mainly in intercellular spaces (apoplast). However, although a role in disease resistance has been heavily implicated, the functioning and activities of these proteins are, as yet, unknown.

Enzyme levels

Levels of enzymes involved in amino acid and amide synthesis do become increased in infected tissues. Elevated levels of respiratory enzymes, enzymes involved in pathways of photosynthesis, phenylpropanoid metabolism, peroxidases, and polyphenoloxidases have been measured. Such increases have been associated with the plants response to mechanical injury, wound repair and with the biosynthetic pathways involved in altered phenol metabolism. Increases in phenylalanine ammonia-lyase (PAL) have been associated with changes relating to necrosis and the hypersensitive response.

Phenylpropanoid metabolism

Phenylpropanoids have a number of important roles in higher plants, acting as signal molecules in plant–microbe interactions, as insect repellents, phytoalexins and wound-sealing compounds. A generalised scheme showing the biosynthetic pathways of phenylpropanoid metabolism is given in Fig. 4.14. Products of these pathways are precursors of plant products, such as flavonoid and isoflavonoid phytoalexins, coumarins, lignins, phenols, and the stilbene phytoalexins.

A great deal of detailed information has now been accrued concerning the pathways, mechanisms of phenylpropanoid metabolism and the nature of the individual enzymes involved (Hahlbrock and Scheel, 1989), most notably in parsley (*Petroselinum crispum*), French bean (*Phaseolus vulgaris*), soybean (*Glycine max*) and potato (*Solanum tuberosum*). Modern molecular biological methods have allowed the structure, mechanisms of activation and gene regulation to be determined for many of these enzymes. It is interesting that much of this work has been carried out using cultured plant cells and protoplasts since these have proved easy to manipulate in experiments and have been shown to retain the biochemical machinery of phenylpropanoid metabolism. This has provided highly uniform material and facilitated the use of molecular probes.

The enzyme phenylalanine ammonia-lyase (PAL) has a key role in these pathways and has been found to be present constitutively throughout whole plant and cell culture development in a number of species. Using immunohistochemical techniques with healthy parsley plants, PAL has been shown to accumulate in cells where oils are normally found, e.g. oil ducts and epidermal cells. It has also been seen to accumulate, after infection, in areas surrounding those cells in which a hypersensitive reaction occurs. Additionally, other enzymes involved in phenylpropanoid metabolism (e.g. 4-coumarate: CoA ligase) also accumulate rapidly around infection sites.

Lignification and cell wall modifications

The phenylpropanoid pathway also generates precursors for the synthesis of lignin and wall-bound phenolics. Lignins and suberin are highly resistant materials laid down during the normal development of a plant. Increased levels of synthesis occur, often very rapidly, in response to wounding and pathogen invasion. There is, however, some evidence that lignin molecules which are synthesised after induction in this way differ slightly in molecular composition.

Hydroxyproline-rich glycoproteins (HRGP) have a major structural role in plant cell walls. There have been many reports of increases in the synthesis of these polymers (particularly extensin) in pathogen-inoculated

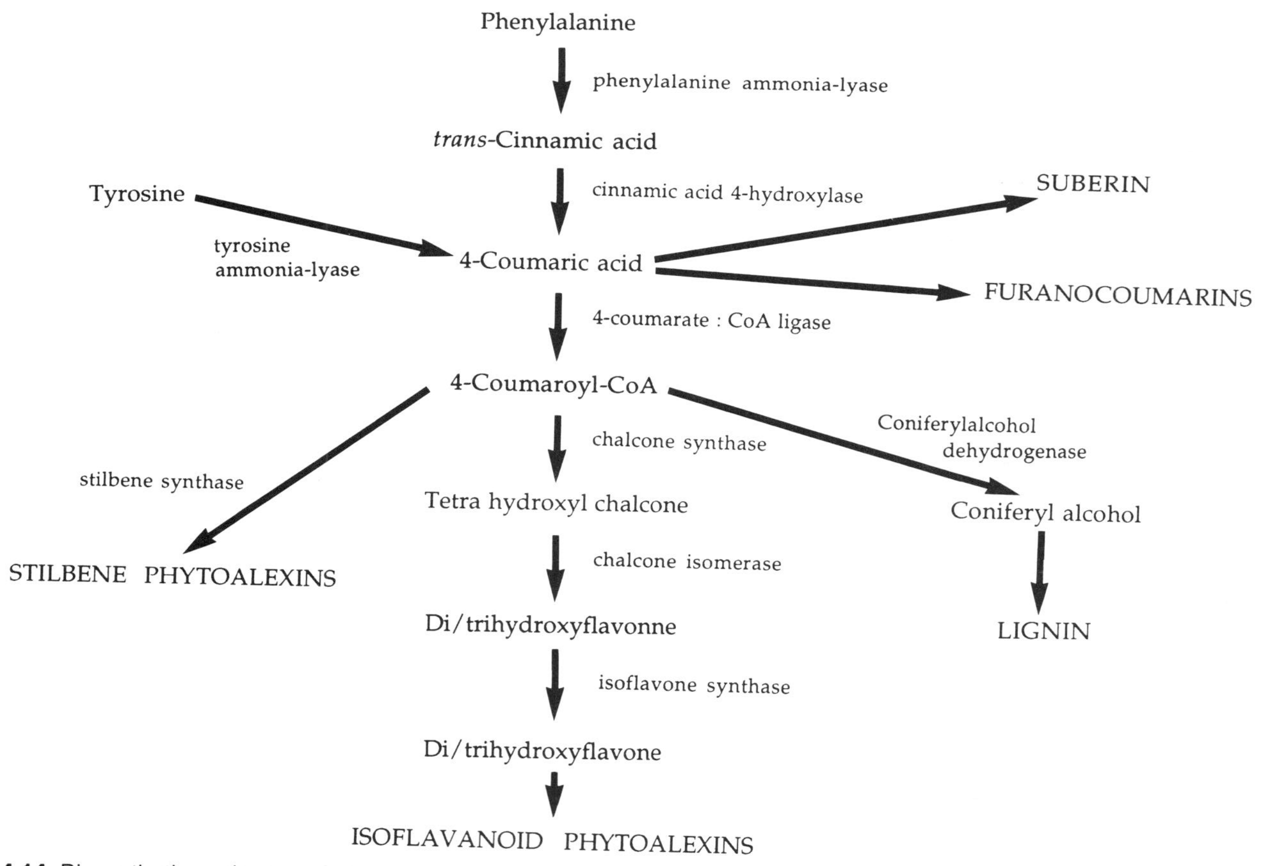

Fig. 4.14 Biosynthetic pathways of phenylpropanoid metabolism.

plant tissues, such as French bean (*Phaseolus vulgaris*), tobacco (*Nicotiana tabacum*), wheat (*Triticum aestivum*), rice (*Oryza sativa*) and carrot (*Daucus carota*). It is clear that increases in levels of associated enzymes correlate with infection and wounding (Stermer and Hammerschmidt, 1987) and this activity can also be induced by treatment with elicitors.

Hydrolytic enzymes

Increases in the activity of chitinase and β-(1, 3)-glucanase have been shown to occur very early in infected plants, ethylene-treated plants and in elicitor-treated plants. Both these enzymes can act to hydrolyse fungal cell wall polymers and are now considered to be components of plant defense systems, by virtue of this antifungal activity. Certainly no substrate is normally present in plant tissues appropriate to degradation by chitinase activity. In uninfected tissues chitinase is thought to be most usually located in cell vaculoes, although there is now evidence for extracellular location in some plants (e.g. carrot and oats), whereas β-(1, 3)-glucanase is found in intercellular spaces. These enzymes may have an important role in releasing elicitor molecules from the outer surfaces of invaders, and may therefore form an early defense warning (Mauch and Staehelin, 1989). It is now clear that these enzymes act synergistically within plant tissues (Mauch *et al.*, 1988) against pathogenic invaders.

Ribonucleases

Ribonucleases (RNases) have been investigated during obligate biotrophic (rust and mildew) infections. RNases in infected plants have been shown to exhibit different substrate preferences from those in uninfected plant tissues (host RNase) and from those of invading fungal mycelium (fungal RNase). Additionally, this type of novel RNase activity was not detected in resistant hosts which therefore suggests that changes in RNases may be related to the expression of resistance in the host or to modification of host RNase by the pathogen (Manners and Scott, 1984; Barnes *et al.*, 1988). It has been shown, however, that levels of RNases are not necessarily related to the severity of the disease and are probably synthesised as a result of infection. Intercellular washing fluids obtained by vacuum infiltration of plant tissues followed by centrifugation (Barna *et al.*, 1989), have been shown to contain RNase activity. Since these washings were apoplastic and not contaminated by cytoplasmic constituents these RNases probably have intercellular localisation in the intact plant and may therefore form a defense mechanism, protecting the host from any nucleic acids derived from pathogenic invaders.

EFFECTS OF INVASION ON PLANT GROWTH

A plant may show very dramatic changes in growth rate and habit after infection. Such alterations are accompanied by changes in ultrastructural characteristics, physiological and biochemical effects. Fungal pathogens have major influences on all cellular and metabolic functions, leading to increased rates of respiration and changing rates of photosynthesis and may also cause destruction of photosynthetic machinery and chlorosis in some tissues. Changes in carbohydrate metabolism, altered patterns of translocation and aberrant protein synthesis may also occur. The pathogen absorbs nutrients and diverts supplies of metabolites for its own use and therefore the invader acts as a great drain on all plant resources and reserves. These effects, even in combination, are not necessarily lethal to the plant but may severely reduce the rate of growth and development. However, as disease progresses it is even more likely that vital functions will be impaired. Extensive cell wall degradation may lead to the fairly rapid death of cells and tissues and colonisation by the pathogen may cause blockage of xylem vessels leading to irreversible wilting and desiccation in plant tissues.

The invading pathogen therefore exerts effects on the plant at cellular, tissue and whole plant levels. All these effects together give rise to morphological changes which are often characteristic of the disease and can be used to identify the particular disease syndrome as it develops. As well as the diagnostic value these symptoms also allow estimations to be made of the impact of disease on host plants. The types of symptom, rate of development and the severity of the disease caused is influenced by a range of factors. The species of pathogen will determine the kinds of symptoms which develop. The virulence of the strain, together with the size of the inoculum received by the plant, will affect the rate of disease development. The stages of growth, location, and type of tissue infected and the degree of inherent resistance of the plant to that pathogen will also affect the outcome of infection. The nutritional status of the plant and levels of reserves it contains, and other environmental factors which affect general plant health and vigour, also have a role in the speed and severity of disease development.

It is not always easy to assess the damage and yield losses caused by a pathogen on a host plant, or a population of host plants, in a field situation, e.g. the influence of a particular pathogen on a crop. Nor is it easy to estimate the extent to which plants are able to tolerate or escape disease development. Tolerance is an interesting concept which has attracted much argument, but the term has often been used very loosely in the literature (Clarke, 1984). The term tolerance describes the ability of the plant to endure infection by a pathogen. However, views concerning the expression of symptoms by that plant are varied. It is clear that a plant may be tolerant of invasion by a pathogen and not develop symptoms characteristic of the disease. Alternatively, a plant may be sensitive to the destructive metabolic effects of the pathogen, showing symptoms, but may be able to compensate for the effects so that the infection may not result in as drastic reductions in growth and yield as might occur in a non-tolerant plant. It is likely, therefore, that the degree of tolerance which is exhibited is related to the plants ability to compensate for any damage caused by the pathogen.

Plants may also escape disease development and remain healthy even though they are not immune. If spores are not present on a plant at a time when it is susceptible infection will not result. For example, an early crop may be harvested before a disease can become established. Additionally, spores may not reach susceptible tissue and germ tubes may not achieve entry to the plant. Agriculturally good standards of hygiene in fields, control of diseased materials, soil sterilisation procedures, crop rotation, etc., are techniques which can be used to encourage disease escape in the field.

Factors which cause stress to plants also aggravate the development of disease. Planting density, competition for nutrients, water stress and wounding alter the susceptibility of the plant to pathogenic invasion. Stresses are also generated in a plant as the result of pathogenic invasion and the physiological effects which ensue. Some of the symptoms of disease are the result of increased levels of growth regulators present in plant tissues, generated either as a response of the plant or produced by the invading pathogen.

GROWTH REGULATOR IMBALANCE IN GROWING PLANTS

Plant growth regulators (growth hormones) are important substances, present in healthy plants, which are involved in the organisation of growth and development which gives rise to the normal growth habit. Any changes in the levels of these compounds which are present, and the balance between different regulators, upsets normal developmental patterns and may therefore result in abnormal growth. Those morphological changes which accompany invasion of a plant by a pathogen are often characteristic of a particular disease interaction and are regarded as symptoms of that disease. Some such changes have been attributed to alterations in the normal levels of growth hormones present in infected tissues.

There are a number of classes of growth regulators. Indolyl auxins, cytokinins, gibberellins and ethylene are involved in growth promotion by stimulating cell division and expansion, increasing activity in meristematic tissues and metabolising nutrients in the plant. Abscisic acid is considered to be a growth inhibitor and is concerned with leaf abscission, general senescence and the preparation of the plant for dormancy. It is interesting to note that each of these compounds apparently has a number of roles in the intact plant, many of the results of these activities overlapping in the effect produced but probably achieved by different biochemical and molecular means. It is also important to realise that these compounds often have co-operative effects. A slight increase in the level of one compound alone may have little effect on the plant but an increase in two or more, or a change in the ratio between the different regulators, may have drastic effects on cells or tissues. The disruption of the balance between these hormones may have profound influences on plant development. All these compounds are active at very low concentrations and affect the plant at all levels, from effects on individual cells, disruption of normal biochemical and metabolic pathways, to changes in tissues and disruption at the whole plant level. The most dramatic of these changes are typified by tissue overgrowths (e.g. galls and cankers) such as elongation, swelling, proliferation of stems and roots (e.g. Witches' brooms and club roots), distortions of leaves (e.g. leaf curls) and stunting effects.

Growth regulators are often formed in plants as a response to stress (e.g. wounding and water deficit). It is not always possible to decide whether measured changes in growth regulator activity are the primary cause of observed symptoms or occur as the result of the plants response to these other factors. Pathogenic invasion may be sufficient wounding to stimulate growth regulator production in a plant, or wilt symptoms, caused by vascular blockages, may cause sufficient water stress to give rise to increased hormone levels.

Fungi have been shown to synthesise growth regulator compounds, often by the same biosynthetic pathways as plants. Many pathogenic species have been shown to liberate these substances *in vitro*, into culture filtrates which have been tested against healthy plant tissues, and are often shown to have effects similar to disease symptoms. Accordingly it is difficult to determine, in a host plant, whether the observed effects are due to growth regulators produced and liberated by the invading fungus or whether production has been induced in the plant by the direct stimulatory effects of fungal activities. The source of increased levels of growth regulators in infected plants has been the subject of much experimentation and argument. Modern methods with improved specificity, and more sensitive means for detecting individual compounds, may provide greater insight to the quantitative and qualitative changes which occur in plant tissues. A greater understanding of the synthesis and supply of growth regulator precursor compounds may also provide an indication of the degree of disruption and alteration to biosynthesis which occurs in infected plants. Some diseased tissues, not apparently exhibiting any morphological abnormalities or other symptoms which might be attributable to plant growth regulators, have in fact been shown

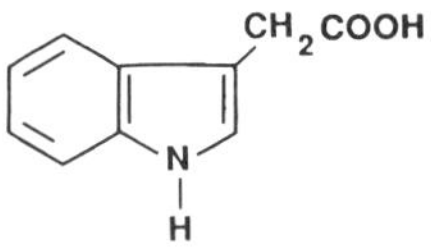

Indole-3-acetic acid

Fig. 4.15 Chemical structure of indole-3-acetic acid.

to contain elevated levels and may well be physiologically disrupted. The mechanisms which trigger or suppress symptom development are largely unknown and it is quite clear that very subtle changes occur in plant tissues.

The significance of plant growth regulators in disease development and whether these compounds have any major part in determining a compatible or incompatible disease reaction between host and pathogen, are not well understood. However, it is likely that hormones are involved in changes to growth patterns even though much of the evidence is, as yet, circumstantial.

Auxins

Auxins are major growth regulators involved in the control of many growth processes in plants, but particularly in cell enlargement and differentiation. The main active auxin found in plants is 3-indole-acetic acid (IAA) (Fig. 4.15). It is synthesised in active leaves and shoot tips and is produced via the shikimic acid pathway from tryptophan. It is also synthesised by fungi and other microorganisms via a similar biochemical route. In healthy plant cells auxins are often found in bound forms and easily make complexes with sugars. IAA may be stored in this way and may also be compartmentalised in cells. It is active in very low concentrations (10^{-5}–10^{-6} M).

IAA is often described as having a wall-loosening effect on plant cells. Increased levels provide plasticity to the structure of the cell wall which allows that cell to expand and elongate. The exact mechanism by which this loosening effect may occur is not yet clear. There have been suggestions that it is a genetic effect and that mRNA and synthesis of enzymes are involved. However, the effects of auxin on cell walls are extremely rapid and much faster than could be the case if such a mechanism was operated. It has also been proposed that auxin causes a rise in H^+ ions in the wall region so that a pH effect loosens the bonding between the wall polymers. In fact this may be the first step in the process, with further changes dependent on enzyme activity and therefore enzyme synthesis (Hadwiger *et al.*, 1970).

Many fungal diseases of plants cause an increase in the levels of auxin in tissues (hyperauxiny), often to very high concentrations. Any such change will bring about an auxin imbalance in the plant. This results in tissue overgrowth, often involving epinasty (downward growth of shoots), hypertrophy (increase in the sizes of cells) and hyperplasia (increases in the number of cells; uncontrolled cell division). This leads to gross morphological abnormalities in infected tissues. Such distortions can sometimes, but not always, be induced by the application of IAA to plants and auxins are therefore heavily implicated in symptom development. However, there are also reported cases of diseases in which very high levels of IAA can be detected in tissues but no changes in morphology are seen, e.g. *Puccinia graminis* f. sp. *tritici* infections of wheat. It is very important to remember that growth regulators often have effects in combination so that the study of any one in isolation may lead to results which are difficult to interpret. The interactive effects between growth regulators are not yet well understood.

Probably the best-known example of increased IAA levels in plant tissues infected by fungi is club root disease of Crucifers caused by *Plasmodiophora brassicae*. This pathogenic fungus invades root tissues by means of motile zoospores which gain access to the root via root hairs. The fungus upsets the plants normal regulatory control of IAA synthesis which leads to root distortion in infected individuals. It is in these abnormal masses of tissue, 'club roots', that the resting spores are eventually formed. Increased levels of IAA have also been found in rust-infected tissues of safflower during the growth of *Puccinia carthami*. A ten-fold increase has been measured in susceptible potato tuber tissues infected by another root-invading pathogen, *Phytophthora infestans*, and also in tumour tissues of corn infected by *Ustilago zeae*, which causes corn smut disease.

A great deal of research has been concentrated on the source of increased IAA levels in infected plants. However, this has proved difficult to determine (Pegg, 1984). Some pathogenic fungi do produce IAA in culture (e.g. *Fusarium oxysporum*, *Phytophthora infestans* and *Taphrina deformans*), and while this ability has often been interpreted to imply involvement, this capacity does not necessarily contribute to the course of disease or the development of symptoms. Some pathogenic fungi also produce toxic compounds which cause symptoms in the host plant which are similar to those caused by increased levels of IAA. Fusicoccin stimulates cell growth in host plants and also causes acidification of cell walls (Marré *et al.*, 1974). Victorin induces rapid cell elongation (Saftner and Evans, 1974). However, although some pathogenic fungal invaders have the capacity to contribute to elevated IAA levels and may be the source of the increase there is no conclusive evidence.

Alternatively, host plants may be stimulated to produce more IAA in response to infection by the fungus. Certainly some fluctuations in levels of IAA are due to changes in levels of precursors which are available. Additionally, disruptions to carbohydrate metabolism and respiration may well affect the synthesis of auxins. Such changes in the metabolic balance of the plant cell may also result in loss of control of IAA synthesis and bound forms of IAA or compartmentalised auxins may be released. The normal degradation of IAA may be prevented in a host plant, by the inhibition of IAA oxidase leading to the accumulation of IAA in greatly elevated levels within infected tissues. IAA oxidase has been shown to occur at higher levels in healthy *Euphorbia cyparissias* plants than in those infected by the rust *Uromyces pisi* (Pilet, 1960). A correlation was also found between the increase in IAA and the decrease in IAA oxidase in healthy plants.

Gibberellins

The gibberellins were first recognised as a 'toxic' substance produced during the infection of Japanese rice seedlings by *Gibberella fujikuroi* (the telomorph of *Fusarium moniliforme*). Infected plants grew very much taller than healthy seedlings and as a result were somewhat more fragile and earned the name 'foolish seedling' disease (Bakanae disease) for the outcome of this interaction. Gibberellic acid was isolated and purified from cultures of the fungus and the hormone properties were recognised. Gibberellins have also been isolated and identified from other fungal species and from higher plants. About 60 are now known and form a group of compounds with similar molecular structure (Fig. 4.16). The gibberellins are diterpenoid acids for which a great deal is known concerning the biochemistry of synthesis, and are active at concentrations of 0.001 $\mu g.ml^{-1}$.

In a healthy plant the action of gibberellins is of great interest. In general terms these compounds are responsible for internode extension, reversal of dwarfism, maintenance of cell division and apical dominance and the induction of enzymes, particularly for starch synthesis and cell wall synthesis. Cells of the aleurone layer, which surrounds the endoplasm in cereal seeds, respond to the presence of gibberellic acid by producing α-amylase and proteinase. In this way, gibberellic acid influences the development of the embryo and has been shown to affect protein and RNA synthesis, changes which are blocked by inhibitors of protein synthesis. Gibberellic acid may have the effect of derepressing a gene which gives rise to mRNA synthesis and subsequently to enzyme synthesis. It is suggested that, in molecular terms, the action of gibberellic acid and auxins may be similar in plants.

In addition to stimulating enzymes involved in starch and cell wall synthesis, gibberellic acid also has a role in growth promotion and maintenance of active apical meristems, stimulating cell elongation and division. The effects are very similar to those of auxins and have been suggested

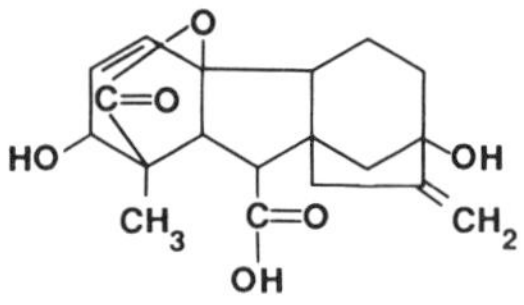

Gibberellic acid

Fig. 4.16 Chemical structure of gibberellic acid.

to enhance the action of auxins and possibly encourage increased auxin synthesis too.

There have been reports that fungal species, e.g. *Verticillium* spp., produce gibberellin-like substances but there are no reports of detection in diseased tissues. In rust infections of creeping thistle (*Cirsium arvense*) caused by *Puccinia punctiformis*, higher levels of gibberellic acid were detected in taller plants during disease development (Bailiss and Wilson, 1967). There was good correlation between the concentrations of gibberellic acid measured and increased growth and symptom development which implicates it in disease development. Additionally, gibberellic acid probably has effects in infections of cassava (*Manihot esculenta*) by *Sphaceloma manihoticola*, which initially causes the formation of necrotic leaf spots and stem canker, but subsequently gives rise to massive elongation of internodes as a characteristic disease symptom. A gibberellin has been identified in culture filtrates of the fungus (Zeigler *et al.*, 1980) and good correlation with disease symptoms has been identified in infected tissues.

Cytokinins

In healthy plants cytokinins are recognised as having a role in the control of the cell cycle, by inducing cell division, and also in the inhibition of senescence. They also affect the mobilisation of nutrients, patterns of nutrient transport and nutrient accumulation. The cytokinins are substituted adenyl derivatives with an isoprenoid side chain (Fig. 4.17). They are a group of chemically similar compounds, the activity of which depends on the properties of the side chain. They are active in plants in very low concentrations (1 $\mu g.kg^{-1}$ fresh weight). Most cytokinin synthesis takes place in active cells of roots and they are translocated through the plant via the xylem, in xylem sap, to distant tissues. Cytokinins are reported to stimulate activity and cell division in dormant buds, far removed from the sites of synthesis.

Cytokinins are reported to work in the plant in conjunction with other growth regulators. Much of our knowledge on this aspect has been gained

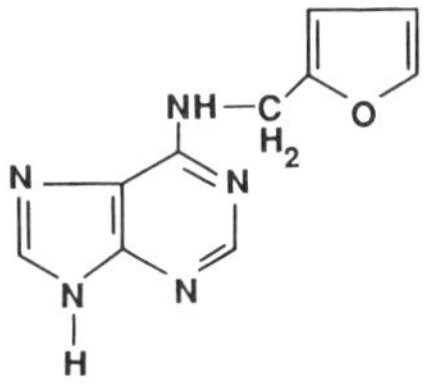

Kinetin

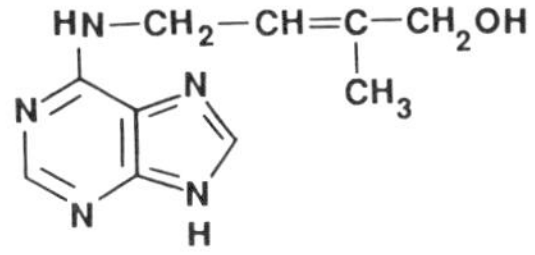

Zeatin

Fig. 4.17 Chemical structure of two cytokinins.

from studies with tissue cultures of plant cells. Cytokinins are required to promote differentiation and cell division in plant tissues and are therefore added to cell culture media to encourage growth of plant cell cultures. It has been shown that a higher ratio of cytokinin to auxin added to undifferentiated callus cultures results in the promotion of differentiation of shoot and bud tissue. However, a low ratio of cytokinin to auxin promotes the differentiation of roots from callus material. When these hormones are added in equal quantities callus tissues remain undifferentiated. It is clear, therefore, that very tight regulatory control must operate within a healthy growing plant in order for the growth habit of individuals to be maintained. Any disruption of this normal balance is likely to result in disorganised growth.

Symptoms develop in plant tissues infected by fungal pathogens which have been attributed to cytokinin effects and in many such diseased tissues cytokinin levels have been shown to be elevated. Additionally, many of the symptoms can be mimicked in the plant by the exogenous application of cytokinins which certainly implies the involvement of these compounds. In general, however, few accurate quantitative data are available and since much of the effect of cytokinins is realised in conjunction with other growth regulators, the situation in an intact plant is complex and not easy to measure or interpret.

In rust-infected tissues (*Uromyces* spp.), levels of cytokinin have been found to be higher than in healthy plants, and similarly for maize smut disease (*Ustilago maydis*), although in this latter example levels of auxin were increased in tumour tissues as well. The morphological manifestations of increased cytokinin levels are obvious in peach leaves infected by *Taphrina deformans* (peach leaf curl), which causes great distortion of the leaf lamina. Leaves become wrinkled and curled due to greater growth of cells in the upper layers of the leaf (palisade mesophyll) than in the lower (spongy mesophyll) layers. In this disease it has been shown that cytokinin levels are raised in infected tissues. It was also shown, by chromatographic analysis (Sziraki *et al.*, 1975), that three forms of cytokinin present in infected tissues were the same as those found in healthy leaves but there was also an additional form, not detected in healthy tissues.

In club root disease of cabbage (*Plasmodiophora brassicae*) gross morphological changes are caused by repeated cell division, resulting in the production of tumorous tissue on roots. Cytokinin levels are greatly elevated in these tumours (×100) and such tissue is self-supporting for cytokinin supplies in tissue culture. Additionally, high auxin levels have been reported in this tissue which increases the host plant responses to elevated cytokinin levels.

In rust and powdery mildew diseases cytokinins appear to have a different role in symptom development. Characteristic green islands are formed around infection sites on leaves. These are areas of active tissue in which chlorophyll and photosynthetic activity are retained, and RNA and protein synthesis are stimulated, in leaves which are otherwise chlorotic and senescent. This is thought to be an effect of increased cytokinin activity at infection sites. However, the evidence is based largely on the fact that exogenously applied cytokinins give rise to similar symptoms in healthy leaves. In these cases cytokinins are suggested to prolong photosynthetic activity in affected cells and to mobilise nutrients within infected tissues, to accumulate around infection sites which act as nutrient sinks in the leaf. Levels of cytokinins in such infected tissues have been shown to be elevated. Spores from some species of rust (*Uromyces* spp.) have been shown to contain endogenous levels of cytokinins. Chromatographic analysis has shown that the cytokinin activity of *Uromyces* spores and mycelium is different from that found in increased levels in infected host tissues.

Cytokinin activity is therefore heavily implicated in the development of disease symptoms in infected plants, although few details of the effects and molecular mechanisms are available. Indeed it has been suggested that the process of infection may induce cytokinin activity in plants.

Ethylene

Ethylene (C_2H_4) has an important role in the regulation of plant growth and is implicated in a number of disease symptoms, e.g. tissue swelling, stimulation of root formation, leaf abscission, chlorosis, epinasty and fruit ripening. It is produced by plants and some species of pathogenic microorganisms, is effective in very low concentrations ($1.0\ ng.l^{-1}$) and is synthesised from methionine in the presence of light.

Ethylene is produced in plants as a reaction to wounding and to stress which has made the evaluation of its role in the development of disease symptoms and pathogenicity of microbes difficult. The use of detached leaves and tissue sections provides material which will obviously be subject to increased levels of ethylene as a response to wounding and so data from such experiments must be interpreted with care. Additionally, abscisic acid stimulates ethylene production in leaves and interactions with other plant hormones also complicate interpretation of experimental data. However, a number of roles are attributed to ethylene.

Increased levels of ethylene are characteristic of tissues infected by fungi, possibly as a consequence of injury on penetration and invasive growth. In healthy plants, ethylene has an important influence on leaf abscission and is present in higher levels prior to leaf fall, probably as a result of its effect on cell wall degrading enzymes. The defoliation symptom caused by *Verticillium* wilt is also attributed to increased levels of ethylene in infected hop plants. Enhanced levels of ethylene also advance the ageing and senescence of plants which may increase the general susceptibility of plant tissues to fungal invasion. In *Verticillium* wilt disease ethylene has been attributed a dual role; acting directly as a toxin on plant tissues causing leaf drop, epinasty and premature senescence and also inducing disease resistance in infected plants. Treatment with ethylene has been shown to reduce the effects of *Sclerotinia* root rot and reduced symptom development in tomato wilt disease caused by *Verticillium albo-atrum*.

Ethylene stimulates the production of a number of enzymes. In some cases these probably have a major influence on plant cell walls, directly affecting wall integrity, extensibility and growth. In healthy plants ethylene is reported to influence the levels of hydroxyproline-rich proteins in cell walls which are important to wall structure. Ethylene treatment increases the levels of cellulase and β-(1,3)-glucanase, which are involved in the degradation of cell wall polymers. The role of ethylene in disease resistance may be attributed to non-specific induction of phytoalexin synthesis and the stimulation of enzymes which are active in phytoalexin production, e.g. PAL (Hadwiger *et al.*, 1970), peroxidase, polyphenol oxidase and pectin esterase. These enzymes oxidise phenolic compounds and are involved in resistance reactions of host plants.

There are also reports that ethylene treatments stimulate the production of antifungal compounds in carrots and the synthesis of pisatin in peas. It can be seen, therefore, that the effects of ethylene on both healthy and diseased plants are varied and complex.

Abscisic acid

Abscisic acid (ABA) has an important regulatory role to play in plant growth and development as a growth inhibitor. In addition to reducing meristematic activity ABA induces dormancy in plants, and inhibits

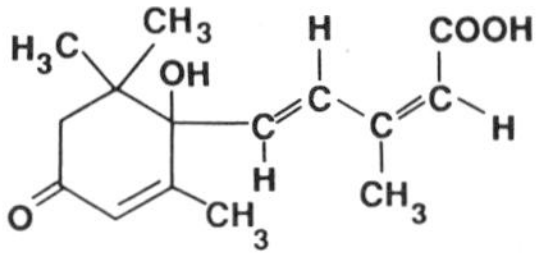

Abscisic acid

Fig. 4.18 Chemical structure of abscisic acid.

bud break. It also has an important role in stomatal closure. Stunting, which accompanies fungal infections of plants, and which is an important symptom, has been attributed to increased ABA levels. Chemically ABA is an isoprenoid, synthesised via the mevalonic acid pathway (Fig. 4.18). It is chemically similar to, and produced by the same biochemical route as gibberellic acid, but gives rise to almost opposite effects on plant tissues. It is interesting that ABA causes chloroplast ageing and eventual degeneration, and that gibberellic acid is primarily synthesised in chloroplasts and plastids.

The role of ABA in disease expression is not clear. It is most likely that changes in ABA are one factor in symptom development and interactions with other growth regulators must be important. Treatments with (exogenous application) ABA and gibberellic acid have been shown to increase the susceptibility of bean tissues to rust fungi (Li and Heath, 1990). Increased levels of ABA have been measured in cotton plants infected with a strain of *Verticillium albo-atrum* which caused defoliation, whereas infections with a non-defoliating strain did not cause an increase in ABA. Additionally, higher levels were found in tomato plants with *Verticillium* wilt disease (Wiese and De Vay, 1970). ABA treatment has been shown to stimulate suberinisation in potato tubers, a response induced by wounding. ABA may have a significant and very precise role in the regulation of wound healing. Exogenous applications have been shown to suppress the activity of PAL in treated soybean hypocotyls, increasing susceptibility to *Phytophthora megasperma* f. sp. *glycinea*. ABA may affect the transcription of plant defense genes (Ward *et al.*, 1989) and may therefore have a very important role in the regulation of fungal–plant interactions.

ABA levels are seen to rise in healthy plants which are subject to wilting. Guard cell turgor is dependent on H^+/K^+ ion exchange but in such wilted plants increased ABA inhibits the uptake of K^+. This prevents any increase in guard cell turgor and maintains stomata in the closed state, so preventing further water loss through stomatal pores. It has been suggested (Pegg, 1981) that loss of turgor in a plant due to wilting (e.g. in vascular wilt disease) will lead to increased levels of ABA in infected tissues. The development of stunting in these tissues may therefore be due to the ABA produced by the plant in response to water deficit. This increase in ABA is the result of wilting and is not the primary cause of symptom development. ABA also stimulates ethylene production in plants so that some observed effects may be due to increased ethylene levels in infected tissues.

EFFECTS OF REDUCED GROWTH ON CROP PLANT YIELD

The structural and functional damage, together with the metabolic disturbances, which occur during the course of infection and disease establishment result in reduction of the potential productivity of affected plants. In agricultural terms this may have very serious economic implications. It is therefore important to assess the degree of damage which occurs to crops under the prevailing conditions. Such information potentially enables the evaluation of any control methods which may have been operated. Additionally, the building of a body of information, from which those factors which influence the outbreak and spread of diseases (epidemiology) may be recognised and predicted, will increase the success of disease forcasting.

Many of the changes brought about by disease result in the development of symptoms and may have far-reaching consequences to the plant and its developmental pattern. Such influences are manifest in different ways, often referred to as the syndrome. Visible effects give some indication of the problems faced by an infected plant and enable preliminary diagnosis of the disease, but some symptoms are more characteristic and easily recognisable than others. Additionally, in some cases the physiological functioning of a plant may be impaired as the result of microbial activity, without the manifestation of symptoms (Smedegaard-Petersen, 1984). Hence, it is not always easy to make an assessment of any losses which may occur.

Infections may cause the plant to grow more slowly and the disruption of physiological functions may result in the development of smaller plants. Many diseases affect the photosynthetic potential of the plant, either by direct physical damage to chloroplasts and membranes or by interference with biochemical pathways. Such effects may limit plant size but will also alter the reserves of the plant and probably therefore the sizes and numbers of seeds and storage organs, such as the roots, tubers and bulbs, which develop. In economic terms it is parameters such as these which are of most concern to the farmer. The overall yield, or total biomass production, of a plant may not be as important in many cases as the production of specific organs or tissues which can be sold commercially or used for future crop production.

The effects of particular pathogens are also heavily influenced and mediated by the health of the plant, its nutrient status, and any levels of natural resistance to the particular strain of pathogen which is encountered. In addition, the timing of the initiation of infection and disease within the developmental sequence of the plant is of primary importance in terms of the economic damage which may be inflicted. Effects occurring at the very early seedling stage may not be recognised in terms of final yield reduction since these plants do not become established and losses may be counteracted by overplanting. Diseases which take hold after the development of tillers and the filling of grain may also not hold any great significance for the farmer. However, influences which occur at, or prior to, the partitioning of assimilates within the plant may have much more serious agricultural consequences.

Means by which the extent and severity of a disease could be assessed rapidly, reliably, reproducibly and easily would be useful. Some systems used require destructive harvests and are therefore retrospective, and some parameters are qualitative and therefore open to interpretation and error. Keys and descriptions have been constructed to increase the reproducibility of observational assessments, for fruits, shoots and leaves. Leaves are responsible for the manufacture of the storage materials which are laid down by the plant. As a consequence any impairment of leaf function is likely to affect the export of carbohydrates and lead to a reduction in yield. Comparisons between the reduction in green leaf area (GLA) caused by foliar pathogens, particularly where these are confined to the leaf lamina, and yield, might therefore be expected to correlate and provide a useful assessment system for the estimation of damage. Investigations have been made for barley (Carver and Griffiths, 1981; Griffiths, 1984) for which biomass production and grain yield have been shown to be closely related. For powdery mildew infections (*Erysiphe graminis*) GLA was found to be proportional to (but not equal to) the area of leaf mildewed during growing stages. However, it is unlikely that the relationship would be similar for all cultivars and all conditions, indeed environmental factors can be seen to exert important influences and affect the comparison of data between experiments (Carver and Griffiths, 1982; Griffiths, 1984). Such relationships allow the assessment of the efficacy of fungicide treatments under one set of conditions for powdery mildew of barley. However, reduction in GLA was shown to be a poor indicator of leaf performance for leaf blotch (*Rhynchosporium secalis*) which gives rise to necrotic spots on laminae. Evidently the appearance of necrotic lesions does not reflect the extent of the functional disruption caused. Similarly for brown rust (*Puccinia hordei*) of barley, in which the green island effect may lead to the retention of greening in affected cells, so that GLA does not reflect the extent of leaf dysfunction. The situation is more complicated for other diseases, particularly those which cause symptoms that are difficult to evaluate or that affect underground organs.

In addition to reductions in yield of commercial crops, pathogens also have other effects. The saleability of produce may be affected by superficial lesions, blemishes or deformations but which have relatively little physiological significance. Even a slight reduction in quality will affect market value, even very small blemishes changing the market grade at which this produce may be sold, and if affected material is to be removed by hand, this may ultimately be a very costly procedure. The storage potential of produce may also be impaired if even a small proportion is likely to harbour organisms which may cause post-harvest destruction.

Mutualistic symbioses 5

INTRODUCTION

Physiological interactions between plants and fungi begin very soon after contact has occurred. In many instances the changes which are induced in plants may be detrimental to the general physiology of the plant, to a greater or lesser degree, or may even be fatal. However, there are also a great many cases where interactions and long-lived associations occur between fungi and plants, from which the partners appear to derive benefit, e.g. mycorrhizas and lichens. In general, these are also interactions which involve the flow of nutrients between the associates (Smith and Smith, 1990) resulting in increased growth, vigour and survival in the natural environment. The partners may be able to survive independently of such associations (facultative) or may be totally dependent on such interactions (obligate) to survive and reproduce.

De Bary (1866, 1879) first used the term symbiosis to describe the living together of differently named organisms (two or more) for part of their life cycles. His concept of symbiosis included mutualistic associations, in which one organism benefits to the detriment of the others involved. Much consideration has been given to this definition and some authors have since considered the term distinct from parasitism, and contrasting with it (Lewin, 1982). Others have, however, retained De Bary's original concept (Lewis, 1973, 1985; Starr, 1975) and it is in that sense that symbiosis is used here.

Much discussion has also been given to the concept of harm and benefit in this context. Such parameters are difficult to describe and even more difficult to measure. Mutualistic associations may result in enhanced survival, nutrient aquisition, reproduction and growth for the component organisms. However, some of these changes are not easily assessed in terms of cost and benefit to the individual. For example, some features of the individual organisms may be lost in the association. Growth rates for the associated organisms may be lower than rates of growth achievable for the component individuals in similar conditions (e.g. lichen thallus proliferation). Law and Lewis (1983) have considered the increased fitness of individuals in association, to constitute mutualistic symbiosis. It is also important to remember that the relationship between organisms may change with time and/or may be different under other biotic and abiotic conditions, so that a relationship may not be mutualistic or parasitic throughout the life cycles of those individuals concerned.

A range of symbiotic interactions can be recognised and have been classified as a continuum (Lewis, 1985). Read (1970) considered interactions between populations of two species using a score system, where interaction resulted in increased population growth (+), decreased population growth (−) or where growth was unaffected (o). Other workers have extended this to consider the potential fitness of the organisms concerned and Lewis (1985) has given score combinations to demonstrate the continuum (Table 5.1). Lewis (1985) uses the term agonism to describe the −/+ combination in which the fitness of one associant is enhanced to the detriment of the others. This includes parasitic relationships. Antagonism is preferred to collectively describe competition, amensalism and agonism.

This chapter considers mutualistic symbioses involving fungal partners, e.g. associations with plant roots (mycorrhizas), with algae and cyanobacteria (lichens) and asymptomatic endophytic symbioses with aerial parts of plants, demonstrating the varying degrees of intimacy, nutritional interdependence and the physiological interactions which occur.

MYCORRHIZAL SYMBIOSES

The term mycorrhiza (literally 'fungus root') was first used by Frank (1885) to describe the long-lived association between plant roots and fungal mycelium. Mycorrhizas are mutualistic symbioses, the co-existing

Table 5.1 The symbiotic continuum expressed in terms of the fitness of the associated organisms (after Lewis, 1985)

Mutualism	+/+
Commensalism	0/+
Nyetralism	0/0
Agonism	−/+
Amensalism	−/0
Competition	−/−

Potential fitness decreased; − −; Potential fitness not affected − 0; Potential fitness increased − +.

partners have mutual dependence on each other. The normal state is mycorrhizal and dual organs are formed with consistently recognisable morphology. Exchanges of nutrients and metabolites occur between living cells and both partners derive benefit from the relationship. Mycorrhizas are probably the most abundant of all symbioses world-wide and are the main organs of nutrient uptake in terrestrial ecosystems (Harley and Smith, 1983).

It is interesting that the majority of all land plants are mycorrhizal (Harley, 1986, 1989). In a European survey, 76% of all plants examined were shown to have mycorrhizal infections (Harley and Harley, 1987). These associations are formed with most species of angiosperms, all gymnosperms, pteridophytes and some bryophytes (particularly liverworts). The presence of the fungal partner influences the absorbtion of materials from the soil, increasing the efficiency of nutrient uptake by the plant. Greater root branching and a rise in the surface area of roots increases the volume of soil which may be exploited by a plant. Additionally, mycorrhizal associations often enhance the longevity of the roots involved. Although some structural changes occur to the plant roots the host is not damaged and little necrosis occurs. Where roots are mycorrhizal plant growth is usually enhanced in comparison with that of uninfected plants.

Representatives from all the major taxonomic groups of fungi (Zygomycotina, Ascomycotina, Basidiomycotina and Deuteromycotina) form mycorrhizal associations. The fungus invades plant root tissues and in general, gains a supply of carbon from the host. Many of the fungi involved are poor competitors in the soil environment, and some are obligate symbionts which are not capable of independent growth either in the natural environment or in culture. In general, however, the fungal partner maintains contact with the soil, sometimes forming an extensive mycelial network in the immediate vicinity which may have a very important role in the nutritional interdependence between the partners.

MYCORRHIZAL TYPES

A range of forms of mycorrhizas occur and have been classified (Harley and Smith, 1983) and grouped together by the structural characteristics at maturity (Table 5.2). This is a deliberately descriptive scheme. The great similarities which link the structural relationships between host and fungal cells have been used to group the associations together, allowing comparisons to be made without implications concerning the physiological nature of the involvements. Very close contact is set up between the fungus and plant root cells, often with intercellular growth of hyphae and in all cases except the ectomycorrhizas, the fungal partner

Table 5. 2 The characteristics of mature mycorrhizal states (simplified from Harley and Smith, 1983)

Characteristic	Mycorrhizal association						
	Vesicular-arbuscular	**Ectomy-corrhiza**	**Ectendo-mycorrhiza**	**Arbutoid**	**Monotropoid**	**Ericoid**	**Orchid**
Fungi septate	–	+	+	+	+	+	+
aseptate	+	(+)	–	–	–	–	–
Hyphae within cells	+	–	+	+	+	+	+
Fungal sheath present	–	+	+/–	+	+	–	–
Hartig net formed	–	+	+	+	+	–	–
Hyphal coils in cells	+	–	+	+	–	+	+
Vesicles in cells	+ (or –)	–	–	–	–	–	–

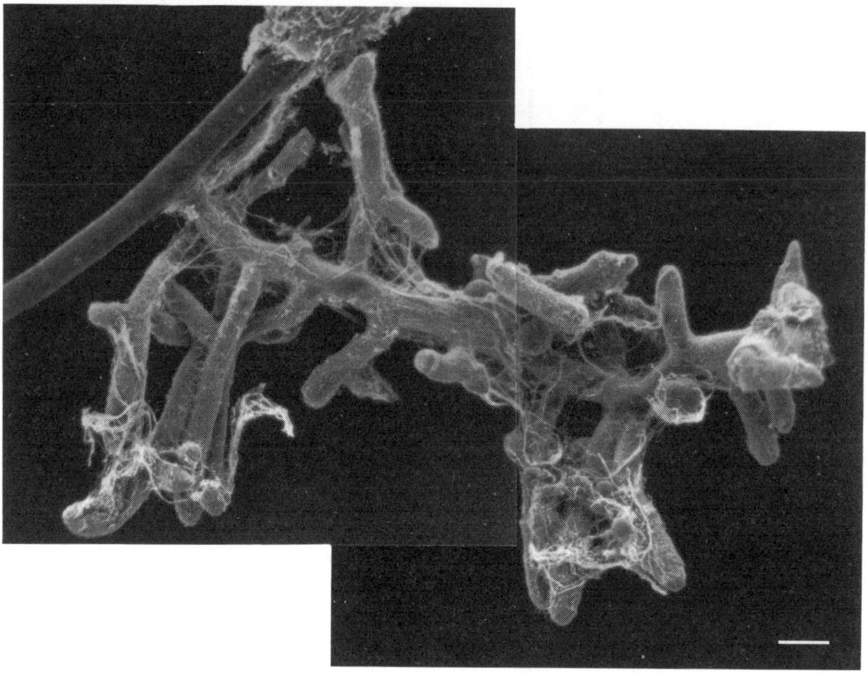

Fig. 5.1 Characteristic branching of mycorrhizal roots of birch. Bar represents 1 mm. Reproduced courtesy of Dr P. Mason and Dr A. Crossley.

penetrates the host plant cells. Further divisions can be made on the basis of the fungal partners involved and whether the fungal component forms a tight sheath or mantle around the outer surface of the root. It is important to remember that morphological characteristics may be different during the development of the association and at senescence. In fact the relationship between the fungus and the plant constantly changes throughout the association. The main types of mycorrhizas will be considered independently.

Ectomycorrhizas

Ectomycorrhizas are very common associations for forest trees and for shrubs, particularly in subarctic and temperate regions. These are particularly important associations for plants which inhabit regions where growth is restricted at some times. Many of the host plants belong to the families Pinaceae, Fagaceae, Betulaceae and Myrtaceae.

Uninfected roots of species which form ectomycorrhizas are usually of two forms. Main roots which extend, elongating into the soil, and

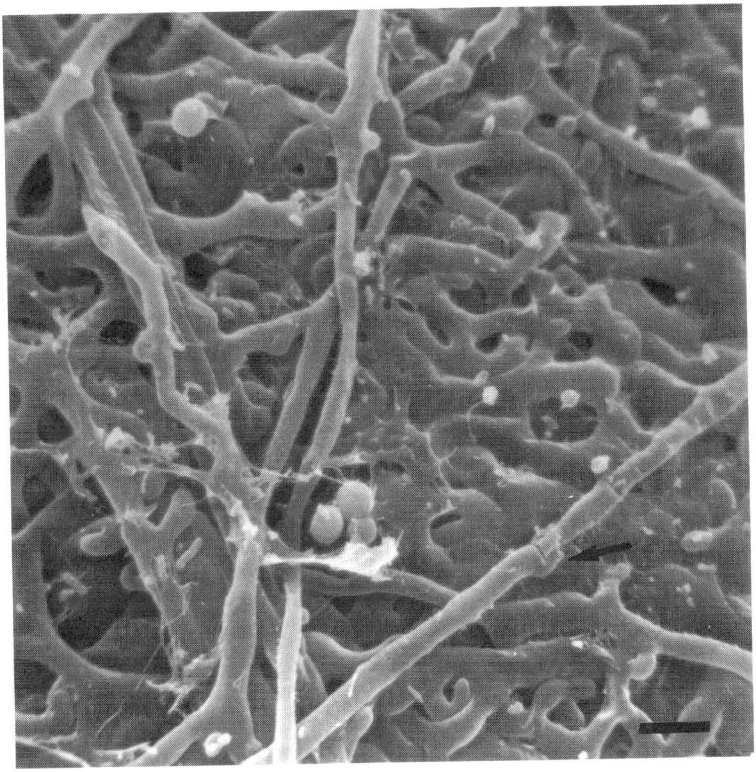

Fig. 5.2 Smooth outer surface of ectomycorrhizal sheath with a few fungal hyphae distinguishable, note the presence of clamp connections (arrowed). Bar represents 10 μm. Reproduced courtesy of Dr P. Mason and Dr A. Crossley.

lateral roots which undergo more limited extension. Uninfected roots may develop root hairs. After mycorrhizal infection when roots have been colonised, a number of morphological changes are seen and the roots take on a different, but characteristic, appearance. The main roots, which are usually longer, continue to elongate and show less morphological change after infection and also become longer lived. Laterals become thickened, are often pigmented, and elongate only slowly. These ectomycorrhizal roots branch in a characteristic fashion (Fig. 5.1). All the ectomycorrhizas branch in a racemose manner, except associations with the genus *Pinus* in which the mycorrhizas branch dichotomously. Functional mycorrhizal roots have many short branches and the lateral roots are completely ensheathed by a layer of fungal tissue, usually 20–40 μm thick (sheath or mantle). The sheath is quite variable and may only be one or two hyphal layers deep, but often occurs in denser, more compact and thicker forms. No root hairs are formed after infection.

The outer surface of the mycorrhiza is smooth (Fig. 5.2), although a few hyphae maintain connection with the surrounding soil, and sometimes extramatrical mycelium may develop.

In the natural environment, it is likely that new roots are colonised as they are formed by hyphae from the sheath of the main roots. Infections of seedling roots are probably brought about by spores or extramatrical mycelium from nearby mycorrhizal roots. These hyphae are attracted by root exudates and grow towards the root aggregating around the outer surface. At first, a very loose network is formed but then hyphae penetrate between the epidermal cells of the root and grow between the cortical cells. All penetration is achieved by mechanical pressure but if any enzymic degradation of cortical cell walls does occur it is highly localised. The fungal sheath is also connected to hyphae which grow between the cells of the outer root cortex. These plant cells are pushed apart by the invading hyphae and eventually become radially elongated (Fig. 5.3) with hyphae growing between them in a network, known as the Hartig net. The invading hyphae are active and remain intercellular, not penetrating the cortical cells. The infection never extends beyond the endodermis into the stele or into undifferentiated tissue, although the root tip is covered externally by the fungal sheath. As the root grows the fungus grows with it, giving rise to and maintaining the characteristic mycorrhizal root forms. The fungal component may contribute 25%, or more, to the dry weight of an infected root. Ectomycorrhizal associations have been shown to increase the growth rate and biomass production of the host plant and also to influence the proliferation of the root system and the development of lateral roots (Table 5.3).

In general, the fungi which form ectomycorrhizal associations are Basidiomycetes with representatives from 25 families, e.g. Agaricales (*Amanita, Boletus, Suillus, Tricholoma*); Russales (*Russula, Lactarius, Cortinarius*); Aphyllophorales (*Thelephora*); Hymenogastrales (*Rhizopogon*); and Sclerodermatales (*Scleroderma*). Some Ascomycetes are also reported to form ectomycorrhizas, e.g. members of the Tuberales (*Tuber* spp., the truffle fungi) and two species of Zygomycetes (Endogonaceae) from the genus Endogone, a group which are more usually noted for forming only endomycorrhizal (vesicular–arbuscular) associations. Although many of the fungi involved will grow in culture they are usually very slow growing and often have quite complex nutrient requirements. They compete very poorly as saprophytes and probably depend on associations with higher plants in the natural situation. The fungi are generally able to form associations with a wide range of host plant species although some are more species-specific (e.g. *Suillus grevillei*). Plant species may be host to a number of associated fungal species, indeed sometimes a tree may be host to several species simultaneously. In particular, species of the genera *Salix*,

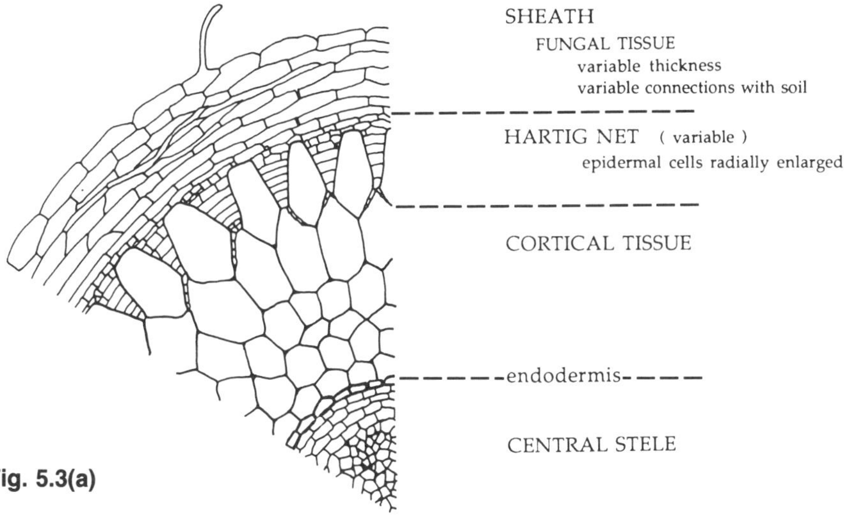

Fig. 5.3(a)

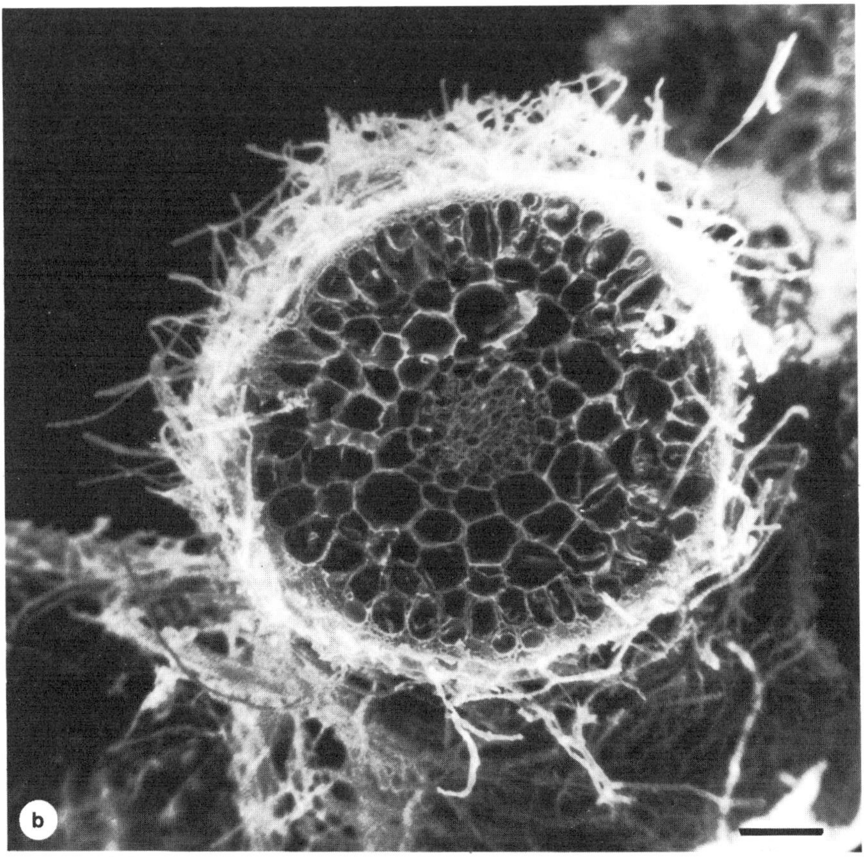

Fig. 5.3(b)

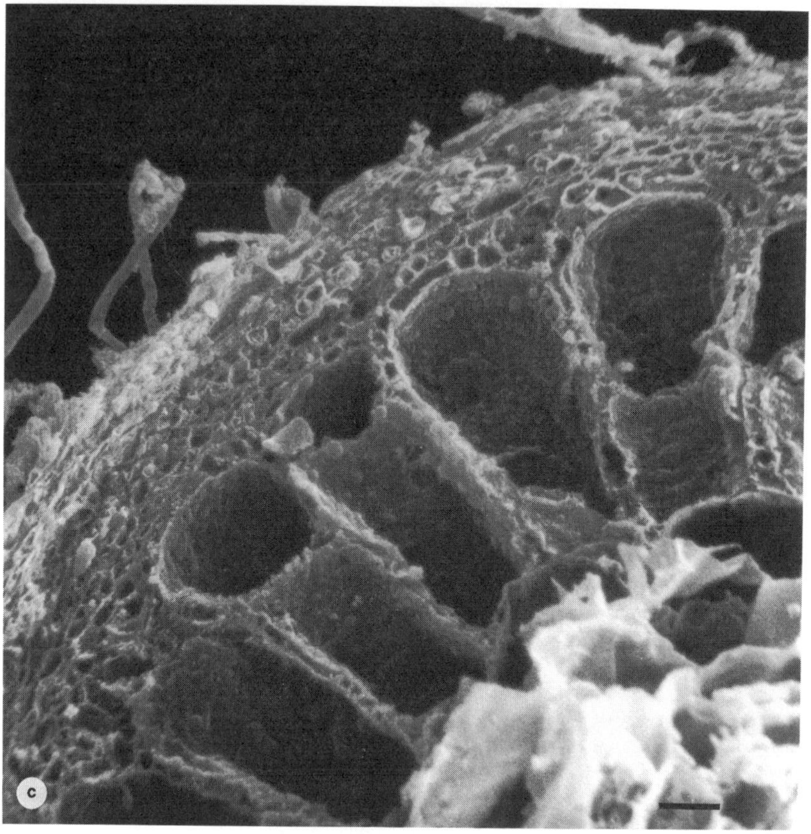

Fig. 5.3 Formation of ectomycorrhizas. (a) (p.273) The morphological relationship between the plant and fungal components. (b, c) Scanning electron micrographs of birch ectomycorrhizas (transverse sections). (b) (p.273) Section through whole root. (c) Higher magnification similar to (b) showing the development of the fungal sheath and Hartig net. Note the elongation of the outer cells of the root cortex. Bars represent, (b) 40 μm and (c) 10 μm. Reproduced courtesy of Dr P. Mason and Dr A. Crossley.

Prunus and *Acacia* form both ectomycorrhizas and vesicular–arbuscular mycorrhizas.

Endomycorrhizas

Those associations grouped together as endomycorrhizas are extremely diverse structurally. In all cases fungal hyphae penetrate into the cortical cells of the host plant root to form a highly intimate relationship. A fungal sheath and/or Hartig net may, or may not, be present (Table 5.2).

Table 5.3 Growth of Sitka spruce (*Picea sichensis*) seedlings inoculated with *Lactarius rufus* in aseptic culture under greenhouse conditions (data from Alexander, 1981)

	Control	Inoculated	Significant difference between means p<
Mycorrhizal infection (%)	0	58.2	0.05
Shoot height (cm)	5.8	9.2	0.05
Shoot dry weight (mg)	47.1	100.2	0.01
Root dry weight (mg)	25.5	74.0	0.001
Total biomass (mg)	72.6	174.2	0.001
Root: shoot ratio	0.59	0.76	ns
Length lateral root (cm)	98.2	219.0	0.01
Total number root tips	60.8	236.7	0.05

Incubation time 14 weeks.
The slight rise in root:shoot ratio in infected tissue may have been attributable to fungal biomass.

In the classification scheme of Harley and Smith (1983) reference is also made to ectendomycorrhizas (Table 5.2). These are often viewed as variants of ectomycorrhizas and yet, also display morphological characteristics of the endomycorrhizas. A sheath of variable development (very often reduced in size), intracellular penetration of cortical tissues and Hartig net development may, or may not, be found. The associations appear to be intermediate forms and their classification is difficult. Ectendomycorrhizal forms are found in *Pinus* and *Larix* species.

Vesicular–arbuscular mycorrhizas

The vesicular–arbuscular forms are the most common and widely occurring of all the mycorrhizal associations. In fact, it has been suggested that up to 25 000 plant species have the potential to form vesicular–arbuscular mycorrhizas (Law and Lewis, 1983), with representatives from almost all taxonomic groups. The vesicular–arbuscular relationship is particularly common in crop plants, herbs and especially tropical trees. These mycorrhizas are the most agriculturally important forms and potentially have great economic significance.

The fungi involved are aseptate, from the family Endogonaceae (Zygomycotina, order Mucorales), and all fall into four genera: *Glomus*, *Acaulospora*, *Gigaspora* and *Sclerocystis*. Classification of these fungi has been achieved using the morphological characteristics of the spores. It is not usually possible to identify the fungus within host tissue since the morphology of that partner may change in association and, in any case, is not comparable in different host plant species. However, the zygospores and chlamydospores are very large (up to 800 μm diameter) and characteristic of the species. These thick-walled, resistant spores can be recovered from soil by sieving and can occur in very large numbers, often accruing as a result of their highly persistent nature. These fungi are obligate symbionts, having no saprophytic ability and not living free in the soil or rhizosphere environment. None has yet been isolated into axenic culture for any length of time.

The name 'vesicular–arbuscular' is derived from the structures which are formed within the host plant tissues by the invading fungus (Fig. 5.4). In general, root morphology is unchanged by the infection. Root hairs are present and the invading fungus forms a very loose hyphal weft over the root surface (not a sheath). After spore germination the developing germ tube is probably attracted to grow toward the root by exudates. Hyphae branch to form a 'fan' near the root surface and hyphae come into contact with the epidermal cells. After contact with the root surface fungal hyphae form appressoria and the fungus penetrates between the outermost layer of cells to grow within and between the cells of the root cortex. Usually there are few penetration points and a relatively small region of the root is invaded (5–10 μm). Penetration is probably mechanical and any wall degrading enzyme activity is limited. Most penetration occurs in newly differentiated tissue. Meristematic tissues are not invaded and the fungus does not enter the stele or any chlorophyll-containing cells of the plant. In the outer cortex, the fungus penetrates host cells and coils of hyphae are found within cells in this region. These are thought to have a role in nutrient transfer. In the outer to middle cortical layers of the plant root the fungus forms terminal or intercallary swellings either in or between host cells. These are termed vesicles and are often large and thick-walled and may distort the host cells as they are formed. Such vesicles are rich in lipids and contain oil droplets and are probably used as storage organs or may possibly have reproductive purposes. Within cells of the inner cortical layers, nearer to the central stele, the invading fungus branches dichotomously and forms finely divided, relatively thin-walled, projections which penetrate into the host cells, invaginating the host membrane (Fig. 5.5). These are called arbuscules and provide a very large surface area of contact between the host and the fungus and are thought to be the regions where most nutrient and mineral exchange takes place. These are often considered to be akin to the haustoria of biotrophic infections. Where the symbionts are in very close association ATPase activity has been shown to be located on the plasma membrane of the partners, providing that both are physiologically functional (Fig. 5.6). The host and fungus are not in direct physical contact but there is an

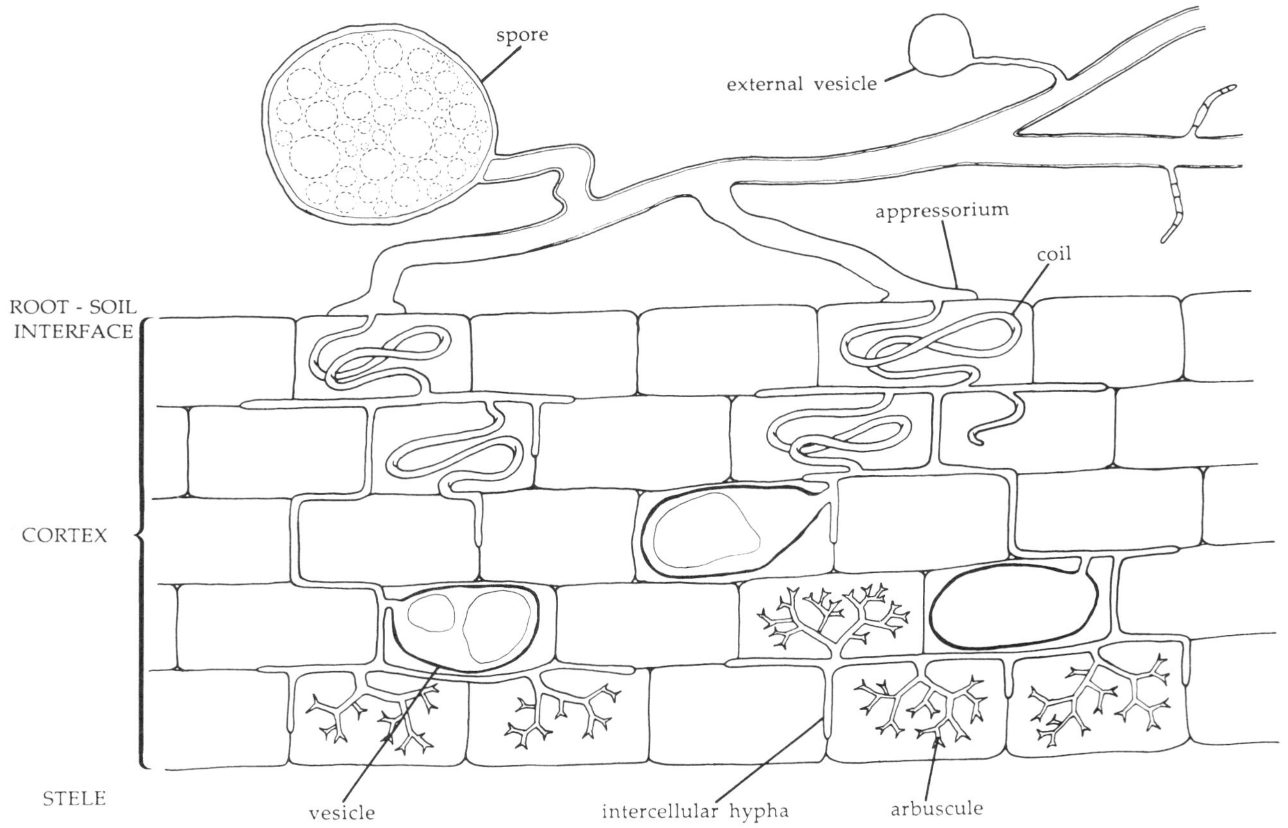

Fig. 5.4 The relationship between the fungus and the host in vesicular-arbuscular mycorrhizal relationships. New arbuscules are formed and older ones become senescent continually during the relationship.

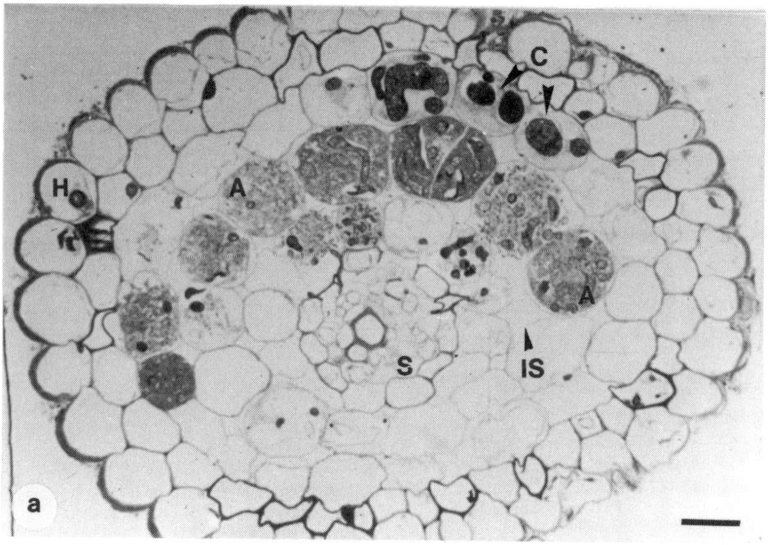
C
H
A
A
S
IS
a

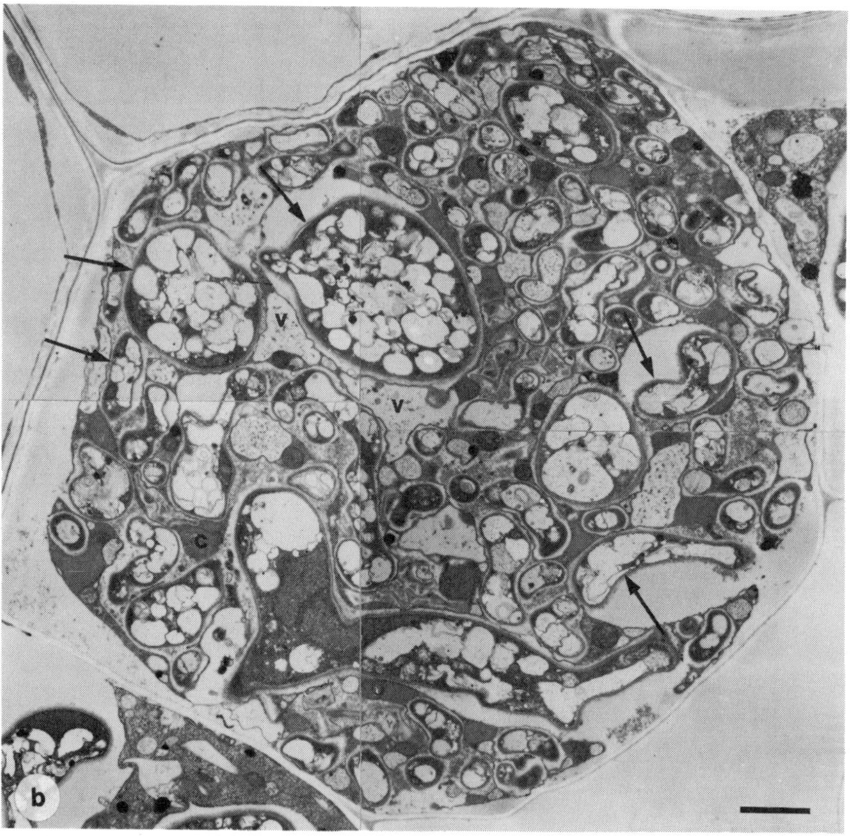
v
v
c
b

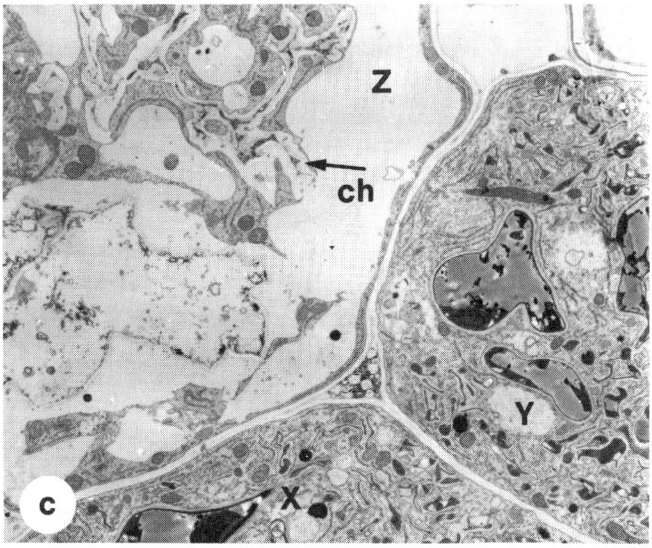

Fig. 5.5 Ultrastructure of symbiont relationships of endomycorrhizas in *Gentiana lutea* roots infected by *Glomus intraradices*. (a) Light micrograph (transverse section) showing linear hyphae (H) in outer cortical cells, with coils (C) and arbuscules (A) in parenchyma cells. No hyphae are visible in the central stele (S) or intercellular spaces (IS). Bar represents 25 μm. (b) Electron micrograph showing a cortical cell containing an arbuscule. Sections of hyphae (arrowed), are surrounded by host cytoplasm (c) which contains many organelles. Almost the whole of the host cell has been filled and little host vacuole (v) remains. Bar represents 0.9 μm. (c) Three parenchyma cells in the root cortex. Cells X and Y contain senescent arbuscules with lipid-rich and collapsed hyphae and Z contains collapsed hyphae (ch) (x3 500). (a, b) Reproduced from Jacquelinet-Jeanmougin *et al.* (1987), courtesy of Dr V. Gianinazzi-Pearson and Balaban Publishers. (c) Reproduced courtesy of Dr V. Gianinazzi-Pearson.

interfacial matrix between the host cell membrane and the hyphal wall. The fungal hyphae do not appear to be morphologically modifed in any way. These arbuscules are relatively short lived (4–15 days) and collapse and degenerate to sporangioles, devoid of cytoplasmic contents. However, new arbuscules are continually formed in other root cells. It is not clear if this degeneration occurs as the result of autolysis by the fungus itself or by host plant digestion. Host cells remain alive and may be invaded again by another hypha. It has, however, been suggested that the host is able to regulate the growth of the fungus (Buwalda *et al.*, 1984).

Up to 20% of the vesicular-arbuscular mycorrhizal root dry weight may be attributable to internal fungal biomass. The fungus remains in connection with the soil environment. Hyphae may spread through the soil, ramifying several centimetres from the infected root surface, and

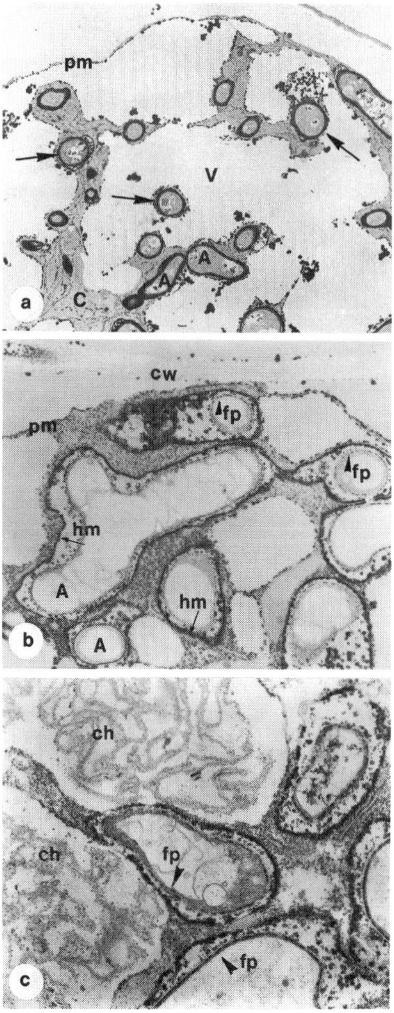

Fig. 5.6 ATP hydrolysing enzyme activity in endomycorrhizas. The electron-dense precipitate of Pb phosphate indicates Mg^{2+} ATPase activity. (a) ATPase specifically localised along host (*Alium cepa*) membrane and in the interface around the fine arbuscule branches (arrowed) as the fungus develops in the host cell. Very little enzyme activity detected along the peripheral host plasmalemma (pm). Host vacuole (V); host cytoplasm (C); arbuscule branches (A) (x5000). (b) Detail of ATPase localisation as in (a). Notice ATPase activity also along fungal membrane (fp). Host cell wall (cw), peripheral host plasma membrane (pm), host membrane (hm) (x15 000). (c) ATPase localised along host (*Acer pseudoplatanus*) membrane and in the interface surrounding living arbuscule hyphae but absent from around dead, collapsed hyphal remains (ch). ATPase also along fungal plasmalemma (fp) (x25 000). Reproduced courtesy of Drs V. Gianinazzi-Pearson, S. Gianinazzi and J. Dexheimer.

both vesicles and spores are formed on this external mycelium, although these fungi are not capable of free-living, saprophytic growth. Spores are often produced in very large numbers. The infection of new roots may occur following contact with hyphae which are already associated with infected roots or by the penetration of germ tubes formed from resting spores stimulated to germinate in the presence of a host root. There is very little host specificity among the partners which form vesicular–arbuscular mycorrhizas. It is likely that early infection is important to the establishment of crop plants and may have a very important influence on productivity and eventual yield. Host plants with such mycorrhizal infections certainly show enhanced growth.

Ericalean mycorrhizas

There are three types of mycorrhizas formed by higher plants of the order Ericales (ericoid, arbutoid and monotropoid). These plants are woody shrubs or small trees which often grow on relatively nutrient-poor and acid soils. Although there are similarities between the forms of these associations they are also very variable.

(a) Ericoid mycorrhizas

The ericoid mycorrhizas occur on plants belonging to the Ericoideae (*Erica*, *Calluna*), Rhododendroideae (*Rhodedendron*), Vaccinioideae (*Vaccinium*), as well as genera in the Epachridaceae and Empetraceae. These plants are small shrubs and trees which grow on heaths and peaty, acid soils.

The roots of these plants are extremely fine and quite simple morphologically. A small stele is surrounded by a few layers of cortical cells (1–3 cells thick). These roots do not have an epidermis and do not form root hairs. Mycorrhizal infections only occur in cortical cells and fungal hyphae never penterate into the stele. The fungal partner grows loosely over the root surface, forming a weft of hyphae, but not a sheath, and achieves penetration into the root cells without the formation of an appressorium. The penetrating hypha lays down a collar of material at the point where it enters the host cell (Bonfante-Fasolo and Gianinazzi-Pearson, 1979). The host plasma membrane is invaginated around the infection hypha and a matrix of pectic substances fills the space between the fungal cell wall and host membrane. After penetration the intracellular hyphae form extensive coils which fill the host cortical cells. These cells are usually infected by hyphae from outside, so that any root usually has a very large number of penetration points (Read and Stribley, 1975). No penetration of meristematic cells occurs. As many as 70% of host cells may be infected by fungus by the time the association is mature. There is no growth of hyphae outwards into the surrounding soil. The association is physiologically active for a short time (3–4 weeks) after which the host plasma membrane looses integrity, followed by degeneration of the fungus (Duddridge and Read, 1982a). This process seems to be linked to the growth and development of new roots.

The fungi which form ericoid associations are not very host-specific within the species range (Pearson and Read, 1975). Most associations are formed with Ascomycete fungi of the genus Pezizella although there are also reports of the involvement of Basidiomycetes too (Bonfante-Fasolo and Gianinazzi-Pearson, 1979). It has been suggested that the fungi involved develop well in the host plant because the internal pH of those cells is near to the optimum (pH 6–7) for the fungus. Certainly growth in culture at pH levels near those of the peaty soils involved (pH 3) was shown to be very limited (Pearson and Read, 1975). Growth of such fungal isolates is very slow and fruiting in culture is rare.

(b) Arbutoid mycorrhizas

Arbutoid mycorrhizal associations are formed with members of the genus Arbutoideae and Pyrolaceae, which are woody shrubs and trees. These have thicker roots than the Ericaceae. Long roots may have very sparse infections, but intercellular hyphae also form a Hartig net. Short roots have a thicker outer sheath and a more highly developed Hartig net between the outer cortical cells. The fungus also penetrates and forms coils within the cells of the outer cortex. The fungal partners are Basidiomycete fungi, e.g. *Cortinarius zakii* and *Cenococcum graniforme*. Fungi which form arbutoid mycorrhizas with these hosts often form ectomycorrhizal associations with other plants (especially conifers).

(c) Monotropoid mycorrhizas

The monotropoid mycorrhizas are specialised forms which differ from other ericoid forms in a number of respects. The family Monotropaceae includes herbaceous plants which lack chlorophyll (achlorophyllous) and which are totally dependent on the mycorrhizal fungi for a supply of carbon, which supports growth. The roots of *Monotropa hypopitys* form a 'root-ball' which is colonised by fungal mycelium. This mycelium (Basidiomycete) also encloses mycorrhizal roots of neighbouring trees. The root-ball overwinters underground to produce flowering shoots (scapes) in spring. During active root growth a fungal sheath and Hartig net are produced. Additionally, close contact with cortical cells is achieved by fungal pegs. A single peg is initially invaginated by the host cell wall and then enclosed by it (Duddridge and Read, 1982b). At the maturity of the association the contents of the fungal peg are released into a sac enclosed by the host plasma membrane. This may provide the host plant with important nutrients and this often coincides with the full development of the plant.

Orchidaceous mycorrhizas

The Orchidaceae includes thousands of species which occur world-wide and which exhibit an extremely varied range of forms. All the orchids are mycorrhizal and form associations with fungi in the early stages of their development. The seeds are very small, often just a few cells, with extremely limited nutrient reserves. It is therefore important to the continued development of the embryos that mycorrhizal associations are formed very soon after germination. Some of the host plant species involved lack chlorophyll throughout their lives and are totally dependent on the active mycorrhiza. The carbon and exogenous vitamins required to sustain plant growth are obtained wholly from the fungal partners.

The fungi involved are members of the Basidiomycotina, often from the genus *Rhizoctonia*. Most of these species grow well and rapidly in culture. The fungi are self-supporting and are able to hydrolyse quite complex nutrient sources (e.g. starch, cellulose and lignin). Many of the fungal species occur on other host plants as quite aggressive pathogens, e.g. *Armillaria mellea* and *Rhizoctonia solani*, and many are active wood decomposers, e.g. *Coriolus versicolor*, *Fomes* spp. and *Marasmius* spp.

Infection of newly germinated seedlings occurs when a fungal hypha penetrates into epidermal cells or epidermal hairs near the suspensor of the embryo. In a compatible infection very little cellular disruption is caused and the fungus grows into the plant cell invaginating the host plasma membrane and is surrounded by host cytoplasm. Hyphal coils (pelotons) are formed in cells, providing a large surface area of contact (Fig. 5.7a). Infection spreads between the cells and the young orchid protocorm becomes heavily colonised. Invagination of the host plasmalemma may be so pronounced that there is very little cytoplasm between it and the tonoplast (Hadley, 1975). In the symbiotic phase fungal hyphae may be surrounded by an encasement layer (Fig. 5.7) apparently of plant origin and has been shown to be continuous with the host plant cell wall. Much membranous material often occurs in close proximity to fungal hyphae. Intracellular hyphal coils are active for a few days before loosing turgor, collapsing, becoming disorganised and degenerating. The host cell may, however, be reinvaded by another hypha (secondary infection), usually arising from a neighbouring cell. There is a very fine balance between symbiosis and parasitism in such associations, which is easily disrupted.

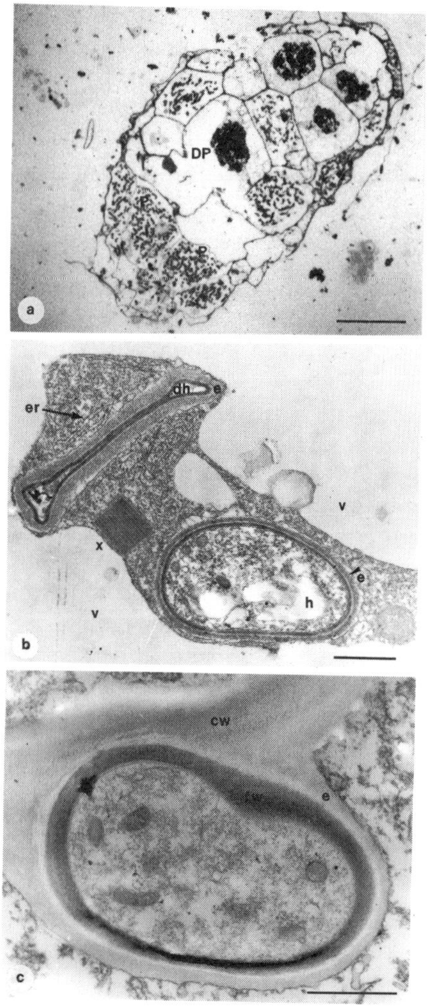

Fig. 5.7 Structure of orchid mycorrhiza; *Dactylorhiza purpurella* infected with *Rhizoctonia* sp. isolate T. (a) Thin section through a protocorm showing outer cortical parenchyma cells containing healthy pelotons (P) and the remains of digested pelotons (DP). Epidermal cells (many collapsed) are not usually infected. Bar represents 200 μm. (b) Hyphae within host cytoplasm showing the encasement layer (e) around a normal hypha (h) and around a disorganised hypha (dh). Endoplasmic reticulum (er) can be seen in close proximity to the disorganised hypha and the encasement layer is increased in thickness. A crystal can be seen in the cytoplasm (x). Vacuole (v). Bar represents 1 μm. (c) Wall of an intracellular fungal hypha (fw) with an encasement layer (e) which is continuous with the host cell wall (cw). Bar represents 1 μm. Reproduced from Hadley (1975), courtesy of Dr G. Hadley and Academic Press.

Under some conditions the plant may confine and destroy the fungus but in others the fungus may parasitise the protocorm. Infection of adult plant roots occurs from soil, by hyphae associated with nearby root systems. This is important for those species which spread by rhizomes and form new root systems when dormancy is broken.

FUNGAL–PLANT CONTACT

In mycorrhizal associations, invading fungal hyphae show very little morphological modification and additionally, few changes occur to plant cells. The fungi probably penetrate host tissues by means of mechanical pressure, possibly aided by very limited enzyme softening in only a few cases. Hyphae always penetrate in regions of the root where cells and cell wall synthesis are active but never into photosynthetic tissues, meristematic regions, vascular or thickened tissues. The host cells are not killed by invasion, but remain metabolically functional (Duddridge, 1987). Hyphal growth may be inter- or intracellular. In the ectomycorrhizas and the arbutoid and monotropoid mycorrhizas a Hartig net develops between the cortical cells. In this region hyphal branching and cell division are slightly modified and this provides a large surface area for contact between the partners. In all forms except the ectomycorrhizas there is some intracellular growth, although hyphae never penetrate the host plasma membrane directly. Matrix material, containing microfibrils and vesicles, always separates the host membrane and the fungal wall. It has been suggested (Harley and Smith, 1983) that after fungal penetration host cell wall synthesis continues and the newly formed polymers pass into the interfacial matrix. This material may therefore provide a source of nutrient for the fungus.

Where intracellular fungal protrusions develop they are relatively short lived. It is not clear whether this is due directly to autolytic activity from the fungus itself, or whether the host plant has any role in this degradation. In the orchidaceous mycorrhizas it has been suggested that the plant may actively degrade the fungal projections to utilise these as sources of nutrients. It has been shown by radioactive labelling (Smith, 1967) that such plants rely on the fungal associations for the supply of carbon and other nutrients, but it is not clear to what extent the plant obtains these materials from such degradation. It has been suggested that the plant lives almost parasitically on the fungus. However, this mechanism may also be a means for the prevention of the uncontrolled spread of fungus within host tissues. In the monotropoid mycorrhizas, the fungal pegs are invaginated by the plant but later burst and the contents are released, similarly becoming available to the plant as potential nutrients. These plants are also dependent on the fungal partner for a supply of carbon and other nutrient materials.

Specificity and recognition

In general terms the fungus–host specificity of mycorrhizal associations is low. However, common partner combinations can be identified and a range of responses obviously occur. The ericoid mycorrhizas are formed by only a few plant species and by fungi which are closely related. Orchidaceous mycorrhizas form within just one plant family, albeit large, and with a number of fungal species from the Basidiomycotina. Within this group of mycorrhizas a range of specificity can be identified. The ectomycorrhizas occur in a limited number of tree species but with a wide range of fungi. Of all the associations the vesicular–arbuscular mycorrhizas are the least specific. The fungi infect host plant species from an extremely wide taxonomic range. The fungal partners, on the other hand, are drawn from a very limited species range. The specificity among mycorrhizal associants is much less than for biotrophic plant infections (Gianinazzi-Pearson, 1984).

It is likely that a range of factors, probably including environmental characteristics, contribute to the mechanism of specificity and recognition in these associations. Little information is available concerning recognition systems although clearly a mechanism must exist. The molecular basis of the compatibility between the plant and fungus is unknown. The fungal partner is either not recognised as a potential pathogen or some system for the suppression of plant host defences must operate. In association, fungal species may form different types of mycorrhizal relationships which are characterised by different morphological features and physiological functioning, and are dependent upon the host species combination.

METABOLITE EXCHANGE

It is well recognised that mycorrhizal infections are beneficial to host plant growth. This effect is most apparent in nutrient-poor soils where infected plants attain a greater biomass and are more vigorous and healthy than uninfected individuals. In nutrient-rich soils the effect is less noticeable and the frequency of infection is often greatly reduced. It is now well accepted that mycorrhizal infections extend the root system, by virtue of the functional mycelial connections into the surrounding soil, making a greater pool of nutrients available to the plant. In some cases extension of the roots themselves may be limited by mycorrhizal infection but the associations usually increase the longevity of infected roots. It is also clear that these associations increase the nutrient status of the host plants and it has been shown, in many experiments, that nutrients are transferred between the partners. It is important to remember that this transfer takes place between living cells and often there is a two-way nutrient flow. It is also becoming well accepted that mycorrhizas have an important role in nutrient cycling within the soil environment.

Transfer of carbon

Ectomycorrhizas

The transfer of photosynthates from host plants to the fungal partner in ectomycorrhizas was first demonstrated, in axenic culture, by Melin and Nilsson (1957). *Pinus sylvestris* seedlings were grown with the fungus *Rhizopogon roseolus* and allowed to photosynthesise radioactively labelled CO_2. The resultant photosynthates were transported to the roots and passed to the mycelium of the fungal partner. Subsequently, Bevege *et al.* (1975) compared the translocation of radiolabelled CO_2 in mycorrhizal and non-mycorrhizal *Pinus radiata* seedlings over 24 h, and showed an eight-fold increase in ^{14}C per unit biomass in the infected roots.

The work of Lewis and Harley (1965a, b, c) with mycorrhizal beech roots (*Fagus sylvatica*) provided important evidence of carbon flow and the functioning of ectomycorrhizas. The soluble carbohydrates detected in non-mycorrhizal tissues, after incubation in ^{14}C-labelled sugars, were found to be glucose, sucrose and fructose. Infected tissues accumulated these compounds too, but in addition, trehalose and mannitol were also found. Both these carbohydrates often occur as fungal storage compounds. Excised mycorrhizas were placed in contact with agar blocks loaded with ^{14}C-labelled sucrose and experiments were carried out to trace the fates of assimilated carbon. The removal of a ring of the fungal sheath, next to the agar block, left only root tissue and hyphae of the Hartig net in contact with the carbon source. Labelled sucrose moved through the mycorrhizal tissues and was detected throughout the excised portion. In the fungal sheath covering apical regions of the root however, activity was detected in the fungal storage carbohydrates mannitol and trehalose. Lewis and Harley (1965c) suggested that ectomycorrhizal fungi absorb sugars from the host and convert these to reserve carbohydrates, in forms which are virtually unavailable to the host plant. In this way, a concentration gradient for carbohydrate is maintained and movements of sugars from the host to the fungus continue. Additionally, the fungus does not then have to share the reserves with the plant but has rendered them unavailable to it.

It is clear that the fungal sheath has an important role in nutrient storage. The fungal hyphae sequester nutrient materials at times when plants are active and are able to mobilise them later in the growing season. Flushes of fruiting bodies are produced above ground in the autumn arising from the subterranean mycelium. The extent to which bidirectional flow of carbohydrates occurs is not clear but in general it is accepted that little carbohydrate flows from the fungus to the plant in ectomycorrhizal associations. Some recycling of amino compounds may occur (Harley and Smith, 1983), especially as much of the transferred carbon must be accessible to cells of the cortex since these are in close proximity to the fungal sheath, but this is not considered part of the carbohydrate flow.

Endomycorrhizas

The fungi which form vesicular–arbuscular mycorrhizal associations are not free living and are incapable of independent growth. Such associations are therefore obligate for these fungal species and it is likely that the plant provides carbohydrate as an energy source. Certainly hyphal growth in soil is extremely limited until infection has been established, after which one of the characteristics of this type of association is the extensive, ramifying, extramatrial, hyphal network. Experiments with *Festuca* spp. showed that ^{14}C-labelled photosynthates were transferred to host plants by vesicular–arbuscular mycorrhizal infections (Ho and Trappe, 1973) and similarly in other plants (Bevege *et al.*, 1975). It has also been shown that radiolabel applied to *Pinus* seedlings was transported to neighbouring plants via extramatrical hyphae (Read, 1984). It is unlikely, however, that the fungal hyphae in these associations have a role in the storage of carbon compounds, although some lipid granules are laid down in fungal tissues.

The transfer of carbon from host to fungus in ericoid mycorrhizas is similar to that which occurs in the ectomycorrhizal associations (Stribley and Read, 1974). However, in monotropoid and orchidaceous associations the transfer of carbon is undoubtedly from the fungal partner to the plant. These plants are achlorophyllous and are not capable of independent growth and development. *Monotropa* makes use of ramifying underground hyphal networks to form connections which provide a source of nutrients. Orchid seeds are barely capable of germination in the absence of a fungal partner and are totally dependent an the association for growth and development (Smith, 1966). It is interesting that the disaccharide trehalose, normally considered a fungal storage carbohydrate, and unavailable to most vascular plants, can be used effectively for growth of orchid seedlings (Purves and Hadley, 1975). It is also likely that, even in the adult state, the photosynthetic orchid does not supply carbon to the fungus.

The means by which carbon compounds are transferred between host and fungus are of great interest and much speculation. It has been suggested that several mechanisms may be involved (Harley and Smith, 1983). The influence of the fungus may alter host membrane permeability and cause leakage of carbohydrates from host cells to the apoplast from where they may be absorbed into hyphae. However, the effect may be directed towards cell wall polymers. The fungus may disrupt host cell wall deposition, causing polymers to accumulate in the apoplast as these are synthesised, or alternatively the fungus may degrade walls as they are formed and then absorb the hydrolysed materials from the apoplast.

Transfer of mineral nutrients

In poor soils the major limiting factors to plant growth are minerals. The development of mycorrhizal associations is a specialised means by which the supply of essential minerals can be increased. Plants depend on this relationship, to a greater or lesser degree. This dependency is determined by genetic characteristics of the plant involved (e.g. root morphology) and also by the nutrient status of the soil. Mycorrhizal dependence has been defined (Gerdemann, 1975) as the dependence of a plant on the mycorrhizal association to produce maximum biomass at a given level of soil fertility. This concept has been modified to give the expression the quantitative measurement (Plenchette *et al.*, 1983) relative field mycorrhizal dependency (RFMD):

$$\text{RFMD} = \frac{(\text{dry wt mycorrhizal plant}) - (\text{dry wt non-mycorrhizal plant})}{\text{dry wt mycorrhizal plant}} \times 100\%$$

It is therefore possible to calculate, for given nutrient levels, the extent to which mycorrhizal potential is realised, under these conditions. Mycorrhizal fungi maintain functional contact with the soil and translocation of minerals provides the plant with a supply, often drawn from regions of the soil beyond the root system.

Phosphate

Phosphorus is a very important mineral for the growth of plants but its availability in soil is usually low. Of the soil content much is present as insoluble inorganic forms. Organic forms are also present but the availability of these to mycorrhizas is controlled by rates of mineralisation in that environment and mediated by the activities of soil microbes (Hetrick, 1989). Soluble forms are absorbed into plants but a depletion zone occurs readily around roots. Phosphorus is often regarded as immobile in the soil and diffusion rates are extremely limiting to movements.

The physiology of host plants and any increases in phosphate influence the development of mycorrhizas. The phosphorus availability within a plant will affect the membrane permeability of roots and a low phosphate level is likely to increase the amount of root exudates released into the soil. It has been shown that plants which exhibit higher rates of root exudation form and maintain mycorrhizal relationships more easily (Schwab *et al.*, 1984). Additionally, where higher levels of phosphate maintain the permeability of the plant membranes the loss of photosynthate can be controlled

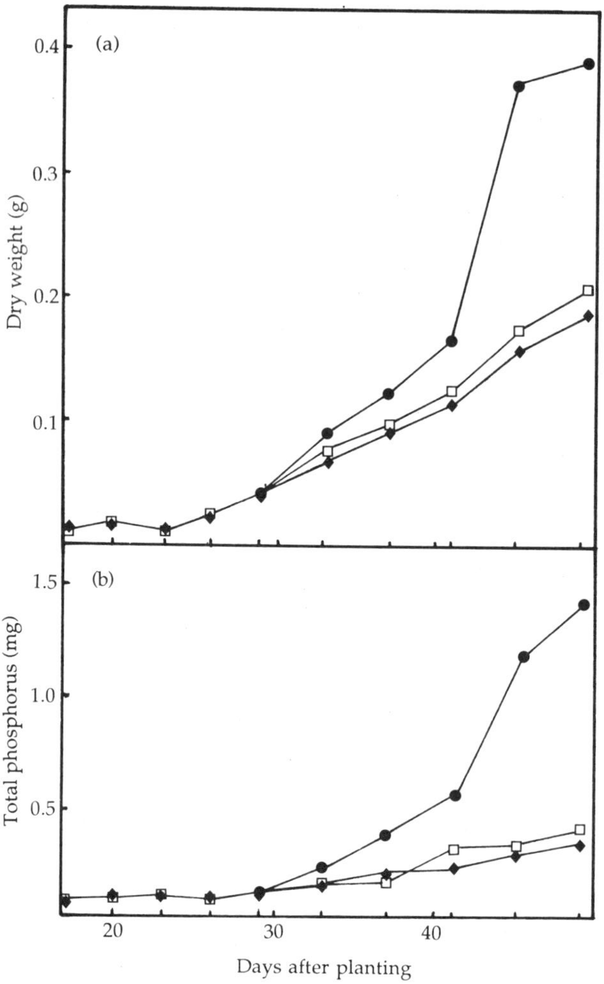

Fig. 5.8 Effects of vesicular–arbuscular mycorrhizal relationships between different fungal partners on the growth of onion plants. (a) Dry weight; (b) phosphorous content. (□) Uninoculated control plants; (●) Plants inoculated with *Gigaspora calospora*; (◆) plants inoculated with *Glomus microcarpus*. The lower effect of *Glomus microcarpus* on plants was attributable to the slower growth and poorer development of this isolate. Data replotted from Sanders *et al.* (1977).

by the plant. Although in that instance there is less exudation and less likelihood of mycorrhizal formation the plant also has less need for such a relationship (Hetrick, 1989).

Mycorrhizal associations increase the phosphorus contents of plants and increase the ability of the plant to obtain phosphorus. This leads to

Table 5.4 Utilisation of two phosphate fertilisers by oil palms in a 32P-labelled sandy clay acid soil* from Soubrè (Ivory Coast), (from Gianinazzi-Pearson and Gianinazzi, 1989, with permission)

Plant treatment	Fertiliser	Dry weight (g. shoot-1)	Fertiliser utilisation coefficient (%)	% P Derived from fertiliser	Specific activity ($^{32}P/^{31}P$)
Non-mycorrhizal	none	0.6	–	–	–
	rock phosphate	1.5	4.3	65	1.7
	triple super phosphate	1.6	5.0	75	1.9
Mycorrhizal	none	2.4	–	–	–
	rock phosphate	3.3	16.8	76	1.9
	triple super phosphate	3.1	13.8	74	2.0

* Soil water pH = 5.2; ppm P (Olsen) = 3.9.

increased plant growth (Fig. 5.8), often as high as several hundred-fold increases in biomass (Menge, 1983). The application of soluble phosphate to soils may give rise to greater effect, in terms of growth, in mycorrhizal plants than non-mycorrhizal plants (Table 5.4). The supplementation of tropical, acid soils with rock phosphate (poorly soluble) gave rise to more growth in vesicular–arbuscular mycorrhizal plants than non-mycorrhizal plants (Gianinazzi-Pearson and Gianinazzi, 1989). The application of triple superphosphate (highly soluble) fertiliser also gave very similar results. This, together with the specific activity and percentage phosphate derived from fertiliser values, suggests that the plants drew on the same phosphate pool in each case but that mycorrhizal plants were most efficient in absorbtion.

After uptake phosphorus is translocated through the fungal mycelium and is passed to the plant. Acid phosphatases have been detected, by cytochemistry and by the use of polyclonal antibodies against phosphatase, in ericoid and ectomycorrhizal fungi, and have been shown to be wall-bound enzymes (Gianinazzi-Pearson and Gianinazzi, 1989). This fungal activity may have a role in the mobilisation of phosphorus from complexed forms in the soil, but the extent and the significance of such cycling has not yet been assessed. Phosphorus is also stored in fungal hyphae as polyphosphates. These have been detected in the hyphae of the ectomycorrhizal fungal sheath which probably acts as a significant phosphate store.

Nitrogen

A low nitrogen supply may be limiting to plant growth, indeed the requirements of a plant for nitrogen may be high (×10) in comparison with the requirement for phosphate (Read *et al.*, 1989). Nitrogen is incorporated in to key enzymes used for the fixation of carbon. Where rates of organic matter turnover are high good supplies will be available but in situations of low organic turnover nitrogen mobilisation is restricted and plant growth may be limited.

Early work of Melin and Nilsson (1953), using radiolabelled nitrogen (^{15}N), showed that ectomycorrhizas passed nitrogenous compounds to pine seedlings. It has been shown, more recently, that ectomycorrhizal associations are able to utilise more complex nitrogen sources. Some ectomycorrhizal fungi are able to use amino acids and some can degrade protein (Abuzinadah and Read, 1986). Ericoid mycorrhizas have also been shown to absorb ammonium efficiently, with glutamine as the major product transferred to the host plants (France and Reid, 1983). These species have also been shown to utilise free amino acids from soils. This is a previously unrecognised nitrogen source for these species and may be particularly important in autumn when a peak of amino acids appears in surface layers of soil (Abuarghub and Read, 1988a, b; Read *et al.*, 1989). It is likely, therefore, that the release of enzymes from mycorrhizal hyphae, to degrade such complex nitrogen sources, will have an effect on nitrogen cycling in the soil environment.

Transfer of other compounds

Further influence on the growth and development of mycorrhizal plants may be attributable to the production of plant growth regulators by the fungal partners. Any such synthesis in the natural situation must represent an increase in these compounds above the normal level for root tissues. Exogenously applied auxins cause morphological changes in roots which are similar to those observed in ectomycorrhizal associations. Auxin synthesis by ectomycorrhizal fungi in culture is depressed by a good nitrogen supply. In situations where ample nitrogen is available ectomycorrhizal associations break down, which suggests the involvement of enhanced levels of auxin in the establishment and maintenance of these relationships. Ectomycorrhizal fungi also produce cytokinins and gibberellin-like compounds which may have some role in regulating the longevity of mycorrhizal roots.

ECOLOGICAL SIGNIFICANCE OF SOIL EXPLOITATION BY EXTRAMATRICAL HYPHAE

It can be seen that there is a relationship between the occurrence of different types of mycorrhizal association and different environmental characteristics. The predominant vegetation in an area and also the associations which are formed by those plants, are heavily influenced by changes in the prevailing climatic conditions at that site and also by geographical location, by altitude and latitude, and by soil conditions (Read, 1984). Much of the effect arises from the nutrient status, pH, and water availability of the soil . Although there is a great deal of overlap between the zones which have been described, the scheme shown in Fig. 5.9 has proved to be a very influential system in determining our interpretation and understanding of the nature of the different mycorrhizal types in ecological and physiological terms.

Ericaceous species are often the dominant vegetation in relatively cold, wet, climates where there is high rainfall and low levels of evaporation. Soil pH is usually low and rates of organic matter decomposition are slow under such conditions. The fungi which form ericoid mycorrhizal associations are poorly adapted to growth at low pH and show little

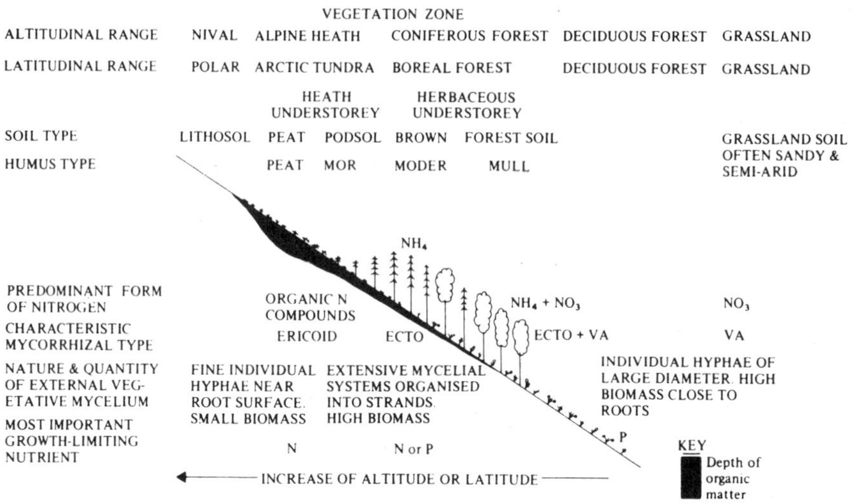

Fig. 5.9 The postulated relationship between latitude or altitude, climate, soil, and mycorrhizal type, and the development of vegetative mycelium associated with mycorrhizas. Reproduced from Read (1984), courtesy of Professor D. Read and Cambridge University Press.

extension in the soil or in contact with it, proliferating predominantly within host tissues. These fungi produce acid phosphatases and are able to use organic phosphates as growth substrates and increase the phosphate status of infected plants. It has also been shown that ericoid mycorrhizal plants have the capacity to use amino acids (Stribley and Read, 1980) and chitin (Leake and Read, 1990) as a source of organic nitrogen, therefore enabling the use of organic compounds which are not normally available to non-mycorrhizal individuals.

Ectomycorrhizal associations occur with trees of the temperate forest ecosystems, such as woody shrubs and conifers. These predominate in warmer conditions where levels of evaporation are higher. Large amounts of mycelium are formed on the outside of roots, as a well-defined sheath and therefore maintain soil contact directly and also by means of ramifying networks of extramatrical hyphae. Recent investigations have shown that disturbance to these mycelial networks in soil leads to a reduction in the nutrient-absorbing capacity of the mycorrhizal association, independent of the degree of colonisation. Additionally, any disturbances also reduce the infective ability of the fungal partner (Jasper *et al.*, 1989; Evans and Miller, 1990). The sheath and mycelial fans extending (±20 cm) from the root, absorb phosphorus, nitrogen and water from the soil which are passed to the plant.

Networks of fungal hyphae become anastomosed to form an extensive and integrated mycelial system with hyphae often grouped together in 'strands'. These scavenge nutrients from the surrounding soil, not yet colonised by the plant roots, and therefore act as an extension to the root system. These hyphae grow through the soil, also infecting nearby plant roots, and eventually interconnecting neighbouring plants. Autoradiographic analyses of soil chambers (Read, 1984) containing *Pinus coronata* seedlings infected with *Suillus bovinus* showed that $^{14}CO_2$ applied to seedling shoots was assimilated and accumulated in the roots of the plant. Subsequently label was transferred quite rapidly to the external mycelium and then accumulated in the mycorrhizal roots of neighbouring plants. The presence of this extramatrical mycelium is therefore, probably highly significant for the rapid establishment of seedlings, particularly under a very dense canopy where light for photosynthesis may be limiting. *Monotropa* species are totally dependent on the nutrients passed from both the fungal partner and also from neighbouring plants via interconnecting hyphae. Although external mycelium is probably much less important to other well-established and mature plant species, it may represent a significant route for the movement of nutrients through a population.

Vesicular–arbuscular mycorrhizal associations occur with herbaceous species and grasses under warmer, dryer conditions where the rates of evaporation and organic matter turnover are higher and soils are less acid. Although there is no outer sheath of fungal hyphae the extramatrical mycelium of these associations is often very extensive. Most of these hyphae are large (20–30 μm diameter) and thick walled and often support much thinner (2–7 μm diameter), more fragile hyphal branches. Extramatrical hyphae provide a means for the plant to exploit soil beyond the normal root zone which is particularly important for the uptake of minerals. Such routes of transport are also important in the absorption of water into the host plant and, additionally, are also a major source of inoculum for other roots.

WATER RELATIONS OF MYCORRHIZAL PLANTS

It is now well accepted that mycorrhizal infections improve the nutritional status of plants as the result of the improvement of the exploitation of the relatively localised nutrient resources in the soil. Recent work has indicated that the external mycelium, which penetrates into the surrounding soil, probably has a very important role in plant water relations, both directly and indirectly. This may be due, in part, to improvements in mineral nutrition but also to the direct water-absorbing capacity of the mycelium, which acts to increase the conductivity of the plant root system and to maintain water flow to the plant even under conditions of water stress.

Vesicular–arbuscular mycorrhizal infections generally affect root morphology very little but do provide good contact with the soil and also extend the root system. Read and Boyd (1986) have suggested that with an average hyphal diameter of 5 μm and a root diameter of 500 μm, 1 m of external hyphal length provides a surface area equivalent to 1 cm of root length and cite Abbot and Robson (1985), who reported 2.5–14 m hyphal length per cm of infected root. Therefore, the surface areas of vesicular–arbuscular mycorrhizal roots is much increased over those of non-mycorrhizal roots and the extension occurs into regions of soil previously uncolonised by those plant roots. The presence of vesicular–arbuscular mycorrhizas has been reported to increase the mean transpiration rates through infected plants as a result of the increase in stomatal conductance. This may be related to changes in leaf physiology and it has been suggested that changes in plant hormone balance may contribute to this effect (Levy *et al.*, 1983). However, increases in hydraulic conductivity have also been observed (Safir *et al.*, 1971; Graham and Syvertsen, 1984) which have been related to the improved phosphorus status of plants with vesicular–arbuscular mycorrhizal associations. Increases in internal phosphorus contents leads to well-maintained membranes, to efficient membrane function and good permeability control in infected plants.

Hardie and Leyton (1981), working with clover, showed that in addition to increased leaf and root conductance infected plants also maintained a lower leaf water potential which enabled very efficient water extraction from soil. Certainly such vesicular–arbuscular mycorrhizal plants appear to be very much more drought resistant than non-mycorrhizal plants. It is difficult however, to separate the direct effects of hyphal water uptake from the effects of increased nutrient status. Attention has been drawn to the need for experiments which discriminate between nutritional and non-nutritional effects, in order to determine the underlying modes of mycelial influence on host plants, to elucidate the mechanisms involved and to understand fully the biological role of mycelium (Read, 1986).

The growth of ectomycorrhizal fungi is inhibited in very wet soils and ectomycorrhizal infections do not develop readily in waterlogged conditions (Theodorou, 1978), although this may be due to reduced aeration (Read and Armstrong, 1972). Theodorou (1978) distinguished two groups of ectomycorrhizal fungi by their response to water stress, in osmotically adjusted soil. Some species grew below −10 MPa (e.g. *Suillus luteus* and *Lactarius deliciosus*) but others were unable to grow below −3.5 MPa (e.g. *Suillus granulatus*). Amongst ectomycorrhizal woody species root density is lower than for herbaceous species. The outer root surface is covered by a fungal sheath, in contact with the soil, which acts as a protective layer under conditions of water stress.

Duddridge *et al.* (1980) carried out some elegant experiments using Perspex chambers in which a plastic divider separated plant roots in dry soil from a layer of moist peat, so preventing diffusion of water through the

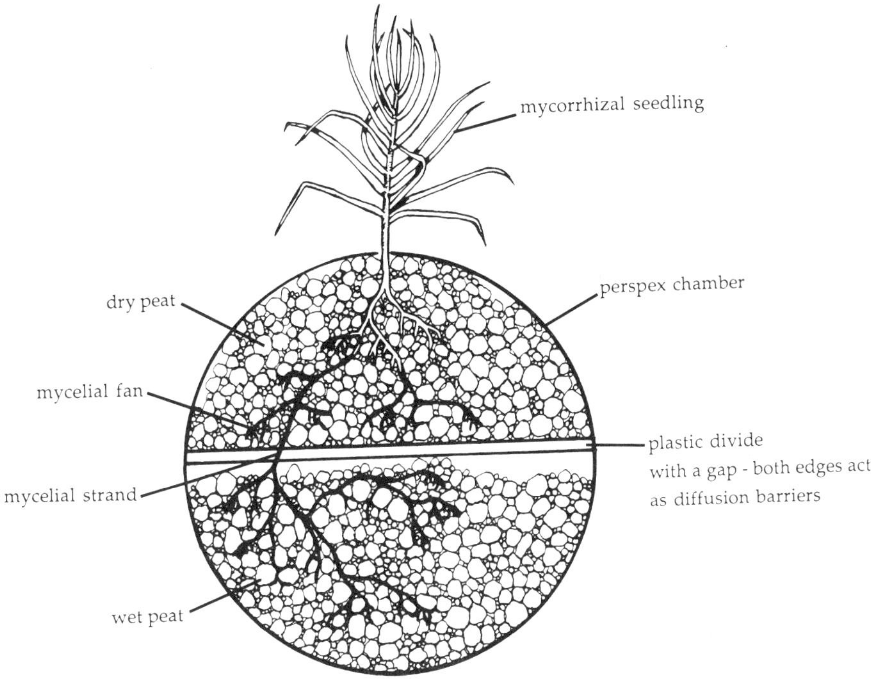

Fig. 5.10 Mycorrhizal seedlings, grown in Perspex chambers, were used to investigate the role of mycorrhizas in the uptake and transport of water (from Duddridge *et al.*, 1980). Peat above the divide was dry but that below the divide was watered. The seedling roots grew only in dry peat but water was obtained through the mycelial strands of the mycorrhiza.

system (Fig. 5.10). Ectomycorrhizal strands developed which established communication between the dry and the wet peat environments. Electron microscope studies established that the central hyphae in the strand had large diameters (10–20 μm) and lacked cytoplasm. These were suggested to be 'vessel' hyphae used for the movement of water and were surrounded by tightly packed, cytoplasm-filled hyphae 5–10 μm diameter. The addition of tritiated water to the system showed that the mycelial fans passed water from the wet peat to the pine seedlings. *Suillus bovinus* transported water over a distance of 20 cm in an hour to host plants (*Pinus sylvestris*). Vessel hyphae are probably major conducting elements in the soil and appear to be very efficient. Severing such strands decreased the water potential and plants died rapidly (Brownlee *et al.*, 1983). This work has shown, by direct analysis, that external mycelium forms a functional extension to the plant root system in ectomycorrhizal infections which is probably of significant advantage to the plant. Read and Boyd (1986) suggested that for ectomycorrhizal systems a value of 10–80 m of hypha are produced per cm of root length. This is a higher value than for vesicular–arbuscular systems.

RESISTANCE OF MYCORRHIZAL PLANTS

It has been suggested that ectomycorrhizal fungi may function as a barrier to the invasion by other root-infecting pathogens (Marx, 1975). Such an effect is not only attributable to the presence of a sheath of mycelium covering, and therefore physically protecting the root surface, but also to the increased competition for nutrient resources at the root surface and additionally to the production of inhibitory compounds. The composition of exudates from such roots is different from those of uninfected plants. Some species of mycorrhizal fungi have been shown to produce antibiotics. There is some evidence that in other mycorrhizal associations too, increased host vigour may account, in part, for a reduction of disease incidence and severity, although such an effect is not universal.

Mycorrhizal infections have also been shown to enhance the ability of plants to survive in the presence of heavy metals at toxic levels (Bradley *et al.*, 1982). During a 12-week experiment high concentrations of zinc and copper were seen to be inhibitory to non-mycorrhizal plants, whereas mycorrhizal individuals showed some growth. Lower levels of metals were detected in the shoot tissues of the mycorrhizal plants and analyses of the root tissues indicated that metals were complexed in mycorrhizas and were not passed to the body of the plant. Fungi which are inherently resistant to such heavy metals may be preferentially selected in contaminated environments and may then form mycorrhizal symbioses.

LICHENS

The vegetative body of a lichen is known as the thallus and is composed of two organisms, a fungal component (mycobiont) and a green alga or cyanobacterium (photobiont). The lichen symbiosis was first recognised by Schwendener (1867) and is one of the most widespread and commonly occurring symbioses. Lichens are long-lived organisms and have characteristics which are quite distinct from those of either partner. The growth of lichens is most favoured by conditions in which minimal physical disturbance occurs and at sites which receive relatively low light intensity and high humidity. Quite high levels of physiological stress are tolerated by lichens and they readily colonise exposed habitats, such as bare rock faces. Indeed slow wetting and drying phases seem to encourage the development of a healthy thallus. Individuals can be found exploiting a very wide range of habitats, in fact the lichens are able to colonise almost any stable substrate provided that adequate illumination is received.

THE SYMBIOTIC RELATIONSHIP

The carbon which is fixed by the photosynthetic activities of the photobiont, passes to the fungus and fully sustains the needs of the mycobiont partner. There is, however, very little evidence for any reciprocal exchange between the fungus and the photobiont. It has therefore been suggested that the relationship between the two partners is a controlled parasitism of the photobiont by the fungus, mediated only by the rapid proliferation of the algal cells (Ahmadjian and Jacobs, 1981; Ahmadjian, 1982). However, characteristics of both parasitic and symbiotic relationships can be identified within the lichens as a group. Since the lichen relationship is a long-lived association it is likely that some degree of mutualism is maintained. Crittenden (1989) has suggested that the photobiont obtains nutrients from the fungal cells by mass flow, or to a lesser extent by diffusion. Alternatively, such materials may be absorbed directly from water trapped in intercellular spaces (Farrar and Smith, 1976), particularly in the growing regions of thalli.

Mycobiont partners

There are over 500 genera and over 13 500 species of lichen-forming fungi (about one in five of the known fungal species). A large majority of the lichenised fungi are members of the Ascomycotina (from 16 orders, five of which are composed entirely of lichen-forming fungi) with only about 20 species belonging to the Basidiomycotina (from several orders of the Hymenomycetes). Most of these species are not found free-living. The taxonomy of lichens is based entirely on fungal characteristics and species names relate to the mycobiont. Within the lichen thallus fungal hyphae may become orientated and, in some cases, differentiated into layers so that many lichens have a distinct structure. In general morphological terms, the fungal hyphae which are present in lichen thalli are very similar to those of non-lichenised species, except for some modifications to hyphae in differentiated regions of highly developed thalli.

Photobiont partners

The range of algal species involved in lichen symbioses is relatively limited, including unicellular or filamentous algae and cyanobacteria from about 24 genera. The majority of photobionts (forming over 70% lichen species) are from the Chlorophyta (green algae), e.g. *Trebouxia* and *Pseudotrebouxia*. Few of these species have been found free-living. These partners are normally surrounded by fungal hyphae and are frequently organised into a layer (10–30 μm thick) covered by a cortex of fungal hyphae. This type of arrangement is termed heteromerous. About 12 genera of cyanobacteria form lichen associations, e.g. *Nostoc, Scytonema* and *Trentepohlia*. In some cases where the partner is a cyanobacterium, the associants are distributed throughout the thallus, which is termed a homiomerous arrangement. In most lichen species only one species of algal symbiont is present; however, in some species a second photobiont species may be found in association. In these cases another partner (occasionally more than one) may be located in cephalodia, which are warty-type protruberances from the main body of the thallus. In fact these cephalodia can sometimes become separated and can exist as independent lichen individuals.

Growth forms

A number of morphological forms are found among the lichens (Fig. 5.11). These range from relatively simple cellular aggregations to more complex and structured arrangements.

Crustose lichens

The simplest morphological types are the leprose forms. These are very simple associations in which fungal hyphae envelop a group of algal cells. Such aggregations have very little structure and form a loose powdery layer directly on the substrate, e.g. *Lepraria* sp. In crustose forms, which have a more organised thallus, the fungal hyphae are arranged as a layer overlying the associated algal cells. These lichens are very closely and tightly attached to the substratum (Fig. 5.11a–c), often almost inseparable from it, and they grow by spreading, radially, outwards over the surface. Such lichens are often found on exposed surfaces of rocks and the bark of trees, e.g. *Lecanora* sp. In some undisturbed habitats complicated mosaics of colonies are found which persist over extremely long periods.

Foliose lichens

Foliose lichens are more structured and have a leafy or lobed appearance, e.g. *Parmelia* spp., *Xanthoria* spp. and *Cetraria* spp. (Fig. 5.11d, e). An upper layer, or cortex, of fungal hyphae covers the photobiont. This region is often regarded as a protective covering for the algal cells. The thallus surface is relatively smooth since the fungal hyphae are embedded in a matrix. This appears to bind the cortex together and may provide mechanical strength (Anglesea *et al.*, 1982). The matrix is probably composed mainly of polysaccharide (notably glucan) and covers the hyphae so that it is not possible to distinguish individual branches. However, a technique for removal of the matrix, using a protease/detergent mixture (Anglesea *et al.*, 1982) has allowed observation of the hyphal network in the cortex

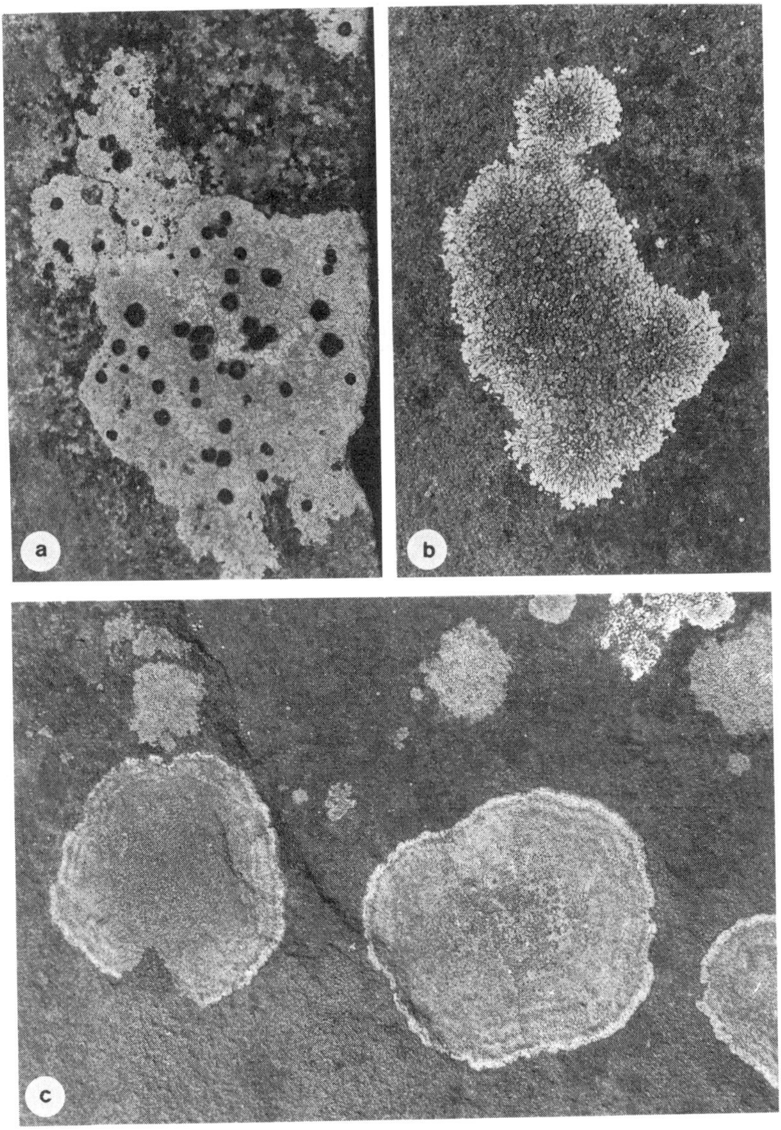

Fig. 5.11 Lichen growth forms. (a–c) Crustose lichens very tightly attached to the rocky substratum. (a) *Lecidea macrocarpa* reproductive structures dark, formed within crust. (b) A very dry specimen of *Xanthoria parietina* in an open habitat, spore-producing structures darker orange, towards centre (arrowed). (c) *Lecanora calcarea*, spore-producing discs formed towards centre of crust. (d, e) Foliose lichens growing on the bark of tree branches. (d) *Hypogymnia physodes*. (e) *Parmelia laevigata*, note soredia on the tips of the lobes. (f, g) Fruticose lichens. (f) *Usnea subfloridana* growing on a tree branch. (g) isolated specimen of *Cladonia portentosa*.

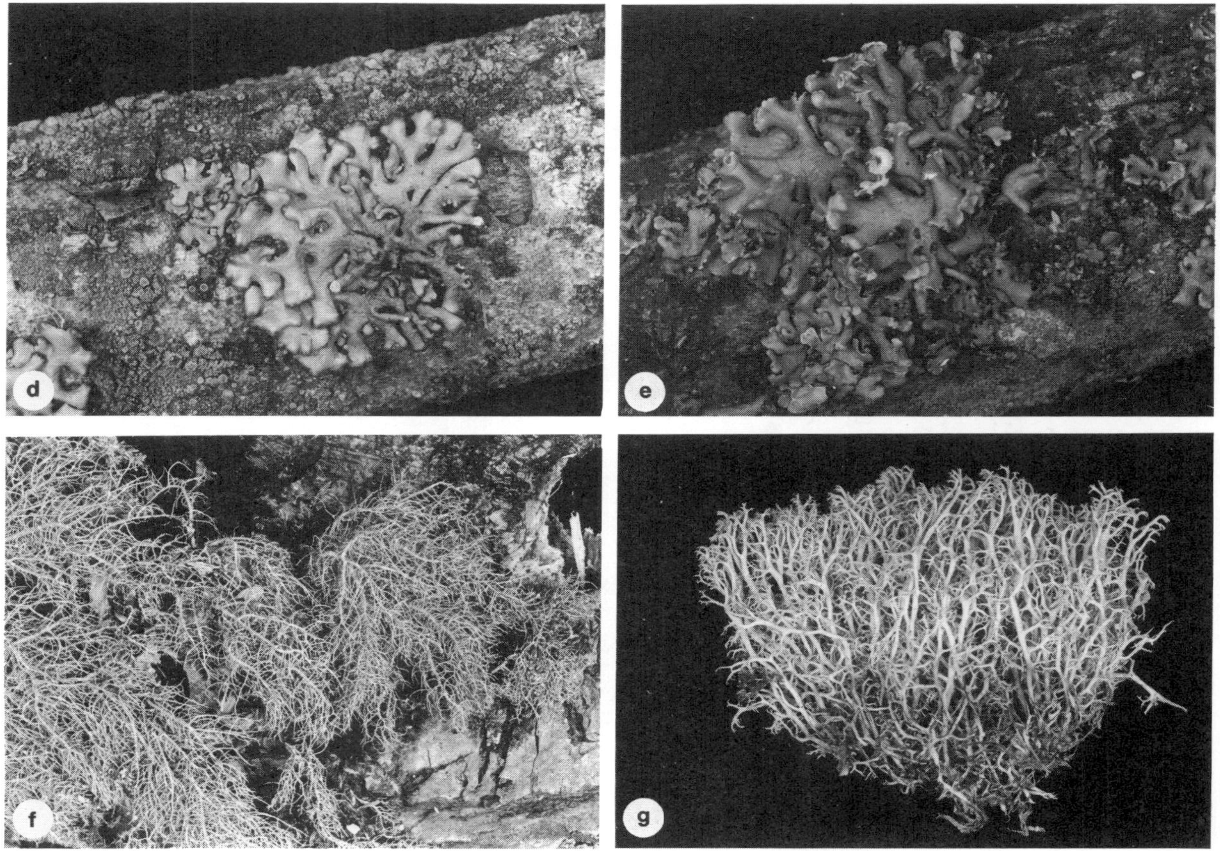
d
e
f
g

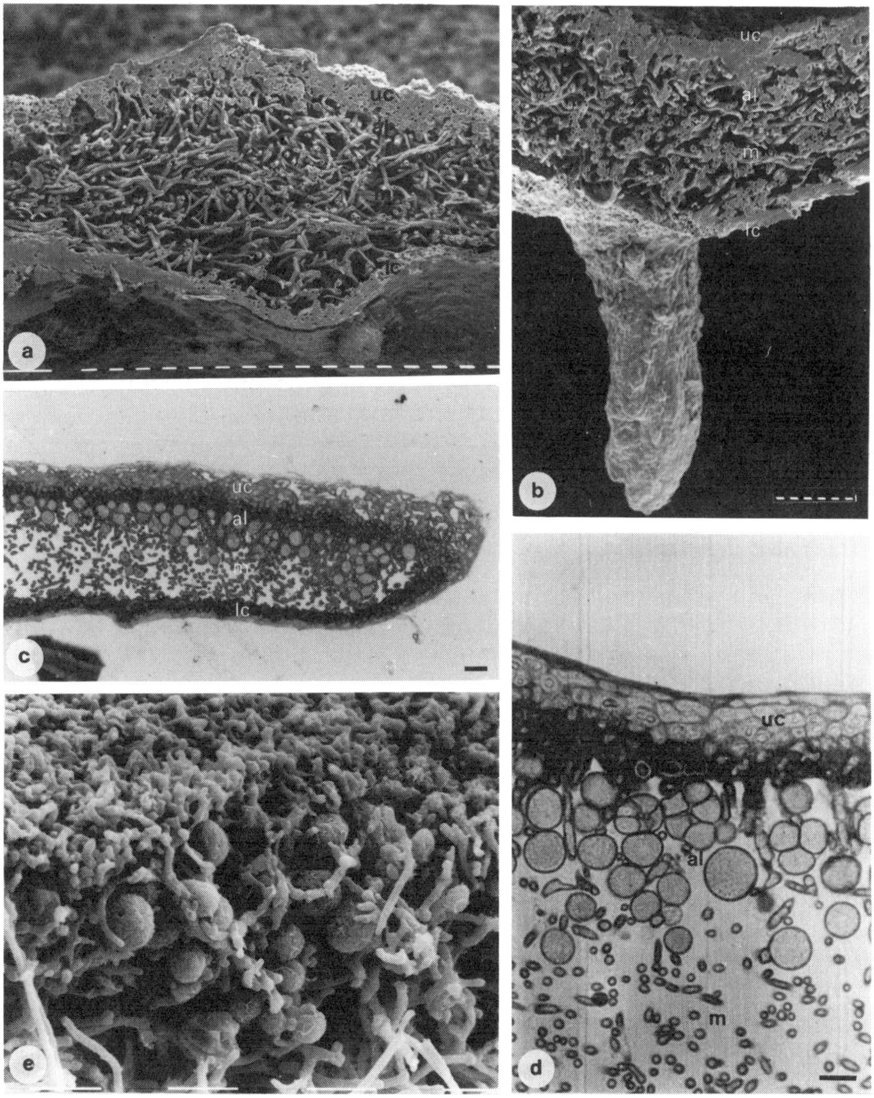

Fig. 5.12 Ultrastructure of foliose lichen thallus. (a) Scanning electron micrograph showing vertical section through *Parmelia perlata* thallus. Upper cortex (uc), algal layer (al), medulla (m), lower cortex (lc). (b) As for (a), also showing rhizine arising from lower cortex. (c) Light micrograph showing vertical longitudinal section through the growing margin of *Parmelia saxatilis* thallus. (d) Light micrograph as in (c) showing upper cortex and algal layer. (e) Scanning electron micrograph showing vertical section of protease/detergent-treated *Parmelia saxatilis* thallus. Bars represent (a) 10 μm; (b) 6.25 μm; (c) 3.5 μm; (d) 7.5 μm; (e) 10 μm. (c, e) Courtesy of Dr D. Anglesea. (d) Reproduced from Anglesea *et al.* (1982), courtesy of Academic Press.

and has shown the relationship between hyphae and algal cells in thalli (Fig. 5.12e). The algae are arranged in a layer underneath the cortex and below that is the medullary region which contains a few algal cells and loosely packed fungal hyphae (Fig. 5.12). Additionally, there may be a distinct lower cortex composed of packed fungal hyphae and it is from this layer that structures arise which attach the thallus to the substrate (e.g. rhizines and cyphellae), but not so closely as in the crustose forms. Such structures may be more or less branched and may be pigmented (Fig. 5.12b). It seems likely that these may have some limited role in the absorption of materials from the substrate but this has not been proven experimentally. These lichens grow across the substrate as lobes, proliferating at the outer margins. In some cases these lobes may be quite fleshy, large and well developed, e.g. *Peltigera* spp., or may be small and scaly or squamulose, e.g. *Cladonia* spp.

Fruticose lichens

Fruticose lichens are more stalked and shrubby, with an erect or pendulous growth habit (Fig. 5.11f, g) and are often highly branched and hair-like. Some species may grow to be quite extensive. For example, the strands of *Usnea* species may reach 5 m in length. The thalli are differentiated radially (Fig. 5.13), and may be round (*Usnea* spp.) or flattened (*Ramalina* spp.), with an outer cortex of fungal hyphae and an underlying algal layer. Beneath this lies a loosely packed medulla containing fungal hyphae with a few algal cells and a central region which may contain a core of fungal hyphae or may be hollow. The thallus surface is smooth and hyphae are embedded in a polysaccharide matrix. Protease/detergent treatment (Anglesea *et al.*, 1983; Greenhalgh and Whitfield, 1987) has shown that algal cells are present at tips of thalli (Fig. 5.13c) and that fungal hyphae grow in parallel towards the end of the thallus, with free hyphal apices at the extreme thallus tip (Fig. 5.13d). Further from the thallus tip hyphae are more branched and form the cortex of the mature thallus (Fig. 5.13e).

Reproduction

A range of diverse vegetative structures, containing both partners, are produced by lichens. Small lobules which arise from the margins of the thalli and often become detached from the main lobes are a very important means of vegetative dispersal. It is thought that these are produced in response to tears in the thallus margin. Soredia are clumps of algal cells which are quite tightly enclosed by fungal hyphae and (Fig. 5.14) which are often aggregated together. These appear to be formed after an overgrowth of algal cells and the aggregates are pushed through breaks in the upper cortical layer, from where some gradually break away and are efficiently

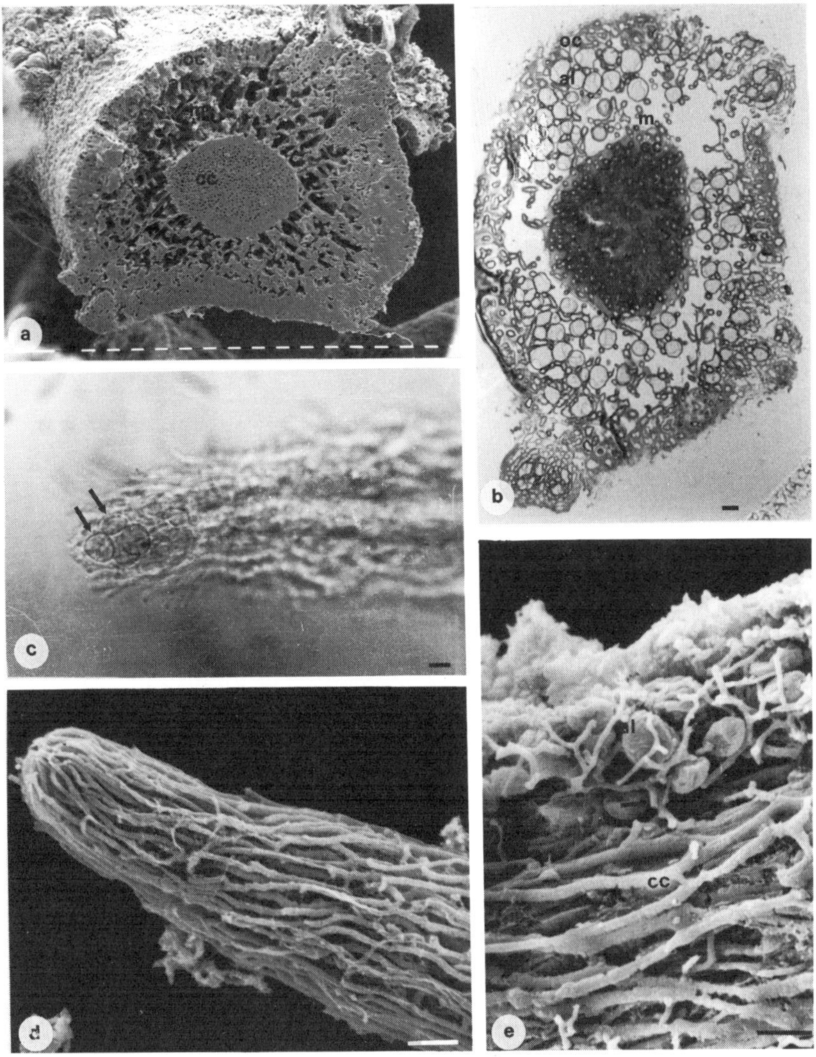

Fig. 5.13 Ultrastructure of fruticose lichen thallus (*Usnea subfloridans*). (a) Scanning electron micrograph showing transverse section of thallus. Outer cortex (oc), algal layer (al), medulla (m), central core (cc). (b) Light micrograph of transverse section of thallus. (c) Light micrograph of protein/detergent-treated thallus, showing algal cells (arrowed) among cortical hyphae at branch apex. (d) Scanning electron micrograph of protein/detergent-treated thallus branch showing parallel hyphae. (e) Scanning electron micrograph section through protein/detergent-treated thallus showing algal layer, central core and connecting hyphae (arrowed). Bars represent (a, e) 10 μm; (b, c, d) 8 μm. (b, c) Reproduced from Anglesea *et al.* (1983), courtesy of Academic Press. (d, e) Courtesy of Dr D. Anglesea.

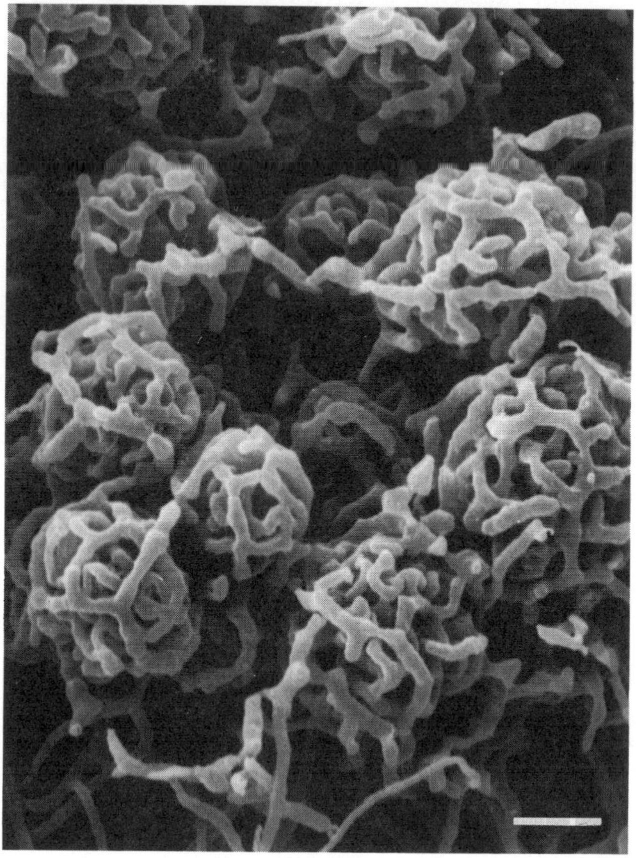

Fig. 5.14 Scanning electron micrograph showing soredia of *Parmelia afrorevoluta*. Bar represents 10 μm. Reproduced from Anglesea (1984), courtesy of Dr D. Anglesea.

dispersed. Many of the most primitive lichen forms are collections of soredia. Isidia are finger-like outgrowths from almost any region of the upper cortex of the thallus which contain both fungal and algal cells. These are easily dislodged and act as vegetative propagules.

Both the associants are also capable of independent reproduction. The fungal reproductive structures which are produced are very like those of similar non-lichenised species and spores are discharged independently of the symbiont. Most fungal partners are Ascomycetes and produce ascospores. Apothecia (open cup-shaped structures) develop on the surface of lichens, in which asci are formed, containing ascospores. Perithecia (flask-shaped structures) are formed, embedded within the thallus and also pycnidia, from which fungal spores are liberated. Basidiomycete partners form fruiting bodies which arise from the surface of lichen thalli and are akin to those of their non-lichenised relatives.

It is only recently that the synthesis of a lichen has been achieved under controlled laboratory conditions (Ahmadjian, 1982; Ahmadjian and Jacobs, 1987). The lichen *Cladonia cristatella* was synthesised under conditions of nutrient limitation and high humidity. The fungal hyphae grew to enclose the symbiont cells, became more branched and produced a gelatinous matrix. Later these aggregations became more differentiated with distinct cortical and medullary regions and the symbiont (*Trebouxia erici*) was confined to a definite layer. The frequency with which such synthesis of lichens takes place in the natural environment is not yet clear, particularly since the symbionts do not have a free-living existence.

PHYSIOLOGICAL RELATIONSHIPS

Experimental work with lichens is notoriously difficult and laboratory experiments are particularly fraught with problems, not least because of the difficulty in maintaining the lichen symbiosis under controlled conditions without one partner becoming dominant. However, some physiological aspects have been successfully investigated and it is very clear that the physiology of a lichen is different from that of the component partners.

Contact between the partners

Contact between the partners in a lichen is close, by virtue of the physical organisation in the thallus, and facilitates the transfer of materials between cells. However, in addition a range of morphological arrangements have been identified, which increase the surface area of contact and the degree of intimacy between fungal hyphae and individual symbiont cells (Honegger, 1984) and vary with changes in environmental conditions. The simplest arrangement is close wall-to-wall contact between algal cells (*Coccomyxa*) and the fungal hyphae in the lichen *Peltigera aphthosa*. Hyphae grow closely around the symbiont surface but no penetration of the algal cells occurs. In *Peltigera* species which have cyanobacterial symbionts (*Nostoc*), hyphae are seen to penetrate into the sheaths of gelatinous matrix which surrounds the symbionts but do not attach directly to the cell walls.

More specialised arrangements are recorded for the wide range of lichens which have trebouxioid symbionts. In well-differentiated lichens (foliose and fruticose) hyphae become attached to the surface of the algal cell wall by an appressorial-like structure which provides a large surface area of contact. The cell wall is not breached by the fungus. There is usually one contact hypha per algal cell. At the time of algal cell division the fungal

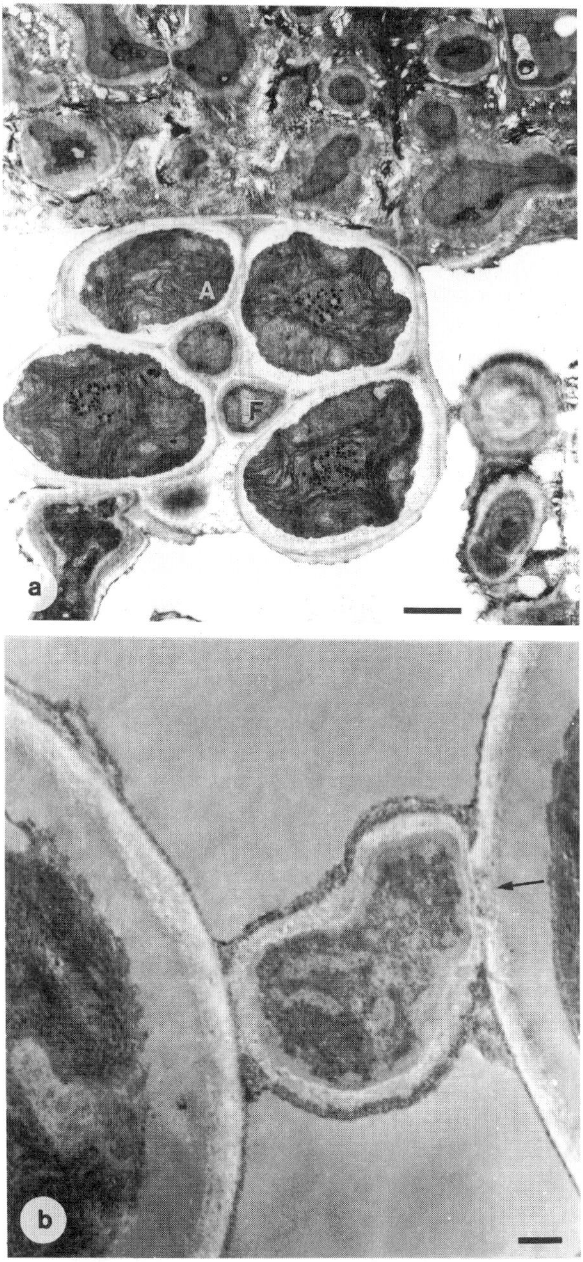

Fig. 5.15 Fungal–algal contact in *Parmelia saxatilis*. (a) Electron micrograph of section through a recently divided algal cell (A). Intrusive fungal hyphae (F) can be seen within the mother cell wall. (b) Point of contact between fungus and alga. Damaged algal wall arrowed. Bars represent (a) 1 μm; (b) 0.1 μm. Reproduced courtesy of Dr D. Anglesea.

hyphae become branched and attach to each daughter cell, growing and pushing the new cells apart (Fig. 5.15). In some instances the fungal hyphae form projections which push into the algal wall, causing it to become invaginated (interparietal), but do not penetrate into the cytoplasm. Relatively undifferentiated lichens from very dry habitats (e.g. *Lecanora*) have been shown to form the closest contacts between the partners. Projections, which have been termed 'intracellular haustoria', breach the cell wall and pass into the algal cells. The cell membrane of the symbiont remains intact.

After active growth the aged and senescent algal cells become invaded by fungal projections and are probably utilised as an additional nutrient supply.

Photosynthesis and carbon transfer

Photosynthesis is carried out within the photobiont cells and supplies the carbon requirements of the fungal partner. In general the chlorophyll content per photobiont cell is very similar to that of non-symbiotic cells of the same species, but the content of the lichen as a whole is quite low because there are few photobiont cells per unit thallus. The rate of photosynthesis is markedly affected by water content. Re-wetting of a dry thallus results in a steady rise in the photosynthetic rate. The optimum moisture level is 40–80% saturation. It has been shown that lichens are able to photosynthesise at lower water potentials than non-symbiotic species in the same habitat.

A large proportion (over 80%) of the carbon compounds fixed by the photobiont pass to the fungus, together with supplies of essential vitamins (e.g. biotin and thiamine) which the fungi lack naturally. Movements of fixed carbon between the partners have been shown in the foliose lichen *Peltigera polydactyla*, using discs cut from active thalli. The discs were exposed to ^{14}C-labelled CO_2 in the light, and were subsequently dissected for analysis (Smith and Drew, 1965). These experiments demonstrated that the transfer of label between the algal layer and the fungal medullary tissues was very rapid. Additionally, the ^{14}C label was present in the fungal tissues as mannitol, which is not metabolised by the *Nostoc* (cyanobacterium) symbiont. Autoradiographic techniques have shown that over 90% of the carbon fixed photosynthetically by the algal symbiont *Trebouxia* in *Cladonia convoluta* is passed to the fungus (Tapper, 1981).

Experiments using an 'inhibition technique' have provided a great deal of information concerning the types of carbohydrates which are passed from photobiont to mycobiont. The suspension of thallus discs in a solution of digitonin results in the release of photosynthetic products to the medium. The treatment inactivates fungal membranes but does not affect the activity of the photobiont. In experiments using a range of lichen species it has been shown that single compounds are released from the photobionts and are made available to the mycobionts, which are specific to the species involved. In cyanobacterial symbionts photosynthetic products are converted to glucose which is then released to the fungal cells. In the Chlorophyceae carbohydrate products are converted to polyols (ribitol, erythritol and sorbitol) prior to release for use by the fungus. These materials are taken up rapidly by the fungus and are converted to fungal carbohydrates, e.g. mannitol and arabitol, which cannot be utilised by the symbionts (Fig. 5.16). This represents a one-way flow to the fungal partner which seems to act continually as a sink for carbohydrates. Collins and Farrar (1978) proposed that the photobionts supply the needs of the fungus by mass transfer in lichen thalli. Fungal polyols act as storage compounds in the mycobiont and can be depleted very slowly over a long period in times of nutrient stress.

It has also been suggested that the fungal partner may stimulate the release of compounds from the symbiont cells. Indeed in non-symbiotic conditions the photobiont cells do not normally release such massive levels of photosynthates. Additionally, a variety of compounds are lost from the non-symbiotic cells rather than the limited range of materials which are passed to the fungal partner in the symbiotic association. The area of contact between fungus and symbiont is limited in comparison with haustorial contacts in other biotrophic infections, which implies that sufficient materials are exchanged efficiently over a small interface. Haustorial-type projections are only present in species which inhabit relatively dry sites. It has been suggested that the fungal partner is able to obtain nutrients by the induction of modifications in the patterns of symbiont wall synthesis, so that materials destined for symbiont walls are continually absorbed by the mycobiont and converted to fungal products.

Water relations

Lichens are able to survive very dry conditions for quite long periods of time and are able to flourish under regimes of wetting and drying. The thalli are able to absorb water vapour from the air efficiently and in sufficient quantities to maintain metabolic activity. A diagrammatic representation of the exchange of materials between the environment and the symbionts is shown in Fig. 5.16. Liquid water can also be taken up very rapidly (Fig. 5.17). Lichens do not have any control over water loss and become dry readily, although some water is bound into molecules within the thallus structure and in matrix materials. A thallus may dry out to 2–5% dry weight but will absorb water extremely rapidly on wetting, often up to 300% dry weight. Total saturation for any length of time results in damage, and metabolic deterioration. Wetting and drying

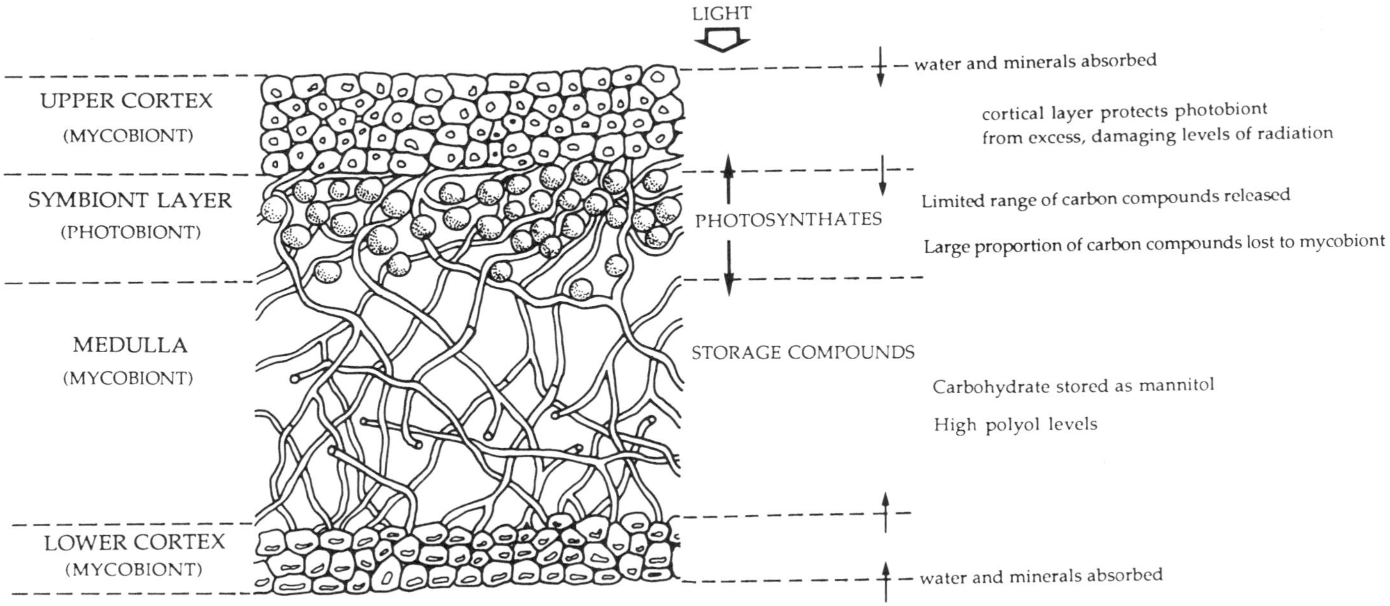

Fig. 5.16 The exchange of materials between the algal and fungal partners in a lichen, and between the thallus and the environment.

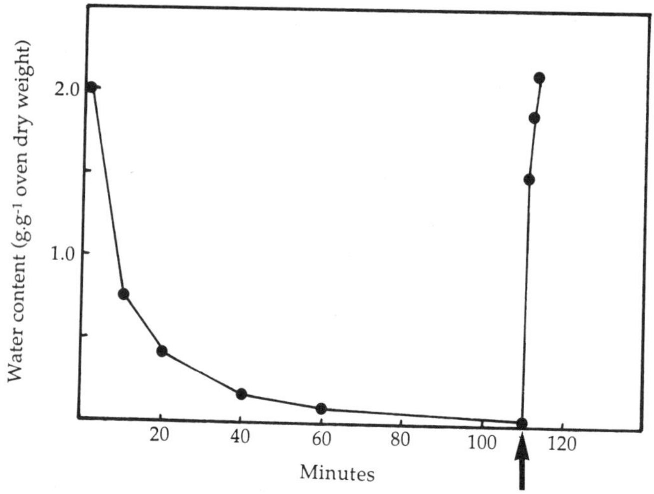

Fig. 5.17 Water content of *Hypogymnia physodes* thallus. Thalli were air-dried and then placed into liquid water at the time indicated (arrowed). On rewetting the thalli absorbed water very rapidly reaching 50% maximum saturation in 10 s and approaching full saturation in 2 min. Data from Farrar and Smith (1976), courtesy of the Trustees of the New Phytologist.

cycles appear to be important and help to maintain healthy activity in lichens (Farrar, 1976).

It has been suggested that the high concentrations of mannitol which are maintained in lichens lower the osmotic potential in the thallus and assist the accumulation of water under conditions of low availability (Table 5.5). It is also proposed that polyols have an important role in the lichens tolerance of wetting and drying. Under conditions of water stress the hydroxyl groups from the polyols may replace water in molecules thereby protecting proteins and maintaining the integrity of membranes (Farrar and Smith, 1976).

Effects of water stress on respiration

Physiological activity is negligible unless the lichen is moist but a very low rate of respiration is maintained in dry thalli. Respiration rates are markedly affected by changes in water content (Smith and Molesworth, 1973). Immediately upon rewetting of the lichen thallus there is a loss of CO_2 accompanied by a dramatic and rapid rise in respiration rate, known as 'resaturation respiration'. There are also losses of some solutes (polyols and phosphates) from the fungal partner into intercellular spaces as a result of alterations in membrane permeability. Carbohydrates may also be lost from the medulla. Some of these components may be reabsorbed after normal function has been restored, although very rapid and extreme

Table 5.5 Concentrations of polyols in *Hypogymnia physodes* under different levels of hydration (data from Farrar and Smith, 1976)

Partner	Polyol	Concentration of polyol			Osmotic pressure (Pa) generated by polyols in symbiont	
		($mg.g^{-1}$)	($mol.dm^{-3}$)	($mol.dm^{-3}$)		
			water content 2.5 $g.g^{-1}$	water content 0.1 $g.g^{-1}$	water content 2.5 $g.g^{-1}$	water content 0.1 $g.g^{-1}$
Alga	Ribitol	10	0.26	6.50	6.33×10^5	1.58×10^7
Fungus	Arabitol	50	0.16	3.50	3.90×10^5	8.53×10^6
	Mannitol	10	0.026	0.65	6.33×10^4	1.58×10^6

These data were calculated on the assumption that the fungal partner accounts for 90% of the thallus and the algal partner 10% (by dry weight), and that water was not preferentially held by either. Osmotic pressure and molarities were calculated on the assumption that water was intracellular.

wetting and drying may cause irreplaceable losses and result in damage to the thallus.

Effects of water stress on mineral uptake

It is likely that the supply of rainwater and run-off which provides leachates from the substratum is sufficient to support the mineral requirements of the thallus. In fact much larger amounts are absorbed than are actually needed by the lichen (Smith, 1975), but the supply is often erratic. Such water is absorbed very efficiently by the thallus and close contact with the substrate is maintained. However, losses which occur on rewetting, after a prolonged phase of drying may be damaging to the lichen physiology. It has been suggested (Crittenden, 1989) that the photobiont may be able to take up some of the solutes lost from mycobiont cells at rewetting and this may therefore represent reciprocal exchange. The photobiont may also be able to absorb some minerals from rainwater taken into the thallus.

Nitrogen metabolism

Lichens with cyanobacterial symbionts have the capacity to fix nitrogen and are also photosynthetic (Galun and Bubrick, 1984). The sites of nitrogen fixation are the heterocysts, enlarged, specialised cells which occur in the chains of *Nostoc* or *Scytonema* cells. In those species which have cyanobacteria as the sole symbionts the amounts and rates of nitrogen fixation are quite large on a cellular basis and sometimes greater than those in free-living cyanobacteria of the same species. In instances where a lichen contains an additional symbiont the cyanobacteria are confined to restricted areas known as cephalodia. From there the cyanobacteria fix nitrogen faster and also transfer a larger proportion of that fixed, to the rest of the thallus. Most of the fixed nitrogen is released as ammonia, which is subsequently converted and stored as glutamate in the fungus although some amino acids are made available for use in the thallus. The rates of protein turnover are generally very low in lichens.

GROWTH RATES

The growth rates of lichens are very low and markedly slower than those of the non-lichenised components. In some habitats, under rather unfavourable conditions growth may be so slow as to be negligible almost throughout the year. In general terms however, crustose lichens may extend across the substrate at about 1 mm per year. Foliose lichens grow more quickly and may achieve an average extension rate of 2 cm per year.

Perhaps the major compensation for the slow growth rate is the extreme longevity of lichen thalli. Many lichens in undisturbed habitats are very long lived. However, the growth rates are extremely variable throughout the year, and are highly influenced by seasonal changes in environmental conditions. Most fluctuations in growth rate relate to the levels of incident light received, fluctuations in temperature and water availability. In some habitats environmental conditions may be unfavourable for growth and proliferation of thalli throughout large parts of the year.

A lichen has a small source of nutrients (photobiont) which supplies a relatively large sink (mycobiont) and productivity may be limited as a result of the low supply. Additionally, environmental extremes may result in very reduced growth for long periods, so that it is not surprising that productivity is restricted.

Lichens are tolerant to quite high temperatures, in comparison with higher plants. Thalli dry out under elevated temperatures but are able to recover from periods of exposure to normal heating by the sun. Very high light levels are more damaging however. To some extent algal cells are protected from high light intensities by the overlying fungal cortex. As the thallus dries out the cortical cells contract in size and the thick walls provide extra protection to the symbiont layer. At rehydration these cells expand and the symbionts receive higher levels of radiant energy. The cortical layers of some lichen thalli are pigmented which also protects the symbiont by changing the quality of the light which is received. Low temperatures are tolerated by lichens providing that the thalli are dry when the temperature is low. Prolonged exposures to very low temperatures do not damage desiccated individuals.

The loss of carbon on rewetting, together with the rapid rise in respiration (resaturation respiration), are the most potentially damaging effects of dry conditions and extremes of temperature. Lichens recover well from these conditions unless the wetting and drying cycles are very rapid and repeated. In such extremes losses may exceed gains and the lichen may be badly damaged. Time is required for the thallus to recover from such effects.

One of the disadvantages of the efficient absorbing capacity of lichens is that aerial radioactive fall-out is also incorporated into thalli. Radioactive strontium (^{90}Sr) and caesium (^{137}Cs) are easily absorbed and accumulated in foliose lichens. In northern regions these lichens are an important part of the diet of reindeer and by this route enhanced levels of radioactivity are passed into the food chain.

Additionally, lichens are very sensitive to atmospheric pollution. As a result the growth of lichens is very restricted around towns and industrial regions. Lichens are particularly sensitive to sulphur dioxide levels and heavy metals which are absorbed from aerosols and accumulate in the thalli. Most sensitivity occurs when the lichens are metabolically active, for example in conditions of high humidity. The non-wettable lichen *Lecanora conizaeoides* is most tolerant to pollution levels and is also a poor competitor. Not surprisingly, this species predominates at sites with low levels of pollution where other lichens are not present.

ENDOPHYTIC SYMBIOSES

Modern usage of the term endophyte in mycology refers to those fungi which live almost entirely within the leaves and stems of apparently healthy host plants, doing so asymptomatically, causing no visible signs of infection. The term endophyte was originally defined by De Bary (1866) to distinguish those species which invade and reside within host tissues or cells from epiphytes, those fungi living on the outer surfaces of host plants. Parasitic antagonistic symbionts which cause visible disease symptoms are more usually referred to as pathogens (e.g. rusts and mildews), even though these may live almost entirely within host tissues (Petrini, 1986). Additionally, although mycorrhizal fungi live both in and on host tissues such associations are not usually included within the endophyte category.

A great deal of interest in endophytes has emerged recently, with the suggestion that such species may be useful as biocontrol agents against insect pests of grasses and with the discovery that such species produce toxins in host plant tissues (Clay, 1986, 1989). There is presently a great deal of discussion concerning the extent to which endophytes are in fact latent pathogens or examples of fungi co-evolving with plants from parasitism to mutualism (Clay, 1988). However, in many cases mutualistic relationships between host plants and fungal endophytes are suspected and in many instances such associations have been identified.

DETECTION AND TAXONOMY

Most recent work has centred around fungal endophytes of grasses, ericaceous species and conifers, although endophytes have been identified in a very wide range of host plant species, including representatives of most major taxonomic groups, including mosses, ferns and liverworts. It has been suggested that the majority of living plants are hosts to endophytic fungi (Petrini, 1986; Carroll, 1986, 1988). An individual plant may be host to a range of endophytic fungal symbionts simultaneously. In morphological terms the fungal hyphae growing in plant tissues are not modified or specialised and it is not possible to identify the species concerned within host tissues. The majority of the endophytic species which have been successfully identified are Ascomycetes and

Deuteromycetes with a few Basidiomycetes and a very small number of Oömycetes.

The isolation and identification of endophytic species involves very careful surface sterilisation of host plant tissues followed by incubation on a range of media to encourage outgrowth of isolates and their subsequent sporulation, so that identification can be carried out. It is often difficult to satisfactorily establish endophytic status for the isolates which are obtained. It is not easy to ensure the exclusion of spores or hyphal fragments which may escape the sterilisation procedure (Petrini, 1986). Some epiphytic species may penetrate host tissues and additionally some endophytic species may also occur as epiphytes under some conditions. Most endophytic species sporulate in culture, although many isolates grow very slowly and a considerable incubation time (sometimes many months) is often required. Mycelium is often located in thickened and lignified tissues which are difficult to examine directly under the microscope, although plating techniques encourage outgrowth on to the surface

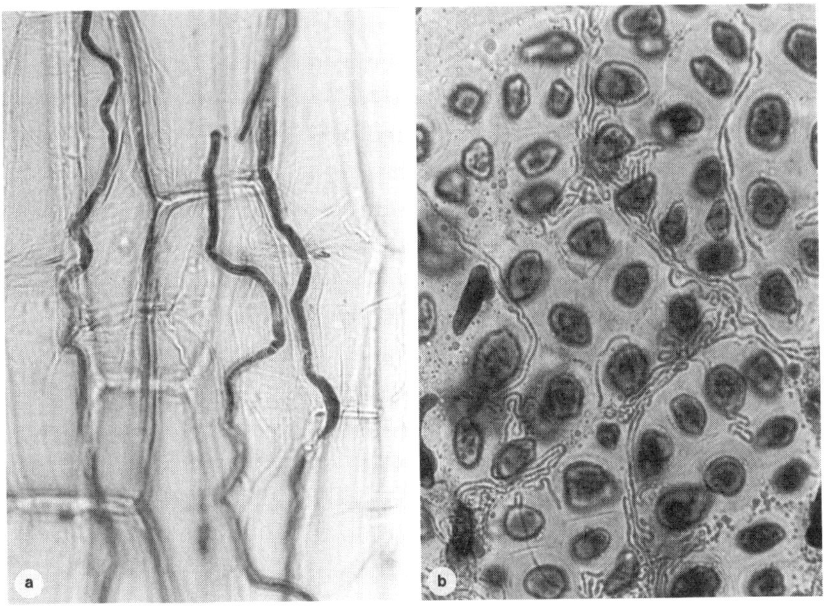

Fig. 5.18 The vegetative growth of endophytic fungi of grasses. (a) Endophytic hyphae of *Acremonium coenophialum* in *Festuca arundinacea* (tall fescue grass) leaf tissue. (b) Hyphae of *Acremonium lolii* in the aleurone layer of seed from *Lolium perenne* (perennial rye grass). Reproduced from Clay (1989), courtesy of Dr K. Clay and Cambridge University Press.

of nutrient medium. Modern methods have also been used to establish the presence of infections in plant materials, e.g. *Epichloë typhina* has been detected in tall fescue using enzyme-linked immuno-absorbent assays (ELISA) (Johnson *et al.*, 1982).

Fungal endophytes of grasses and sedges have received a great deal of attention recently and many species of grasses have been shown to be infected. Many of these fungi belong to the family Clavicipitaceae (Ascomycotina). Five genera from the tribe Balansiae have been found (*Atkinsonella, Balansia, Balansiopsis, Epichloë* and *Myriogenospora*) and about 30 species have been recognised as endophytes. It is interesting that the Balansiae also included species which are grass parasites. Wheat pathogens have been shown to be present within healthy plants throughout the growing phase of plants and have been isolated from apparently healthy tissues, e.g. *Didymella exitialis* and *Phaeosphaeria nodorum*.

Within host tissues vegetative fungal hyphae grow between host cells, usually running longitudinally through leaves and stems, parallel to the long axis of the plant. These hyphae are unremarkable in form (Fig. 5.18a). The anamorphs, or asexual forms, of these endophytes do not sporulate on the host plant and are passed to new plants vegetatively, by growth into ovules and seeds (Fig. 5.18b). Spore-bearing stroma of the teleomorphs, or sexual forms, on the other hand, may be found on host plant leaves and aborted inflorescences (Fig. 5.19). Such endophytic infections inhibit flowering and result in host sterility.

A great many endophytic fungi have also been isolated from ericaceous host plant species. For example, almost 200 fungal species have been isolated from *Arctostaphylos uva-ursi*, e.g. *Coccomyces arctostaphyli* and *Cryptocline arctostaphyli*. *Vaccinium myrtillus* plants have been reported to be host to *Pezicula myrtillina* and *Phyllosticta pyrolae*.

Fungal endophytes of conifers are common although many belong to little known species, probably because these fungi are very inconspicuous and are rarely collected. Some species live almost entirely within a host plant cell (Fig. 5.20). The hyphae are often swollen and distorted with constrictions at the septa, and may fill the whole cell lumen. No further proliferation may occur for long periods of time, often for several years, and not until leaf tissue is aged and/or senescent. It has been seen that many such endophytic species are associated with conifer needles from a single host. Some of these fungi are regularly associated with host species. For example, *Abies* species are often hosts to *Cryptocline abietina* and *Phyllosticta* spp.; *Juniperus communis* is often host to *Anthostomella formosa* and *Kabatina juniperi*. It has been suggested that mutualistic endophyte associations may be as common and widespread as mycorrhizal associations (Carroll, 1986). Rates of endophyte infection increase with ageing of plants and plant populations. It is also possible that

Fig. 5.19 Reproduction of endophytic fungi of grasses and sedge. (a) Stroma of *Epichloë typhina* infecting *Dactylis glomerata*. (b) Stroma of *Balansia oblecta* infecting *Cenchrus echinatus*. (c) (p. 320) Stroma of *Balansia cyperi* infecting the serious weed species *Cyperus rotundus* (purple nutsedge). On the right a healthy inflorescence showing expanded panicle of spikelets. On the left an aborted inflorescence (becoming covered with mycelium). (d) (p. 320) Stroma of *Atkinsonella hypoxylon* infecting *Danthonia spicata*. (a) Reproduced from Clay (1989). (b–d) Courtesy of Dr K. Clay.

some plant species may not occur naturally without endophytic fungal infections.

Host specificity

The degree of host specificity which operates in the endophytic fungi is not yet clear. Some species are commonly occurring and may be isolated from various host plant species and from different locations with differing environmental conditions. In general terms, the geographical occurrence

Fig.5.19c–d

of endophytes is related to the distribution of host species. In some cases almost all individuals in a plant population may be infected by endophytes. *Cladosporium* spp., *Nodulisporium* spp. and *Pleospora* spp. are common. Some endophytes, however, do not show such a wide species range and are often isolated from plants of the same family or closely related families. Other species are only rarely detected.

The degree to which endophytes are tissue- or organ-specific is also not yet clear. Some species are most commonly isolated from similar tissues, particularly the endophytes of conifer needles (Carroll and Petrini, 1983). In other cases the occurrence is less distinct. However, only limited surveys have been carried out to date.

ENDOPHYTIC MUTUALISM

Effects of endophyte infections

Many endophytes live almost entirely within the host plant tissues, often without causing any visible signs of infection. Fungal hyphae penetrate between plant cells or may also grow intracellularly and must obtain nutrient materials through this intimate contact with the host. The occurrence of specialised feeding structures has not been reported in these fungi. Hyphae are sometimes quite wide in diameter (10–15 μm) when in association with plant cells and may be distorted, irregular or bulbous in form (Fig. 5.21). In some instances considerable amounts of fungal biomass are supported in host plant tissues. In physiological terms relatively little is known about the endophytic interactions between host and fungus and it is not easy to see how a host plant may benefit from such a relationship. Endophytic associations do not lead to the development of disease symptoms

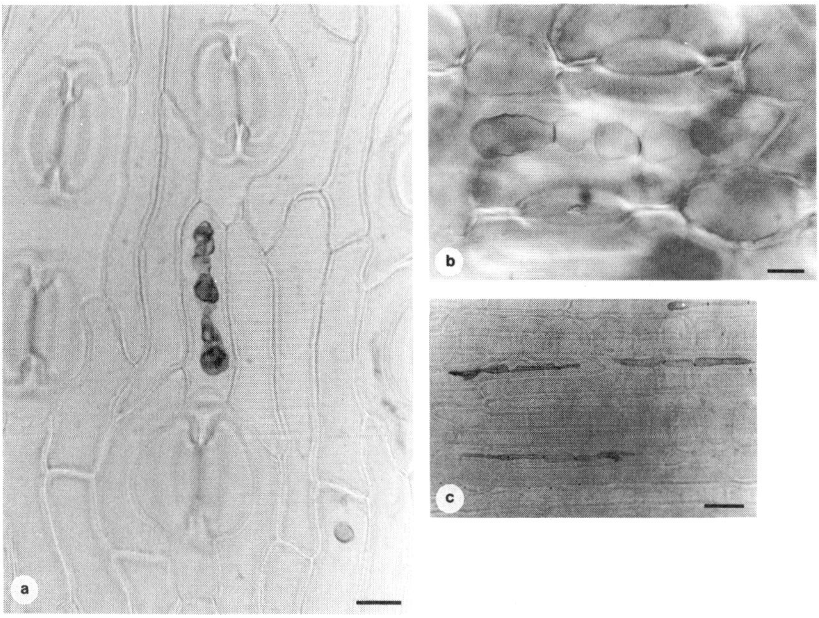

Fig. 5.20 Endophytic infection of conifers. (a) Intracellular hypha of *Choroscypha chloromela* in an epidermal cell of coast redwood (*Sequoia sempervirens*). Bar represents 10 μm. (b, c) Intracellular hypha of *Rhabdocline parkeri* in epidermal cells of Douglas fir. Bars represent (b) 10 μm; (c) 50 μm. Reproduced courtesy of Dr J. Stone.

but do result in some morphological and physiological changes in host tissues which increase the survival and vigour of the plants concerned. Such physiological enhancement would be likely to increase the capacity of a plant to resist disease. It has also been suggested that endophyte-infected plants are more tolerant of water stress and recover more quickly than uninfected individuals (Belesky *et al.*, 1987), although it is not clear quite how photosynthetic rates are affected by the presence of endophytes (Clay, 1989). There are suggestions that endophytes may produce plant growth regulators, which may alter the normal developmental pattern of the host plant (Porter *et al.*, 1985). Reports of secondary metabolite production by endophytes, e.g. alkaloids and antibiotics, which affect a range of herbivores, have attracted a great deal of attention. It has been suggested that endophytes provide the plant with a chemical defense mechanism.

Effects on insects

One of the main reasons for the recent increase in interest in endophytic fungi has been the realisation that endophyte infections have marked

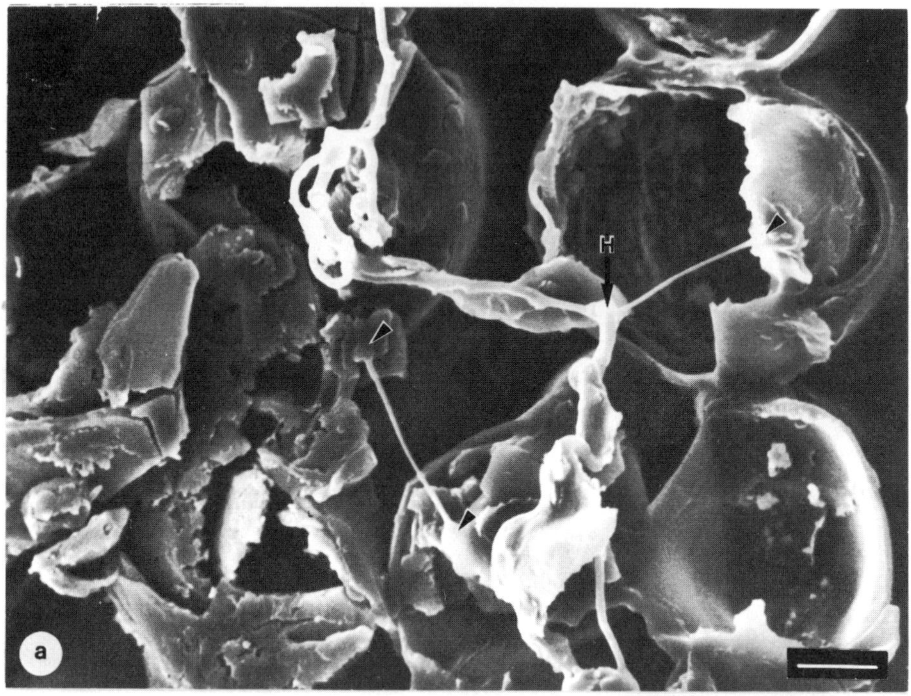

Fig. 5.21 Endophytic infections of symptomless needles from *Picea abies* Karst (Norway spruce). (a) Scanning electron micrograph of a section through mesophyll tissue from a 6 month-old needle. Hyphae (probably of *Lophodermium piceae* (Fokl) Höhn) of different thickness arise from the same branch point (H) showing that they belong to the same mycelium. Areas of contact between fungal hyphae and host cells are indicated (arrows). Bar represents 10 μm. (b) Scanning electron micrograph of a tissue section from a needle over 6 months old. Intracellular swellings (arrowed) are visible along the hypha (H). Mesophyll cell (M), needle surface (S). Bar represents 10 μm. Insert shows higher magnification (bar represents 4 μm) of the region indicated. Intracellular swellings are closely associated with mesophyll cells. (c) Transmission electron micrograph showing a section through an epidermal cell of an asymptomatic needle colonised by endophytic hyphae (H). Epidermis (E), hypodermis (HY), needle surface (S). Bar represents 2 μm. (d) Intercellular hypha (H) containing two endocells (EC). Swellings in endophytic hyphae are probably caused by the occurrence of endocells, or clusters of hyphae. Mesophyll cell wall (MW), electron dense sheath (SH). Bar represents 0.6 μm. (a, b) Reproduced from Süsker and Acker (1987), courtesy of Dr J. Süsker and the National Research Council of Canada. (c, d) Reproduced from Süsker and Acker (1989), courtesy of Dr J. Süsker and Springer-Verlag.

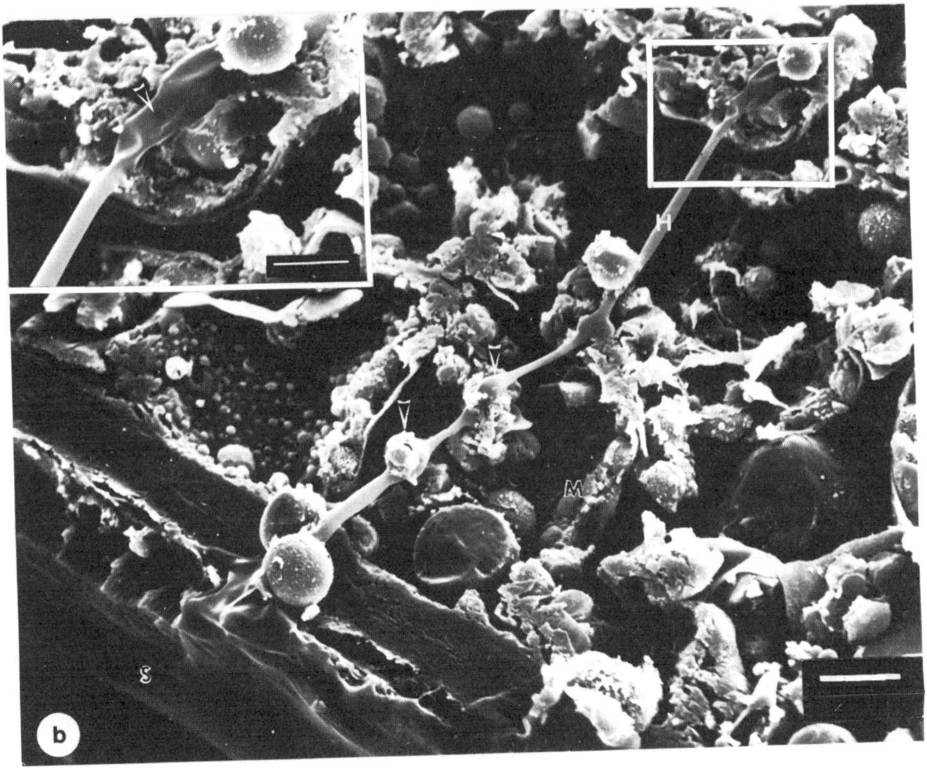

effects on the grazing of insect pests. The presence of endophytes makes plant tissues unacceptable and unpalatable to insects such that infected tissues are avoided. Additionally, infected tissues may give rise to toxic effects on the insects causing poor larval growth and development, reduction in reproductive capacity or death of individuals and so having more far-reaching effects on insect populations.

A range of species of insects have been reported to be negatively affected by endophyte-infected grasses, including crickets, aphids, armyworms and flour beetles. Much of this information has led to the suggestion that endophyte fungi may be suitable biocontrol agents for the protection of grass species (Clay, 1989). A range of field and laboratory feeding experiments has shown that lower levels of insect damage can be correlated with endophyte infections in plants. In no-choice feeding experiments endophyte-infected grasses were toxic to insects and led to a reduction in the rate of growth and development.

An interesting example of the effects of endophytes on insects is provided by the interaction between elm bark beetles and elm trees. The

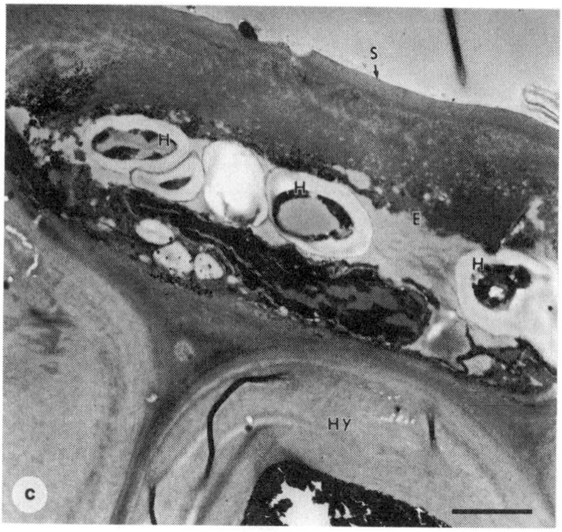

Fig. 5.21(c)

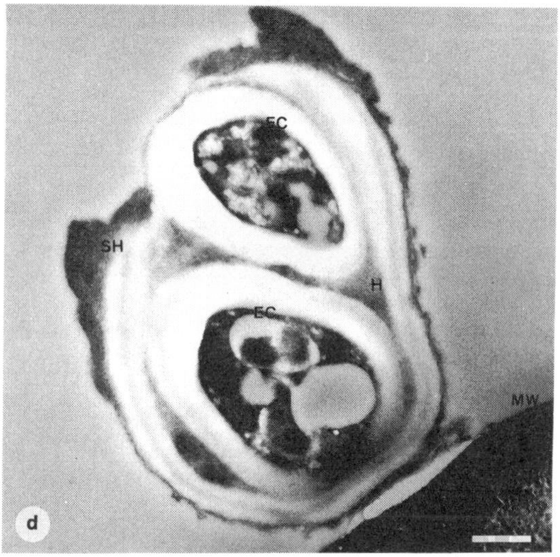

Fig. 5.21(d)

beetles attack elm trees by burrowing into the inner bark tissues and so infecting the trees with spores of *Ceratocystis ulmi*, which causes Dutch Elm disease and has been responsible for the death of many elm trees. A correlation has been demonstrated between the presence of the endophyte *Phomopsis oblonga* and the demise of the bark beetles. Feeding on endophyte-infected wood led to a reduction in the reproductive capacity and a decline in the beetle populations (Webber, 1981).

A number of endophytic fungi isolated from needles of Canadian fir trees were shown to produce secondary metabolites in culture which were toxic to spruce budworms, both decreasing growth rate and increasing mortality of larvae (Miller, 1986). It is not clear to what extent such toxins are produced *in vivo* or the degree of protection which is afforded to the conifers from this source. The presence of the endophytic fungus *Rhabdocline parkeri* in Douglas-fir needles (Carroll, 1986, 1988) increased the mortality of gall-forming midge larvae (*Contarinia*). Toxic metabolites from the fungus were implicated as the responsible agents.

Effects on other herbivores

Although toxicoses induced in domestic herbivores have been known and related to grazing fodder for many years, it is only recently that the observed effects have been correlated with the presence of fungal endophytes in pasture grass populations. Endophytes in grasses responsible for poisoning mammalian stock have now been positively identified. Major examples of such problems have been encountered in grazing areas. In the southern USA the effects of fescue toxicosis have been known for many years. Cattle grazed on infected tall fescue (*Festuca arundinacea*) develop toxic effects including poor weight gain, increased body temperature, lameness, gangrene and limb loss. The effects develop particularly in the summer during hot, dry conditions when the livestock are under most physiological stress. A further example is the condition known as ryegrass staggers, typified by muscular spasms and even, in severe cases, an inability to stand. These symptoms have developed in sheep in New Zealand, and have been correlated with grazing on endophyte-infected *Lolium perenne*. These effects incur substantial economic losses annually.

Although such conditions have been attributed to the presence of fungal endophytes the exact causes are not entirely clear. The endophytes of grasses are Clavicipitaceous fungi and are therefore likely to produce alkaloids, indeed the symptoms of the induced toxicoses are related to those of ergot poisoning. It has now been shown that alkaloids are present in the tissues of these host plants. However, alkaloids have been reported to occur in healthy plant tissues only extremely rarely. Fungi belonging to *Balansia* species do produce alkaloids in culture. It is likely that these are the cause of the toxic effects in mammals.

Plant resistance to microbial pathogens

An increase in host plant vigour will probably enhance the plants inherent ability to resist disease. It has also been suggested that the presence of endophytic fungal infections may protect a host plant from some potentially virulent pathogens (Carroll, 1988). Infected perennial ryegrass plants were apparently protected, to some degree, from *Puccinia coronata*

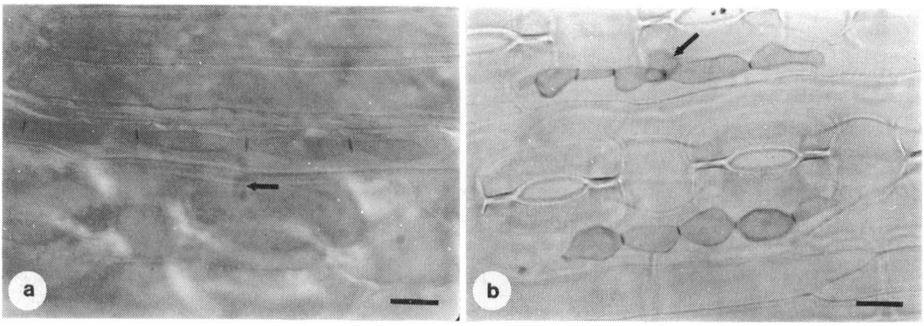

Fig. 5.22 Intracellular hyphae of *Rhabdocline parkeri*, late in endophytic infections of Douglas fir. (a) Hyphae in epidermal cell. A haustorium (arrowed) is visible in an adjacent guard cell. (b) Hyphae in subsidiary cells. A haustorium (arrowed) is visible in the guard cell above the upper hypha. Bars represent 10 μm. Reproduced from Sherwood-Pike *et al.* (1986) courtesy of Dr J. Stone and the National Research Council of Canada.

since the development of rust pustules was less, per unit leaf area, than for uninfected plants (Clay, 1989). It has also been suggested that some endophytic species may be antagonistic towards other fungal species. Such antagonism has been demonstrated in culture, and culture filtrates from a grass endophyte, *Acremonium coenophialum*, have been shown to have inhibitory effects on the growth of potential grass pathogens (White and Cole, 1985).

Endophytic strategies

Although investigations concerning endophytic associations are limited, the available evidence suggests that endophytes have evolved from plant pathogenic fungi. Many of the endophytes which have been described are very closely related to virulent pathogens and occur on the same, or related, host plants too. For example, on grasses *Acremonium coenophialum* is related to the pathogen *Epichloë typhina* and occurs on a similar range of host plants. *Rhabdocline parkeri* is a frequently isolated endophyte of conifers and has been shown to infect Douglas fir (Sherwood-Pike *et al.*, 1986; Stone, 1987, 1988). In autumn germinating conidia become attached to the needle surface by mucilage and produce penetration pegs by which the plant tissues are invaded. Hyphae form intracellularly within individual epidermal cells and remain there for a long time (2–5 years), not proliferating further until needle senescence. As the host tissues age, the fungus then invades surrounding tissues, forming structures somewhat akin to haustoria (Fig. 5.22). Sporulation occurs after needle abscission. *Rhabdocline parkeri* is related to the pathogens *R. wierii* and *R. pseudotsugae*, and in fact all these species may infect the same tree simultaneously. The distinctions between endophytic and pathogenic relationships are often not clear however. Some pathogenic species may show a long period of apparently endophytic growth in a host plant before symptom development occurs. Additionally, the growth requirements of endophytic species indicate the utilisation of a limited range of materials as substrates, a characteristic which is often associated with pathogenic species.

Two strategies of endophytic mutualism have been described (Carroll, 1986, 1988). In grasses fungal species have been identified which do not leave the host plants at reproduction and do not produce any external fruiting bodies. Spread of the fungus is achieved by vegetative growth of hyphae into the ovules of the host so that dispersed seed is already infected by these fungi (Fig. 5.18b). This has been termed constitutive mutualism (Carroll, 1986). Such endophytes infect aerial parts of the plant systemically, commonly developing a large biomass of fungal mycelium in host tissues. These species do not appear to harm the host plant but often produce toxins which may have important deterrant effects on grazing herbivores and may therefore provide protection to the plant. In some instances the presence of these endophytes within ovule tissues results in host plant sterilisation and therefore the cost of this association to the plant is high. However, endophyte-induced sterility may result in more vigorous vegetative growth. Grasses may produce more vegetative tillers enhancing competitive ability in some ecological situations. It is very likely that the metabolic and physiological drains on host plant reserves and energy supplies are also high, since a large fungal biomass is supported by the plant.

An alternative strategy, adopted by some endophytic species, is an inducible mutualism, in which a looser relationship is formed with the host plant (Carroll, 1986, 1988). In these associations the distribution of the endophyte in plant tissues is patchy and probably affected by ageing in the host. These endophytes normally inhabit senescent host tissues and only penetrate metabolically active regions when the plant is stressed. Herbivore wounding of tissues leads to endophyte invasion of active regions. The toxins produced by endophyte activity then give rise to destruction in the herbivore population. There may be little benefit to the host plant directly, although it has been suggested (Carroll, 1988) that the host population may benefit, albeit often in the long term, as a result of the reduction in herbivory. The metabolic cost to the plant is relatively small.

Biotechnology in the study of fungal–plant interactions 6

INTRODUCTION

Since the onset of plant cultivation, even in its simplest form, selection systems have been imposed on chosen species in addition to those exerted by the environment. Those individuals presenting specific traits, such as high yields, greatest palatibility, attractive morphology and general hardiness in particular prevailing conditions, have been nurtured and grown on in successive crops. In this way crop improvement has been practised over many centuries. As a result many genetically different lines and cultivars have gradually arisen. More recently the value of plant breeding has been recognised and more intensive methods, accelerating this process, have been employed, combining useful genetic characters and giving rise to varieties with greatly increased yields, improved quality, harvestability, storage potential and appearance.

A combination of factors have contributed to the increased production, quality, distribution and availability of present-day produce. Improved agricultural practices and crop husbandry, mechanisation and greater use of agrochemicals together with more awareness and understanding of plant genetics have increased the efficiency of world food production. Breeding for lines which are more tolerant of stressful environmental conditions and resistant to pathogen invasion has provided a valuable contribution. Conventional plant breeding techniques have therefore proved to be extremely useful and have been widely applied to great effect (Simmonds, 1979; Bingham, 1981; Dixon 1981).

However, other modern techniques are now available which provide means for the potential modification and improvement of plants. The manipulation of individual plant cells, groups of cells or particular tissues is proving extremely useful in many areas. Tissue culture systems allow the rapid generation of plant lines which would otherwise only be obtained after very lengthy breeding programmes, and which can be easily carried out on a large scale. Somatic cell hybridisation, through protoplast fusion, enables the transfer and exploitation of useful genetic characteristics between individuals and even between species. Such techniques may circumvent incompatibility barriers which often operate in intact plant material and which cannot be overcome by traditional methods. Novel variation can be introduced to cell lines through tissue culture (somaclonal variation), and such material can be rapidly bulked for commercial purposes.

Additionally, biotechnological and molecular biological approaches may now be applied. The genetic modification of organisms and the application of modern techniques to study cellular and molecular processes are now expanding our understanding of interactions between organisms and allowing a greater degree of control over biological processes. In theory at least, very specific changes may now be made to living cells and tissues using genetic engineering techniques. Specific genes may be transferred between plants of different cultivars, lines and species, from microbes to plants and from plants to microbes using vectors, such as plasmids, to integrate new characters in to the plant genome. Although much of this work is currently hindered at the commercial level by the lack of regeneration of plants from some transformed cell lines, such manipulations of the genome may, in time, allow a much greater control over plant functions.

The application of molecular techniques to plant and fungal tissues often requires a great deal more background understanding of the control mechanisms, which operate in such biological material, than is currently available. The development of new technologies is gradually breaking down barriers which previously existed between disciplines and extending our knowledge at the cellular and subcellular level.

PLANT BREEDING FOR RESISTANCE TO DISEASE

There are a range of ways in which the occurrence of fungal infection of plants can be minimised in a crop. Good hygiene and careful practice, such as removing dead and infected material from the site, the use of clean seed and crop rotation, may avoid disease at a site. The use of fungicides as a preventative or direct control measure, and some biological control systems, also prove to be economically viable. However, it is indisputable that an inherent level of resistance to pathogen invasion or establishment in the crop plant is the most useful and cost-effective means for controlling disease.

Plant breeding techniques can be applied to levels of resistance in the same ways as breeding for other characters and are largely based on the testing and selecting of plant lines. It is, however, a lengthy process and may take over 10 years to develop a marketable variety through a breeding programme. For the plant breeder, resistance to disease is usually one of a number of characters which is required in a new plant variety. High yield, growth rate, vigour, quality and so on, are obviously important and so resistance must be added to the objectives of a breeding programme. However, susceptibility to a disease, which could not be easily or adequately controlled by other means, would always be unacceptable to the grower even if the new cultivar were very high yielding.

Sources of genes for resistance can often be identified in native plants, wild varieties or in related plants. In areas where the incidence of a disease is severe the selection pressure on plants is high and strains carrying greater resistance are likely to be present. Such resistant plants can then be used in breeding programme crosses to introduce resistance in to cultivated varieties. It is also important to consider the genetic control of resistance (oligogenic, polygenic, etc.) in the plant and the degree of variability which is likely to occur in the pathogen.

The methods used for any particular programme will depend on the breeding system of the plant concerned (inbreeding or outbreeding) except in instances where plants for cultivation are normally propagated vegetatively and successful pollination is therefore not required. Many cross-pollinated plants are self-incompatible and cannot easily be used as the basis for a breeding programme. This applies to quite a wide range of crop plants, such as rye, maize, runner beans (*Phaseolus vulgaris*), apples, plums, dates, ryegrass, tall fescue, potato, sunflower, sugar-beet, members of the Brassicae, carrot, celery and onion for example. In instances where cross-pollinated species can be selfed vigour is usually lost as a result of such inbreeding. Self-pollinating plants which exhibit resistance can be used in breeding programmes and resistant varieties may be derived from an individual plant. Inbreeding does not reduce the vigour or yield of such species. Some important crop plants are self-pollinating, e.g. wheat, oats, rice, soybean, peas, cotton, tobacco, lettuce and tomato.

Breeding methods require that levels of disease resistance are screened to allow comparison between the levels of resistance achieved. Assessments must be carefully carried out and can often be labour intensive and time consuming. A viable, standardised inoculum is required and trials must be conducted under controlled, uniform conditions in order to avoid interactions with the environment which might interfere with the interpretation of the results. It is therefore necessary to understand the mechanism of host invasion and disease establishment to reliably test for resistance. Screens are designed to distinguish between the incidence of disease (the percentage plants infected), the severity of the disease (the proportion of the total plant affected) and may also allow the identification of differences between lesion size and type, which may reflect the reaction of the plant. The main methods used in the breeding for disease resistance are considered below.

CONVENTIONAL METHODS USED IN BREEDING FOR RESISTANCE

Mass selection

Mass selection is the simplest method used for breeding for resistance. Seed from the most highly resistant plants is used to grow future generations. Plants with high levels of resistance, and other characters where these are important, are selected from a self-pollinating population and the progeny are then bulked together for a further growth cycle and subsequent selection. The individuals are heterozygous and therefore there is always likely to be some degree of variation between them. This method is straightforward to carry out but the level of resistance increases only slowly through each successive generation. In cross-pollinated plants there is no control over the pollen source so that recurrent selection for resistant and desirable plants operates slowly over successive generations.

Pure line selection

Pure lines (or pedigrees) are selected from the progeny of highly resistant homozygous plants. These are tested for resistance and other characters and those individuals with desirable features are then multiplied. In this way the genetic properties of new lines can be well defined. For self-pollinating species off-types which carry useful characters arise occasionally, probably as a result of mutation or hybridisation within the crop. Such plants can also be selected and multiplied, to great effect. This is a widely used method of breeding for resistance.

Pure line selection is more difficult with cross-breeding plants. Selected individuals must be artificially selfed or interpollinated, tested for resistance and the most satisfactory plants used for further breeding. Interpollination between such individuals gives rise to composite crosses.

Back-crossing

Desirable characters from parental varieties can be combined by crossing plants, enabling useful traits to be transferred to the progeny. Those resultant individuals are tested for resistance and then back-crossed to a standard cultivar. Eventually, over successive generations of back-crosses the level of resistance becomes stabilised and is expressed within the useful cultivar. This system is more easily controlled with cross-pollinated crops and has the advantage that several forms of resistance may be transferred simultaneously. Back-crossing is used to introduce resistance into a susceptible cultivar but it is often a lengthy process taking many years. If the resistance to be bred in is a dominant character then back-crossing will retain all traits of the parent plus that resistance. If the resistance is a recessive character then the first generation of back-cross progeny (F_1) must be selfed to enable the selection of resistant plants. This method is not used in cases where resistance is a multigene character.

F_1 hybrids

Inbred lines with high levels of resistance can be produced from cross-pollinating species by selfing or sib-mating (brother × sister crosses) which have been selected for resistance. First generation (F_1) hybrids of homozygous lines carrying different genes for resistance are then used. Two separate lines are selfed (inbred parent lines) and are then crossed. The F_1 hybrid progeny are heterozygous for most alleles, which may enhance the level of disease resistance. A new F_1 hybrid must be produced each year for new crop planting since the seed from any progeny cannot be guaranteed to retain the resistance. F_1 hybrids of Brassicae, carrots, onions and tomatoes are used agriculturally. However, incompatibility barriers, which operate in some species, may require the use of specialised techniques such as bud pollination.

Mutations

Mutations in the chromosome complement of plants do occur naturally and can give rise to useful variants. However, the process is random and very slow (1 in 10^4 to 10^6 alleles per locus). It is possible to induce mutations in selected material by the use of chemical agents, e.g. cochicine, or ultraviolet radiation. A range of chromosome deletions and additions may occur. Colchicine treatment disrupts the movement of chromosomes at cell division and results in changes in the chromosome number of progeny. Diploids ($2n$) and triploids ($3n$) contain two or three complete sets of chromosomes, respectively. Plants with three or more sets of chromosomes are termed polyploids. Autopolyploids contain multiples of haploid or diploid chromosome numbers. Aneuploid plants contain multiples of the normal chromosome number with either extra or fewer chromosomes, as in trisomics ($2n + 1$) and tetrasomics ($3n + 2$) which contain the diploid chromosome number plus one or two chromosomes respectively and monosomics ($2n - 1$) which have the diploid number less one chromosome from a pair. Many different changes are possible and although large changes are likely to be lethal alterations in chromosome number may enhance levels of resistance.

Mutations may also be induced within genes. Exposure to X-rays, ultraviolet light and chemicals such as diethyl sulphate, sodium azide and nitroso-compounds may be applied. Such treatments can introduce quite marked changes in individuals and mutagenised plants have been incorporated into breeding programmes to great effect, particularly with ornamental plants.

Vegetative propagation

The use of cuttings, bulbs and grafts is common in the production of crop plants, especially for the propagation of soft-fruits, herbaceous species and some grasses. Plants generated in this way are heterozygous and the same genotype is maintained through the newly propagated plants. Any F_1 plants derived from such stock must be tested for resistance independently and such plants are not normally used for cultivation.

THE USE OF RESISTANT CULTIVARS

Resistant varieties are very useful for disease control but must be carefully planted in order to avoid problems which arise from the so-called durability of that resistance. It can be seen that there is no need for a detailed understanding of the genetics or the mechanisms of disease resistance in a host, in order to breed resistant lines and to develop resistant cultivars for agricultural use. The application of positive selection pressure to develop resistance is the main requirement. However, dangers do arise from the planting patterns used with any cultivars which are developed.

The durability of resistance in a particular variety is modified by interactions between the host and the pathogen. It is important to understand that the breakdown of resistance in a crop is not as the result of changes in the host or in host resistance but is a reflection of adaptations in the pathogen (Chapter 3). The presence of resistant plants increases the selection pressure on the potential fungal pathogens in that area and eventually new strains, or races, appear (by mutation) which are able to overcome the host resistance presented. The speed at which this occurs depends on the level of variability, and mutation rate in the pathogen but may occur relatively rapidly for some species (particularly for those biotrophs which produce many generations in a year) and in some conditions.

Vertical resistance (oligogenic or race-specific) is conferred by a few major genes and gives rise to complete resistance (immunity) to specific pathogen races. Such resistance is relatively easy to manipulate in a breeding programme and can be useful in some instances. For example, vertical resistance can be very effective in annual crops which have a relatively short life-span. It is also effective when directed towards races of the pathogen which do not spread very easily or which are relatively genetically stable, i.e. in which there is little genetic change. Additionally, lines carrying vertical resistance are useful if they are planted in controlled, confined areas. However, one such variety planted over a large acreage may unduly raise the selection pressure on the pathogen population at that site. If a new pathogen race develops which can overcome those particular resistance characters expressed then the crop will succumb to the pathogen and infections of severe epidemic proportions may occur.

In general terms immunity to disease is not normally required. Horizontal resistance (polygenic, non-specific) provides plants with incomplete protection but such resistance is more durable in agricultural situations. This non-specific resistance operates against all races of the pathogen and will be maintained where the pathogen is present since there is always positive selection pressure on the potential host plants to retain it.

The widespread planting of a genetically uniform cultivar has advantages in terms of the uniformity of ripening, harvesting and quality of a crop and may be required for commercial purposes. However, the generation of a new race of pathogen will eventually result in the infection of resistant plants within such a stand. Infected individuals then act as an inoculum source and in a genetically uniform crop the pathogen may subsequently spread in epidemic proportions. The same effect may occur with the influx of an imported or air-borne pathogen. More genetically heterogeneous crops are less likely to suffer in this way. Although some plants may be immediately susceptible it is not likely that all will become infected. A lower inoculum potential is therefore generated. Interspersed plants which are more resistant also act as barriers to the rapid spread of the disease.

It can be seen therefore that the planting of mixtures of cultivars, which are similar in their characteristics but which are derived from a number of lines containing many resistance genes, is sound agricultural practice. Commonly used varieties must usually be replaced after 3–5 years after widespread planting, to guard against the development of epidemics. Multilines are also used. These are isogenic lines, derived by back-crossing individuals to a common parent and differing in the vertical resistance which is carried. Such multilines have been developed very successfully for the control of rust diseases.

THE STUDY OF DISEASE RESISTANCE USING TISSUE CULTURE TECHNIQUES

Tissue culture techniques exploit the capacity of every individual plant cell to regenerate replicas of the plant from which it was derived. Meristematic cells, parenchyma and cambial cells in particular, retain a high degree of developmental plasticity and are able to form phenotypic copies of the original plant under appropriate cultural conditions (totipotency). Only highly differentiated and specialised cells (e.g. vascular elements or highly suberised cells) do not retain the capacity to alter their pattern of development (de-differentiation).

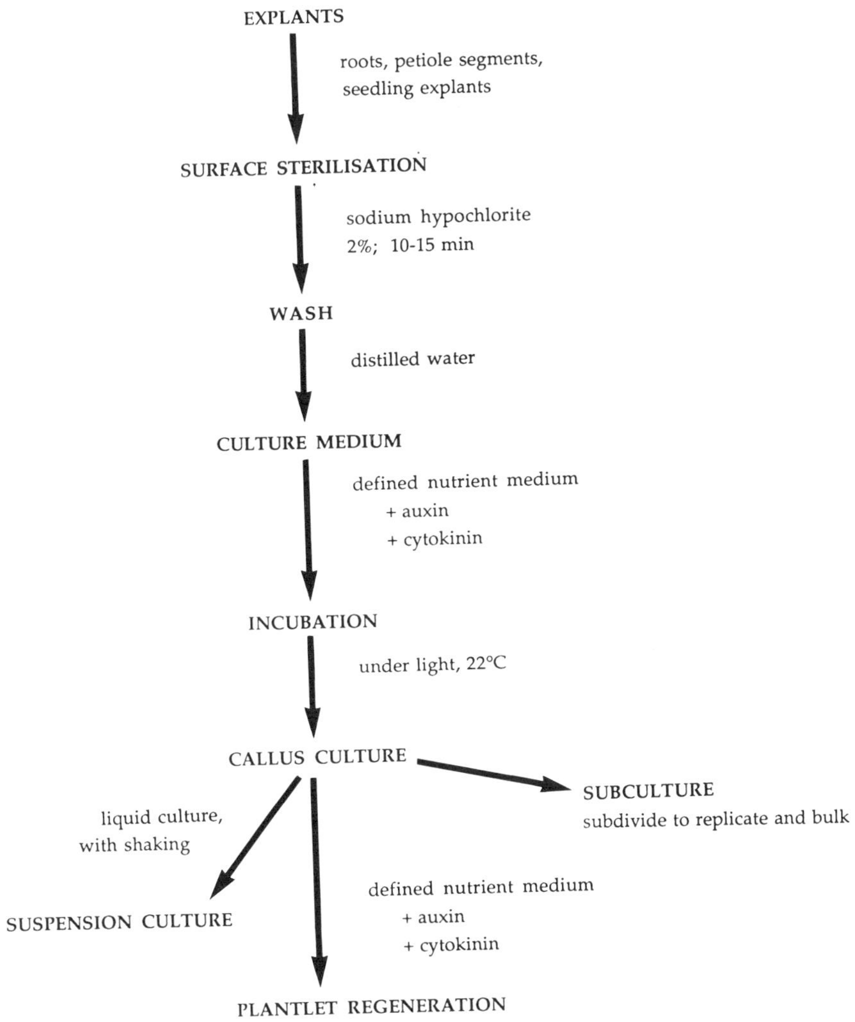

Fig. 6.1 Generalised outline protocol for the establishment of plant tissue cultures. The generation of callus tissue is induced on culture medium by the presence of growth regulator compounds. Callus material may be used to initiate cell suspension cultures. Manipulation of the ratios between the growth regulators allows regeneration of plantlets from callus cultures.

The manipulation of the cultural and environmental conditions under which such cells are grown, either as tissue masses or as individual cells, provides material for the study of cellular processes, interactions with other cells and the biochemistry of metabolism. A range of cultural techniques, often in conjunction with genetic changes and manipulations, has enabled the introduction of novel variation to crop plant species and the development of new cultivars.

PLANT CULTURE TECHNIQUES

Callus cultures

The wounding of an intact, growing plant frequently results in the initiation of cell division and the formation of an undifferentiated mass of cells, or callus, at the wound site as part of the protective mechanisms of the plant. Antimicrobial compounds often become localised into such tissue also contributing to that defense system. The growth of callus tissue is normally limited *in vivo*.

Callus tissue may be grown over a much longer term *in vitro*, providing that appropriate culture conditions are maintained. Plant tissue explants placed on suitable solid or semi-solid culture medium may be induced to form callus tissue (Fig. 6.1) which may be grown for prolonged periods (Brown, 1990). The most appropriate plant tissues are petiole segments, leaf discs, buds, root tips and freshly germinated seedling tissues (Fig. 6.2). Suitable explants must be surface sterilised to eliminate any potentially contaminating microbes. The growth rate of isolated plant tissues is much slower than that of associated microorganisms and cultures may quickly become overrun unless strictly aseptic conditions are maintained. After washing explants can be placed on to culture medium.

The most commonly used plant tissue culture medium is that devised by Murashige and Skoog (1962) and known as MS medium. This contains basic inorganic macronutrients (N, P, K, Ca, Mg and Fe), inorganic micronutrients (Zn, Cu, Mn, Na, Cl, Mo, Co, Al and B), a carbon source (usually sucrose), organic nitrogen (glycine) and vitamins (e.g. nicotinic acid and pyridoxine, thiamine) and is usually used as either a liquid medium or alternatively as a solid or semi-solid medium in an agar base. This is a rich medium in comparison with many of the substrates used to isolate and maintain fungal cultures. In addition to these constituents, growth regulator compounds must be added to encourage proliferation of the cells. A high auxin to moderate cytokinin level will maintain growth of undifferentiated callus cells on such medium and sterile vials are normally incubated at 20–22°C under continuous light. Callus tissue may develop from the ends of explants in contact with the medium or, in very active tissue, may generate throughout the whole explant. Fresh medium

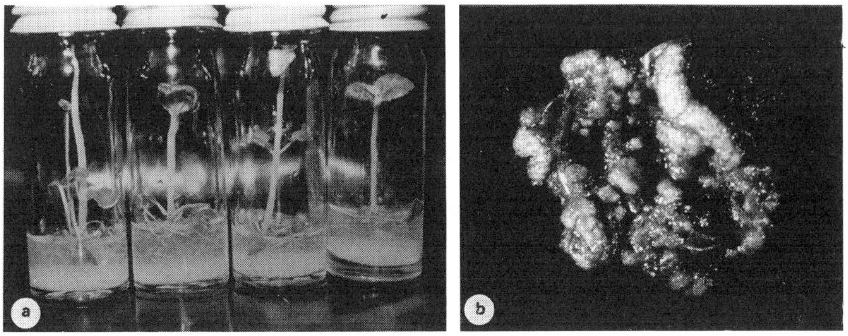

Fig. 6.2 The initiation of plant tissue culture systems. (a) Seedlings of *Coleus blumei* germinated aseptically on MS agar medium (20 mg.l^{-1} sucrose) for use as an explant source for tissue culture generation. (b) Leaf disc explant from *Coleus blumei* generating callus tissue in culture, on B5 salts medium supplemented with auxin (naphthaleneacetic acid, 1 mg.l^{-1}) and cytokinin (benzylaminopurine, 2 mg.l^{-1}). Reproduced courtesy of Dr K. Ibrahim and Dr H. A. Collin.

is required at intervals for continued growth and blocks of tissue must be subcultured. Callus material may be bulked by subdivision at this stage.

Cell suspension cultures

The transfer of rapidly growing and dividing, undifferentiated callus tissue (friable material is particularly suitable) into a liquid culture system, incubated with shaking, results in the break up of the mass. Individual cells or small aggregates of cells continue to grow and divide as a suspension of single cells. The growth of such cultures resembles that of bacteria or yeast liquid cultures and can be manipulated in very similar ways. Plant cell suspension cultures must be subcultured towards the end of the logarithmic phase of growth to prevent staling in the medium. The most finely divided cells, with few aggregates, are normally preferred for subculture to maintain as homogeneous a suspension as possible.

The addition of glutamine and serine to the cell suspension medium is important to encourage good, active growth. Theoretically individual cells could be cultured but for most higher plants cell cultures show a density dependence and in most cases cultures do not grow unless more than 10^4. cm^{-3} cells are present. It has been suggested that cultured cells 'condition' the medium. The use of such conditioned medium in culturing programmes results in rapid growth of cells in that culture and the shortening of the lag phase. Conditioned medium is prepared as a filtrate from cell suspension cultures, most usually from the same species, previously grown for a few days. Materials which have leaked from these cells then act to encourage the rapid establishment, development and proliferation of the newly inoculated cell suspensions. However, individual tobacco mesophyll cells have now been regenerated into plants with about a 50% success rate (Koop and Schweiger, 1985).

Cells from such suspension cultures can also be transferred to solidified medium for culturing. In the presence of a suitable hormone balance callus cultures will be derived from such cells and plantlets can also be regenerated.

Embryogenesis and the regeneration of plants

Callus tissue can often be induced to redifferentiate, producing roots, shoots or embryos in culture and subsequently whole plants can be regenerated. The inclusion of the plant growth hormones, auxin and cytokinin, is required for the establishment of callus cultures. Manipulation of the ratios between the growth regulator compounds can promote root or shoot development depending on the balance between them.

High auxin (indole-3-acetic acid, IAA) to cytokinin (kinetin) levels favours the formation of roots from callus and high cytokinin to auxin favours shoot formation from callus (Fig. 6.3). In some cases it is possible to encourage root growth on callus tissue which has shoots ready formed (or vice versa) and hence plantlets can be grown (Fig. 6.4). Alternatively however, it is possible to induce the formation of embryos directly on the

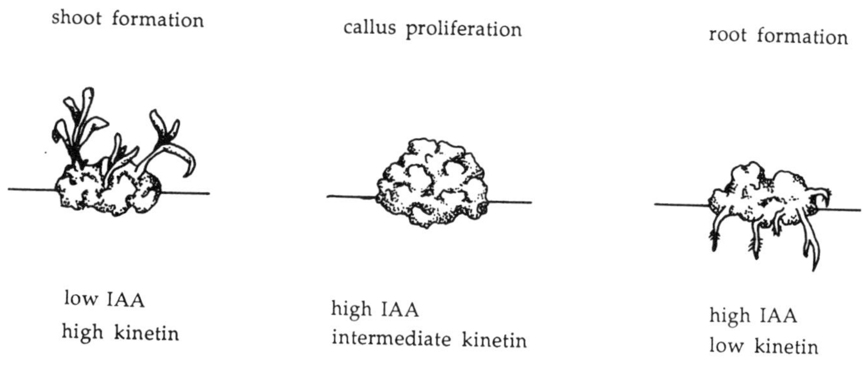

Fig. 6.3 The influence of growth regulators on the morphogenesis of tissue explants from dicotyledonous species.

callus, from which plants can be generated more easily. It is important to remember that although the main requirements for tissue differentiation are the supply and balance between growth regulator compounds, each genotype reacts differently to cultural regimes. In general therefore, a definitive protocol must be developed for each species to be cultured.

Somatic embryogenesis can also be brought about by manipulation of the culture conditions. Growth of callus tissue in the presence of 2, 4-dichlorophenoxyacetic acid (2, 4-D), as the auxin source in the medium, followed by transfer to a medium free from 2, 4-D but supplemented with glutamine as a nitrogen source, results in the formation of many embryoids in the culture. Similarly, embryoids can be induced to develop in cell suspension cultures. The embryoids form through a developmental sequence (Choi and Sung, 1989; Monnier, 1990a, b) which parallels that in normal zygote development, following natural fertilisation. Small globular embryoids are first to develop followed by further differentiation to heart-shaped and torpedo forms (Fig. 6.5). The bipolar torpedo forms elongate and give rise to small plantlets which may be grown up on culture medium, transferred to soil and eventually potted on as normal plants.

Some vascular plants, particularly dicotyledonous species, can be regenerated easily by manipulation of the culture conditions and adjustment of the growth hormone levels in the medium. Other species, especially monocotyledonous cereals and woody perennials, are extremely resistant to regeneration and are more difficult to manipulate in culture. Additionally, many of these species become progressively recalcitrant with successive subcultures. A great deal of research effort has been directed towards these problems, some of which ultimately limit the use of tissue culture systems for some species. However, new techniques and new approaches to solving such problems are being developed.

Pollen culture

Pollen development *in vivo* proceeds with the formation of a germ tube, each pollen grain posessing a single copy of the genome complement (haploid). The culture of pollen cells allows the use of this haploid material for the generation of haploid plants. This is a particularly useful technique for plant breeding and for mutation research.

Anthers may be placed directly on to solid or liquid nutrient medium, on to which pollen will be released after dehiscence. Haploid embryoids can then be generated, on medium containing growth regulator compounds. Alternatively, immature anthers may be ground and filtered to release microspores which can be cultured in a similar way. A major advantage is provided by the very large numbers of microspores which are available for culture from each anther. The nutrient and cofactor requirements

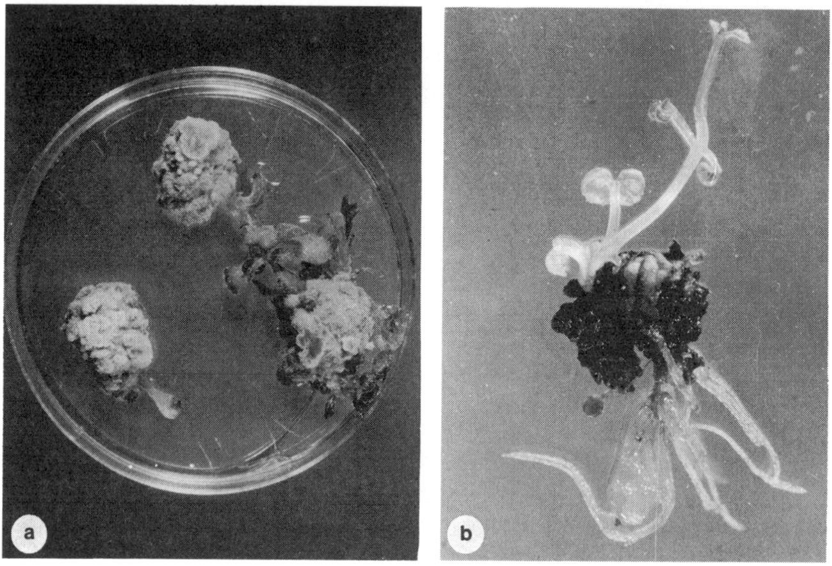

Fig. 6.4 Plantlet regeneration in tissue culture. (a) Shoot development on callus tissue (1 month) of *Nicotiana tabacum.* (b) Plantlet regenerated from callus culture of *Salivia splendens.* Roots and shoots have been formed on the same explant; 40 days after transfer to regeneration medium. Reproduced courtesy of Dr K. Ibrahim and Dr H. A. Collin

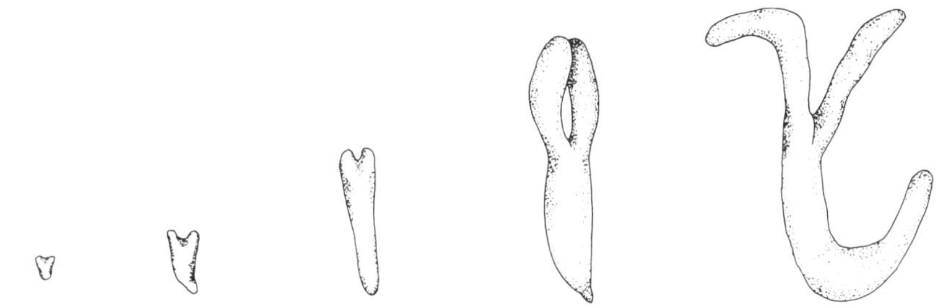

Fig. 6.5 The sequence of somatic embryo development from callus culture.

are very variable between species and conditions must be manipulated to ascertain the most appropriate technique in each case. Ovule tissues may be cultured in the same way although it is often difficult to obtain cells at the most useful stage of development in this case.

Protoplast technology and somatic hybridisation

Protoplasts are membrane-bound, cytoplasmic units, which can be released from plant cells as the result of wall polymer degradation by appropriate enzyme mixtures. Such structures retain their totipotency and, under appropriate conditions, have the capacity to regenerate a new cell wall. Since they are bounded only by a plasma membrane, protoplasts are osmotically fragile and can be used as a gentle and non-disruptive means for the preparation of cell-free fractions from plant tissues. Studies concerning the properties of plant membranes have been carried out using protoplasts. Protoplasts can be induced to fuse with protoplasts from other species, giving rise to hybrids (by the combination of two genomes), partial hybrids (by partial genome transfer) and cybrids (by transfer of organelles) which have very important roles in plant breeding. Additionally, protoplast membranes can be induced to take up small particles, including DNA and RNA fragments, organelles, microbial cells and viruses. This capacity has been exploited to investigate the effects of the inclusion of these sorts of materials on the intact protoplasts and any plants subsequently regenerated.

Protoplast isolation is achieved by the removal of the cell wall. This

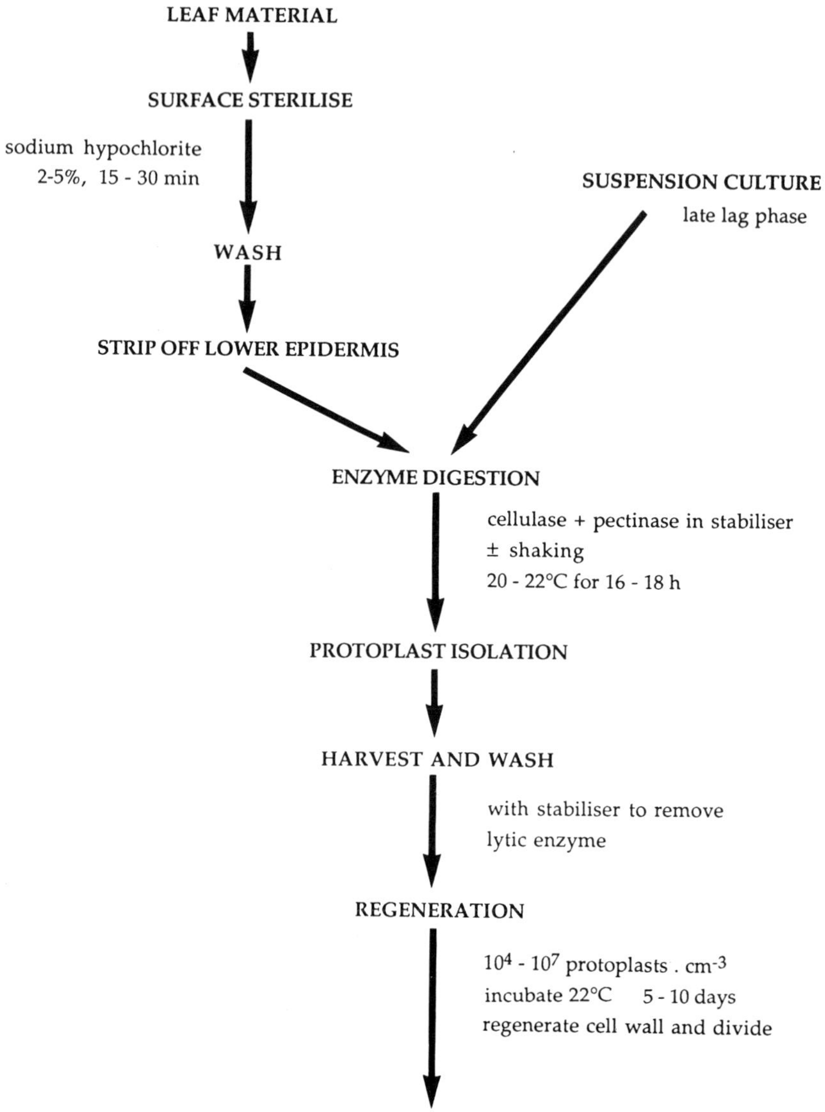

Fig. 6.6 Outline protocol for the isolation of plant protoplasts from intact leaf tissue or from cell suspension cultures. Protoplasts are released after lytic digestion of the cell wall, by appropriate enzyme mixtures, into osmotic stabiliser solution to maintain membrane integrity.

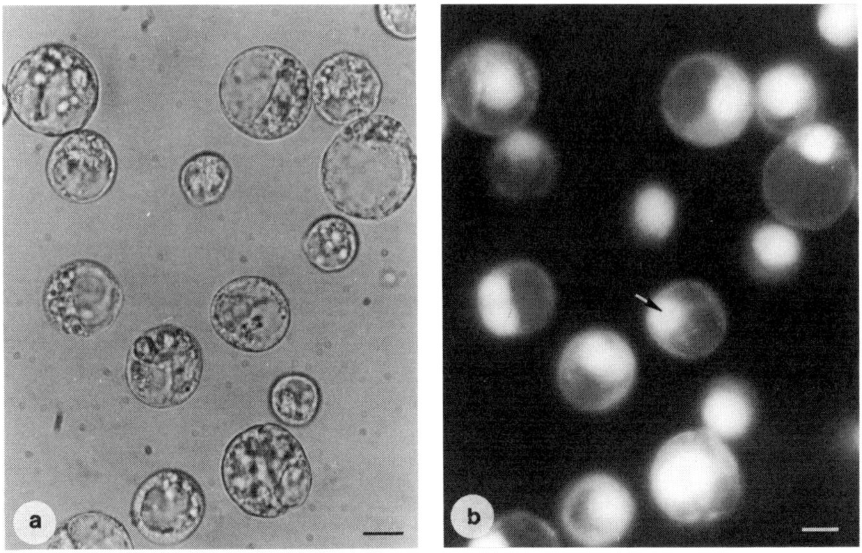

Fig. 6.7 Protoplasts isolated from cell suspension cultures (10 days old) of celery (*Apium graveolens*). Protoplasts were isolated by enzyme digestion (2% cellulase and 1% Macerozym R10) into 0.6 M mannitol stabiliser. (a) Light micrograph; (b) corresponding fluorescence light micrograph with the fluorochrome DAPI to show DNA. Nucleus arrowed. Bars represent 10 μm. Reproduced courtesy of Dr P. T. Lynch.

is normally carried out by enzyme dissolution of the wall polymers, usually incorporating a mixture of enzymes for maximum efficiency. Protoplasts can be released by the mechanical disruption of the cell wall (by grinding of appropriate tissue), although this is more disruptive and damaging to the tissues than enzyme treatments. Some protoplasts may be mechanically damaged and some may be adversely affected by compounds released from other damaged cells. This results in a lower yield of viable protoplasts and so physically gentler, enzyme-mediated, methods are normally adopted. Protoplasts can be released from a range of plant tissues although most usually, physiologically active sources are selected. Different plant species and different tissues may require different treatments for rapid and efficient wall removal and each must be investigated individually to ascertain the most suitable system. Much research effort has been directed towards protoplast techniques and liberation systems have now been defined for many species (Potrykus, 1988a, b; Davey and Power, 1988).

Tissue for protoplast liberation must be surface sterilised and chopped or shredded to enable enzyme penetration. Alternatively, the lower epidermis may be stripped from leaf tissue to improve enzyme access (Fig. 6.6). It is

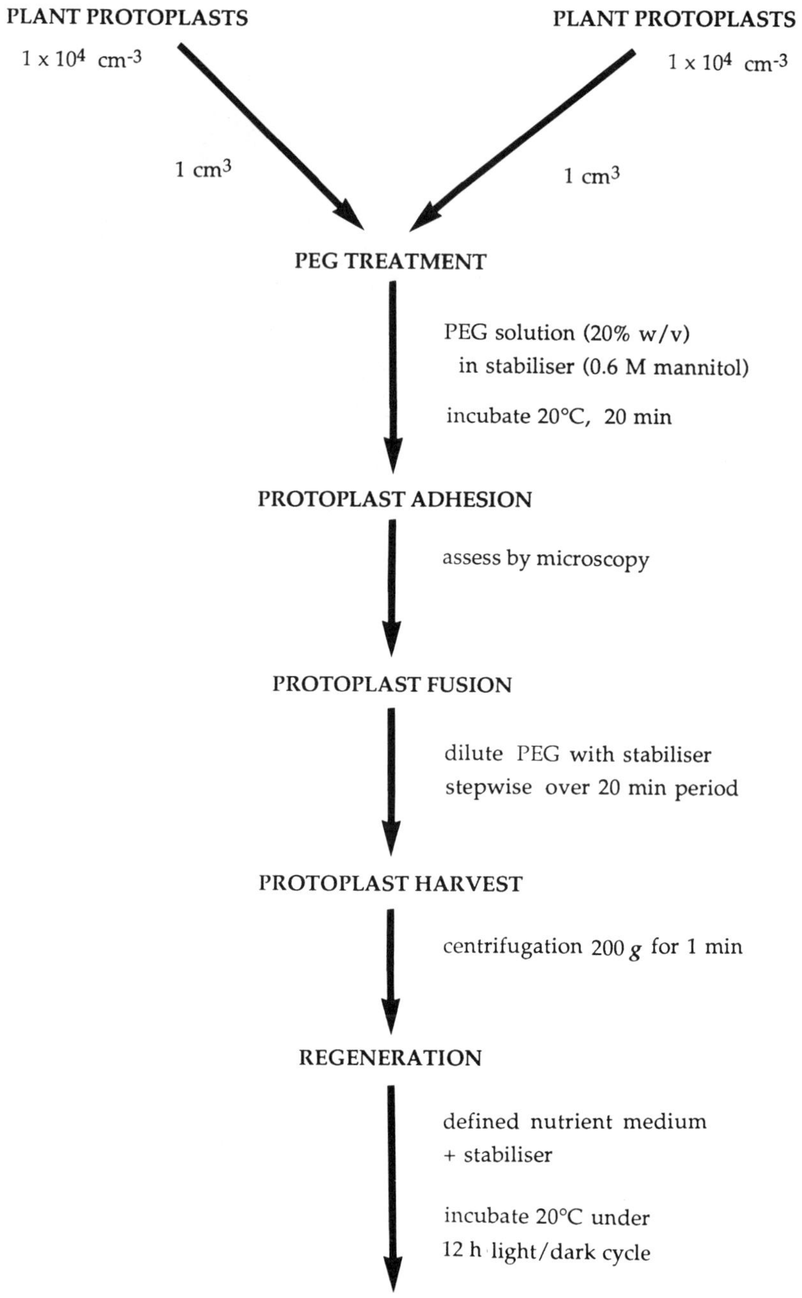

Fig. 6.8 Protocol for the chemically induced fusion of plant protoplasts, using polyethylene glycol solution.

usual to use young, actively growing tissue, to obtain the most efficient protoplast release. Enzyme mixtures normally make use of cellulase and pectinase activities to macerate wall polymers, in an overnight incubation at 20°C. Enzyme preparations such as Macerozyme R-10 and Pectolyase Y-23 are commercially available and give rise to relatively rapid rates of protoplast release. Many of the enzymes used for this purpose are of microbial origin (e.g. Driselase is a mixture of pectinase from a Basidiomycete fungus and cellulase from *Trichoderma viride*). Protoplasts are released into an osmotically stabilised solution to prevent lysis, and may be harvested efficiently by gentle centrifugation. Washing in fresh stabiliser solution removes lytic enzymes. Protoplasts may be separated from contaminating wall fragments by filtering or by layering on sucrose medium. Protoplasts can also be liberated very efficiently, using basically the same methods, from cell suspension cultures (Fig. 6.7). Liberation is achieved most easily if the suspension cells are in an active stage of growth (late lag phase / early exponential phase).

For subsequent regeneration protoplasts are placed on to agar medium, containing stabiliser solution at a density of 10^4–10^7. cm^{-3}. Provided with a suitable carbon source and essential nutrients, as used for tissue culture, protoplasts regenerate a new cell wall after 1–2 weeks. The viability of protoplasts can be tested using vital stains. The synthesis of a new cell wall may be followed by cell division and subsequent colony formation. As contacts between new cells become established differentiation may occur and eventually, under suitable conditions, protoplast-derived plants may be regenerated. However, although it is possible to liberate large numbers of viable protoplasts from some species these cannot always be regenerated and as a result this stage has been a major block to the exploitation of this method in those cases. Regeneration has been achieved for a range of economically important plant species, e.g. tobacco, potato, carrot and sunflower, and yet there is still a major block for other species, most notaby the cereals (barley and rye) which have not yet been regenerated into plants.

Once the cell wall has been removed protoplasts may be used for genetic manipulation and studies concerning somatic hybridisation by fusion of different parental protoplasts.

Chemically induced fusion

Isolated protoplasts are naturally sticky, tend to aggregate in suspension and have been shown to fuse spontaneously during incubation. Chemical treatments can be used to increase the frequency of fusion events. Fusion can also be induced to occur in the presence of high Ca^{2+} and high pH (9–10), but the most commonly used chemical fusogen (Lazar, 1983) is polyethylene glycol (PEG).

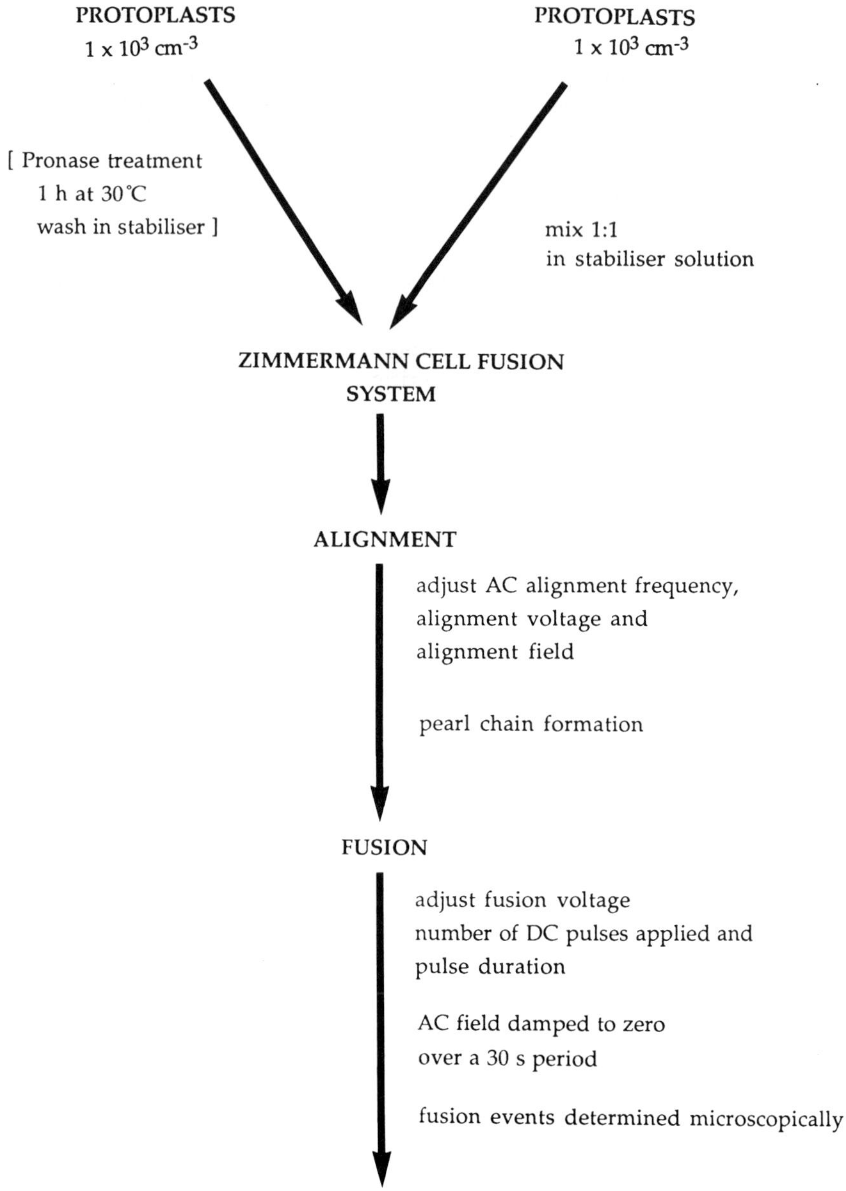

Fig. 6.9 Protocol for the electrofusion of plant protoplasts, using a Zimmermann cell fusion system. Protoplasts are aligned in an AC electrical field prior to the application of a number of DC pulses. Fusion between protoplasts ensues thereafter.

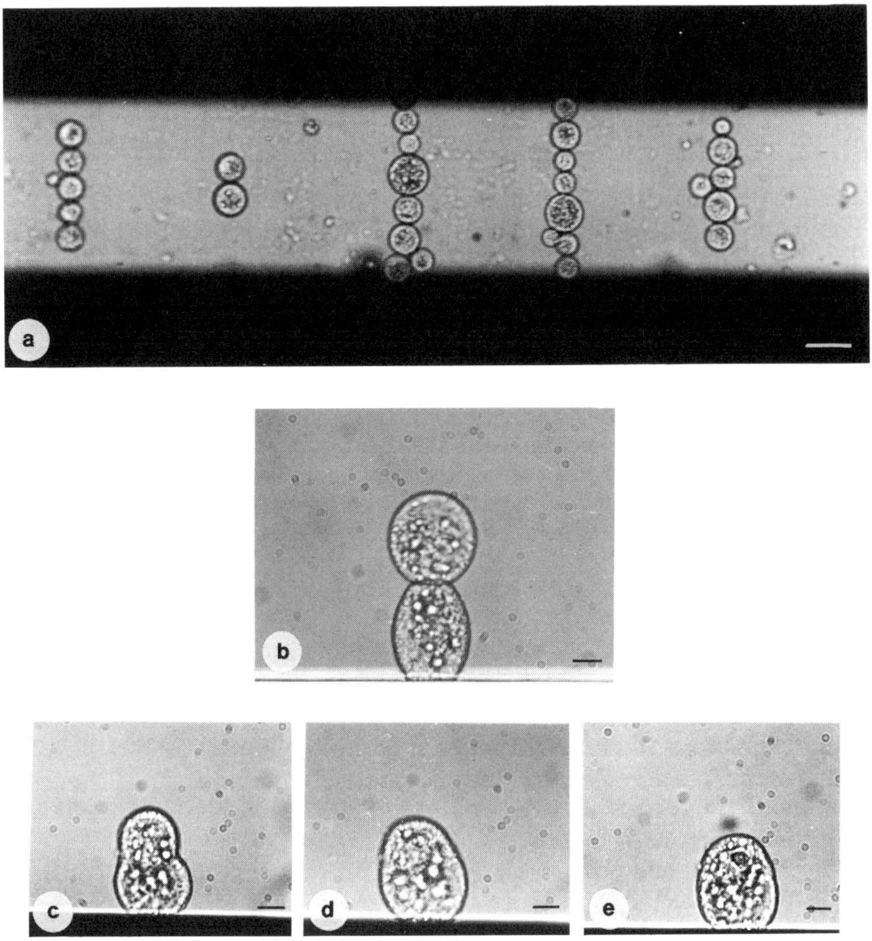

Fig. 6.10 Electrofusion of plant protoplasts (light micrographs) from celery (*Apium graveolens*). (a) Celery protoplasts aligned in pearl chains between the electrodes of a Zimmermann cell fusion apparatus (top and bottom), while exposed to an AC electrical field. (b) Pair of celery protoplasts aligned with close contact between the plasma membranes. (c, d, e) Sequence of micrographs to show fusion between a pair of aligned celery protoplasts in the immediate post-fusion period; (c) immediately after fusion pulses, alignment field present; (d) fused protoplasts rounding up into a single sphere 5 min after fusion pulse; (e) single sphere formed 10 min after fusion pulse. Bars represent (a) 50 μm; (b, c, d) 10 μm. Reproduced courtesy of Dr P. T. Lynch.

The stimulation of fusion by PEG is a two-stage process (Kao and Michayluk, 1974). Mixing protoplasts (in a 1:1 ratio) to be fused with PEG in solution (Fig. 6.8) causes enhanced aggregation and adhesion of protoplasts to their neighbours. It is the subsequent dilution of the stabilised PEG solution, either stepwise or at once, which results in fusion between closely appressed membranes and the mixing of the cytoplasmic contents. PEG is suggested to cause slight dehydration of the protoplasts, and crinkling of the membrane. The addition of Ca^{2+} has been reported to increase the incidence of endocytosis in protoplasts (Harding and Cocking, 1986).

One-to-one fusions between protoplasts isolated from some tissues may occur at relatively high frequency in some systems, but in general the level of fusion is usually a few percent (1–10%). However, large numbers of protoplasts can be treated and screened very easily and this may be adequate for many purposes. Additionally, it has been suggested that chemical fusion agents are toxic and therefore damaging to cells. This may limit the amount of regeneration which can occur from treated protoplasts.

Electrofusion

Higher fusion frequencies have been reported for electrically stimulated fusion events (over 80%), than for chemically induced fusion between plant protoplasts, without subsequent loss of viability or the capacity to regenerate. This is a relatively new technique (Zimmerman and Scheurich, 1981) which has recently been applied to plant protoplasts and which is now finding much favour for plant breeding work.

Electrofusion also occurs as a two-stage process. Protoplast suspensions to be fused are usually mixed in a 1:1 ratio in a suspension medium with low conductivity and placed into a chamber with two parallel electrodes (0.1 - 1.0 mm apart). A non-uniform, alternating current (AC) alignment field (Fig. 6.9) is then applied. The polarised protoplasts move towards regions of high field strength, adhering to the electrodes and becoming aligned between them in so-called 'pearl chains' (Fig. 6.10). This effect is called dielectrophoresis. The membrane interface between adhering protoplasts becomes very closely appressed. The application of a short, single pulse, or a number of pulses of direct current (DC) causes the breakdown of the aligned membranes. This results in fusion between the protoplasts and subsequent rounding up (Fig. 6.10) in the period immediately following pulse application (Zimmermann *et al.*, 1985) as the AC field is reapplied and gradually damped to zero.

Electrofusion combined with manipulation of the electrodes permits the parentage of heterokaryons to be very precisely defined (Koop and Schweiger, 1985; Morikawa *et al.*, 1988). Additionally, electrofusion can be determined and observed microscopically (Zachrisson and Bornman, 1986) so that very precise control can be maintained and electrofusion parameters accurately determined for each protoplast preparation. The disturbance of membranes during electrofusion is reported to be restricted to the zones of contact between them (Arnold and Zimmerman, 1984), which may help to maintain protoplast viability and is an advantage over chemical fusion treatments in which the whole protoplast membrane is affected.

Bulk fusions of protoplast preparations can be carried out in larger volumes (up to 0.5 cm^3) using a helical chamber or a flow cell. Alternatively it is possible to fuse defined pairs of cells which have been individually selected (Spangenberg *et al.*, 1990). Selected protoplasts are placed into a small droplet of fusion medium under mineral oil and adjustable platinum electrodes are manually positioned prior to the application of electrofusion currents. This highly manipulative technology allows the treatment of each cell pair as individuals and precise treatments may be made, tailored for each pair, as necessary. About 50 protoplast pairs can be treated in an hour and callus clones have been established, for *Brassica napus*, as a result. However, the frequency of regeneration achieved so far has been low (15–20 per 1000 fused pairs).

In some cases prior treatment of protoplasts with Pronase E is carried out. This encourages higher levels of fusion, also without other detrimental effects and is used for protoplast preparations that are difficult to fuse. This is a technique employed particularly if the protoplasts to be fused are of different sizes (Salhani *et al.*, 1985). The effectiveness of Pronase treatment probably lies in proteolytic removal of proteins from the membranes. In this way the lipid domains in the membrane are extended. Membrane proteins tend to prevent resealing after electrical treatment. Pronase treatment is often carried out in the presence of Ca^{2+} which binds to and enhances the stability of the membrane. Differences in fusability between protoplasts have been related to differences in size, membrane characteristics and cytoplasmic properties of protoplasts which may affect their electrical characterisitics. Electrofusion parameters must be assessed and optimum treatments determined for each protoplast type.

Mutation

Mutations are permanent changes in the genetic composition of the plant or cell genomes which are passed on to progeny. Protoplasts and cell suspension cultures provide ideal material for mutagenisation, since they are easily manipulated in culture. Mutations do occur naturally but levels may be raised by the use of chemicals or irradiation treatments although the frequency of recovery is usually very low, and mutants can easily be recovered from cell cultures.

BREEDING FOR RESISTANCE USING TISSUE CULTURE TECHNIQUES

Manipulations of plant tissue cultures and the use of protoplasts for somatic hybridisation have potentially important roles in agriculture and plant breeding. However, many of these methods are still being developed and are largely laboratory procedures. Although there are many problems associated with tissue culture methods it can be seen that there are also many advantages, not least the savings to be made in time and space, which are attractive to the plant breeder. Although there are blockages to the use of some tissues for regeneration of plants from some species, methodologies are now available for many plants. Techniques, such as electrofusion, are now being more widely adopted and will obviously be used more routinely in the near future, although more research is needed in the interim. However, these methods have already been commercially exploited in some areas.

Micropropagation

Tissue culture has been most widely exploited for the rapid mass clonal propagation of tissues. This technique is now used very successfully for large-scale commercial production by a number of companies around the world. Micropropagation is essentially a means for the vegetative propagation of plants, carried out *in vitro*, under aseptic conditions. Many different plant species have been cultured in this way, using bud or shoot tip cultures to propagate large numbers of individuals (Evans, 1989) with less variability than after regeneration from callus cultures. Explants are usually established, in aseptic conditions, on suitable culture medium under high illumination. After growth these are transferred to fresh medium to encourage multiplication of the tissue (Fig. 6.11). Shoots are subsequently placed in rooting medium or transferred to suitable, sterile compost for plant production. Potting on into fresh soil is the last stage in the procedure. Axillary buds, shoots or apical meristems are the most usual explants, or alternatively somatic embryogenesis on callus cultures is practised to provide large numbers of new individuals very efficiently. In suitable conditions it is possible to maintain proliferation over several years without loss of viability.

This technique has been exploited particularly for ornamental plants since production can be timed for specific markets (e.g. Christmas) and for woody perennials which are both more difficult and slow to propagate under normal conditions. Hybrid plants can often be propagated more easily via this route than conventional techniques. In addition, there are useful applications in the production of disease-free plants. It can be seen that plant pathogens are often transmitted via seed and plants are essentially never free from infection during the natural lifecycle. The technique of meristem excision and micropropagation has been most widely applied

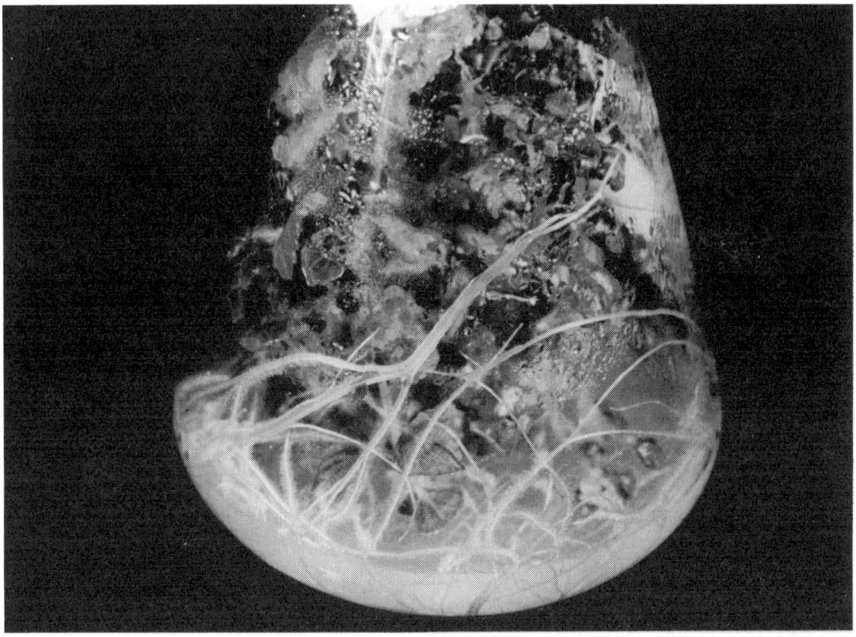

Fig. 6.11 The micropropagation of *Coleus blumei* plantlets from seedling shoot tip cultures, grown on MS salts medium. Reproduced courtesy of Dr K. Ibrahim and Dr H. A. Collin

to plants infected with viruses. Meristem tip cultures can be generated free from viral contamination and the benefits from this have been seen in a range of vegetable and flower crops, e.g. chrysanthemum, carnation and *Pelargonium*. The application of these techniques may be useful in the eradication of fungal species which live wholly as endophytes and which have detrimental effects on crop plant yields (Chapter 5).

Somaclonal variation

Plants regenerated from cultured tissues or cells may show variation in phenotypes in comparison with the original plant from which that material was derived. Such variation is inherited by progeny and often occurs at very high frequency. This has been called somaclonal variation (Larkin and Scowcroft, 1981) and has been exploited to good advantage in a range of crop plants, e.g. tobacco, rice, maize, potato, carrot, wheat, tomato and alfalfa. Plants derived from tissue culture may show variation in terms of plant height, yield of storage tissue, number of flower heads, flower shape, size of leaves, crinkling of leaves, alterations in pigmentation and many other morphological variations. Many such changes have little or no importance in commercial terms but some have greater significance. However, some variants have been shown to possess higher levels of resistance to disease (Larkin and Scowcroft, 1981). Increased uniformity over that of the parent material is also a useful character agriculturally and changes, such as alterations in the timing of ripening may be of useful significance to the farmer in terms of disease escape for some crop plants. In some cases (sugar-cane) it has been possible to regenerate plants which showed resistance to more than one fungal disease simultaneously, from cultures of a susceptible parent.

A range of genetic changes give rise to this somaclonal variation, chromosome rearrangements, crossing-over between chromatids, chromosome deletions and duplications. Polyploidy is common in cultured tissues. Long-term tissue cultures often show more drastic changes to the chromosome complement. Epigenetic changes also occur in such tissue although these are normally less useful because they are not inherited.

Gametoclonal variation

Haploid plants have been derived from a number of important crop plants, e.g. coffee, potato, tomato, wheat, rice, alfalfa and peanut, by conventional breeding or by pollen or anther culture. The frequency of production is often much greater in the cultural systems once these have been developed. Gametoclones are variant plants which arise from such anther cultures either directly or via a callus phase.

Haploids are very useful as homozygous lines and for genetic analysis. Doubled haploids (brought about spontaneously or by chemical treatment, e.g. colchicine) can be used to determine the qualitative and quantitative effects of gene dosage. Haploids are also used in mutagenesis work because recessive alleles are expressed and not masked, preventing detection, as may occur in diploids. Haploids are therefore useful in the detection of minor genes for resistance.

Somatic hybridisation

Plant protoplast fusion techniques have enabled the production of somatic hybrid plants within a range of genera, e.g. *Nicotiana, Solanum, Brassica, Petunia* and *Medicago*. Protoplast fusion allows crosses to be made which would not normally be possible by conventional methods. Intergeneric hybrids have also been produced (usually aneuploids), including *Arabidopsis/Brassica, Atropa/Nicotiana, Petunia/Necotiana,* although in most cases the resultant plants were seriously disadvantaged or infertile. The main block to the widespread use of protoplasts in crop breeding is the lack of regeneration in some species, although protoplast fusion has been achieved. The use of protoplast fusion to transfer resistance to some virus diseases and to insect pests has been reported.

A range of fusion product combinations will result from fusion treatments and can be selected by the use of nutritional markers, cultural differences or by manual selection under the microscope. The fusion of mesophyll protoplasts, which contain chloroplasts and are therefore green, together with colourless protoplasts derived from cell suspension cultures, has allowed the identification of fusion products under the microscope, e.g. *Arabidopsis thaliana / Brassica campestris*. It is also possible to 'mark' the parent protoplasts to enable the more rapid detection of fused individuals. Protoplasts have been marked by the use of specific non-toxic, fluorescent markers (e.g. fluorescein and rhodamine) to label one of the parental populations so that visible selection is possible. In a fluorescence-activated cell sorter (FACS) photocells detect the fluorescence and different fused products are sorted electrostatically. Heterokaryons carrying dual label can be detected and isolated.

Protoplasts are also useful, since they can be easily and gently disrupted, for the isolation of DNA and mRNA for gene cloning experiments and for the purification of enzymes and biochemical analyses. In addition they also provide cytoplasmic units suitable for the uptake of isolated DNA fragments, organelles (Wallin, 1984) and chromosomes (Griesbach *et al.*, 1982) which may carry genes for disease resistance, and also have an important role in genetic engineering techniques.

Screening and selection for resistance in tissue cultures

Screening systems are required to enable the selection of potentially resistant cells and tissues from cultures. The application of a screen relatively early in the culturing process can save a great deal of work. Once tissue types showing increased resistance have been selected it should no longer be necessary to maintain all other samples in addition. However, there is always the possibility that useful tissue will be undetected and discarded unless a very efficient and precise screening system is used. Additionally, the expression of disease resistance is often related to the age, physiological status and other factors in the plant, and might therefore be missed (Jones, 1990). There is great potential in the use of tissue culture techniques but it is not always easy to realise this in practical and commercial situations.

The use of pathogens in selection

The use of the pathogenic agent for selection of resistant tissue is a very direct approach to this problem and has been used in a number of instances for fungal pathogens. Good success has been achieved with viruses (Daub, 1986). However, the inoculation of callus cultures with viable fungal spores (Sacristan 1982, 1985) was not a successsful means for selecting resistant *Brassica* cultures, but apparently more success was achieved in the selection of pearl millet plants (Prasad *et al.*, 1984) with resistance to downy mildew (*Sclerospora graminicola*) from cultures of pre-infected tissues. Culture conditions can influence the growth of the fungal pathogen and the response of the cultured cells, altering the outcome of such screening so that these systems must be carefully defined and controlled.

An important consideration for such a scheme must be the full exposure of cells for screening, to the selection agent. This is not easy to achieve for fungal pathogens in either a callus system or a cell suspension culture. It must be remembered that in culture it is single cells or disorganised tissues that are available and therefore only screening for cellular resistance can be achieved. It is possible that some useful tissue-based resistance mechanisms cannot be expressed in this situation. For example, increased cuticle thickness or altered leaf structures which would prevent pathogen penetration in an intact plant. Additionally, lack of cellular resistance does not necessarily indicate the lack of whole plant resistance (Meredith, 1984; Daub, 1986). Comparisons between resistant and susceptible tissues in culture have confirmed that in some cases resistance is expressed in culture and in others it is not.

The inoculum tested is also important. The use of hyphal fragments gives no information concerning resistance or susceptibility to spore penetration. The application of large numbers of spores may swamp the plant tissue in the culture so rapidly that any resistance response might not be recognised. It is often difficult to manipulate dual cultures so that overgrowth of one or other partner does not occur before resistance/susceptibility can be ascertained.

The nutritional status of the plant tissues (either callus or cell suspensions) is also an important factor and may affect the response to pathogen challenge as a result of alterations in the biochemical pathways which operate. The sensitivity of cultured tissues to toxin production has been shown to vary with temperature. Additionally, the ability of a pathogen to produce toxins in plant tissues may also vary with both nutrient status and temperature (Jones, 1990).

The use of pathogen toxins in selection

The resistance of cultured cells to fungal toxins has been shown to correlate well with the incidence of disease resistance in regenerated plants. Cultured cells can be more evenly exposed to toxins when these are incorporated into the culture medium. Callus pieces placed on a solid surface (Fig. 6.12) or single cells in liquid culture may be used. These tissues have no cuticle and therefore penetration of toxins is not a problem. Protoplast

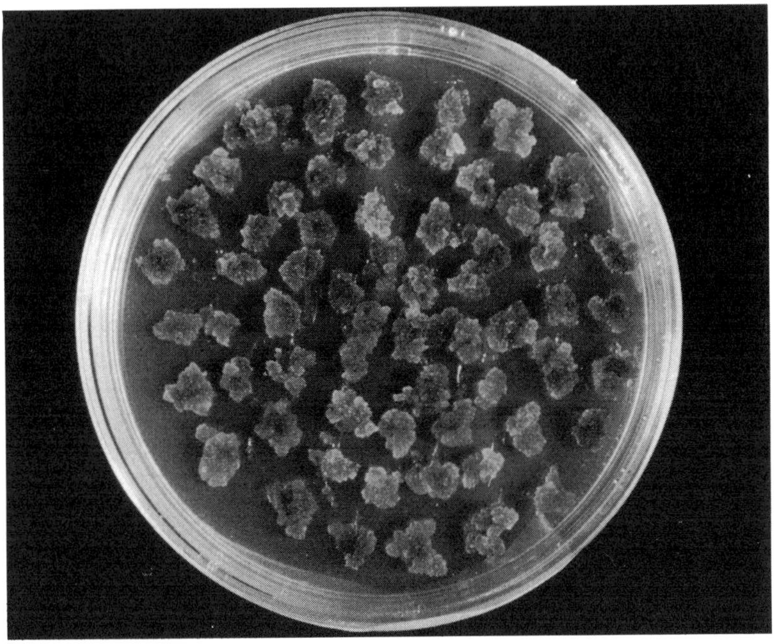

Fig. 6.12 The use of plant callus tissue for screening and selection of new variation. Small masses of *Coleus blumei* callus tissue (14 days old) are used to give good contact with the culture medium which contains selecting agents. Reproduced courtesy of Dr K. Ibrahim and Dr H. A. Collin

preparations can also be used although such a selection system would be of little value for plant species from which protoplast regeneration cannot be achieved. Toxin preparations are particularly useful as screens since the action on the host plant is usually at the cellular level. Additionally, where resistance has been found it has often, but not always, been passed on to progeny.

Some studies have been carried out using culture filtrates from the fungal pathogen for which resistance was sought, as selecting agents. For example, Hartman *et al* . (1984) found increased resistance in regenerated plants of alfalfa selected using culture filtrates from *Fusarium oxysporum* f. sp. *medicaginis*, which was transmitted to progeny. In other studies purified, or partially purified, toxins have been used as selection agents, e.g. sugar-cane cell cultures were selected using partially purified HS-toxin for resistance to *Helminthosporium sacchari*. The increased levels of resistance in selected regenerants were inherited by progeny (Larkin and Scowcroft, 1983). However, several toxic molecules, often chemically related, may function together or independently to produce symptoms in intact plants and this must be taken into consideration with regard to experimental design (Chapter 3). In some instances no resistance to the toxin preparations was found in the cell cultures tested.

The use of toxins in selection systems, and the use of culture filtrates in particular, must be approached with some caution however. Culture filtrates may contain other materials which are toxic to cultured plant cells but which do not have any part in disease. Additionally, the role of toxins in the establishment of disease has not been fully determined in many cases and indeed such compounds may act to aggrevate the symptoms rather than being the primary cause. Some toxic effects are directed towards whole, adult plant systems (e.g. effects on photosynthetic mechanisms) and resistance could not, therefore, be expressed in cell culture. The use of toxin preparations may not always be an appropriate strategy. It has been suggested that large amounts of tissue should be screened in order to generate sufficient potentially resistant lines (Meredith, 1984).

Selection for toxin resistance in cultured tissues is most usually carried out by subculture on medium containing sublethal levels of toxin. It is important to maintain tissue viability and regenerative capacity so that screening usually involves several passages through toxin-supplemented media. Hartman *et al.* (1984) have pointed out that while successive passages on selective media may improve the detection of resistant tissue there is also an increased risk of introducing chromosome abnormalities. A compromise selection strategy has been suggested (Jones, 1990), for which toxin-resistant variants are selected using medium containing sublethal levels of toxin, alternating with subcultures on toxin-free medium (Fig. 6.13). Although such systems have been shown to increase the frequency at which disease-resistant regenerants can be obtained, this is by no means universal. Oilseed rape variants with resistance to *Alternaria brassicicola* (MacDonald and Ingram, 1986) were recovered at similar frequencies from unselected and toxin-selected tissue cultures.

THE USE OF TISSUE CULTURES FOR THE INVESTIGATION OF CELLULAR INTERACTIONS

Tissue culture techniques have a great attraction for studies of fungal–-plant interactions at the cellular level, and in relatively undifferentiated tissues. While it is true that the response of cultured plant cells may not be entirely akin to that expressed in the whole plant, a great deal of information can be obtained from studies with such systems. A great attraction of tissue culture systems is the ease with which axenic material, of reduced complexity, may be generated in a fully and precisely controlled defined environment. Many of the variables that operate in experiments with greenhouse or field-grown plants can be eliminated. In addition, the cells in question are aseptic and unchallenged by any microbial flora and are undamaged by any biotic agents. Tissue cultures may be grown

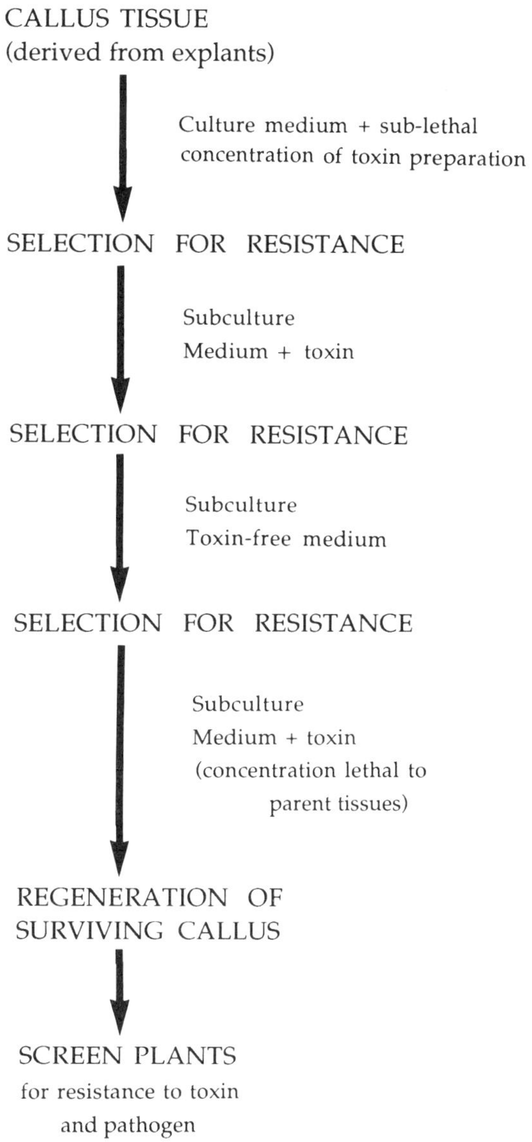

Fig. 6.13 Generalised protocol for the screening of callus tissue for the isolation of toxin-resistant plants (simplified from Jones, 1990).

on a sufficiently large scale to enable biochemical measurements to be made, and for the detection of *de novo* synthesis.

Elicitor studies

Such uniform material is therefore, potentially useful for investigations concerning the mechanisms and the biochemical reactions by which resistance responses are elicited in plants (Helgeson, 1983; Kessmann and Barz, 1987). Plants show complex responses to fungal pathogens at tissue, cell and cytoplasm levels (Ralton *et al.*, 1985). Cellular responses include changes to cell wall structure, mechanical strength and resistance to enzyme degradation. Cytoplasmic responses include phytoalexin synthesis and accumulation, and changes in membrane function. Tissue cultures provide materials suitable for investigations into these processes.

The presence of biotic elicitors (Ward, 1986) such as fungal cell wall compounds, components of culture filtrates, proteins and glycoproteins, acts as a signal (Ebel, 1986) for the induction of phytoalexins in plant cells. Biosynthetic enzymes are then formed *de novo* in plant cells, as a response. The application of such elicitor compounds to plant cell cultures, which can be achieved homogeneously in such systems, has been shown to cause changes in many aspects of cellular metabolism (Barz *et al.*, 1988). Changes in levels of enzymes involved in wall polymer synthesis, polyphenols, lignin synthesis and phytoalexin synthesis have also been shown (Chapter 3). Additionally, other cellular changes occur such as respiration, peroxidase activity and protein metabolism, together with alterations in membrane integrity.

Investigations concerning the interactions between chickpea (*Cicer arietinum*), and the fungus *Ascochyta rabiei* (Deuteromycotina) which causes Ascochyta blight have allowed some of the associated biochemical events which occur to be pieced together (Barz *et al.*, 1988). It is clear that the interactions are complex and although much information has been accrued the situation is still not fully understood even in this well-examined host–pathogen combination. Cell cultures derived from plants with differential capacity for phytoalexin production retained this ability in some cases, although this was not so for all tissues tested. Consequently it is concluded that some cell cultures are useful for such studies (Helgeson, 1983; Barz *et al.*, 1988; Halbrock and Scheel, 1989; Jones, 1990). Increasing our understanding of the mechanisms and control systems for phytoalexin synthesis is important since such defense systems have implications in the field of plant protection.

Dual culture systems

The direct inoculation of callus cultures with fungal spores or mycelium for long-term culture is known as dual culture. Dual cultures of fungal and plant tissue were orginally set up to investigate the relationships between obligate biotrophs that could not be grown in axenic culture and their hosts. In addition to screening for resistance such cultures provide opportunity to carry out biochemical analyses during the course of host–pathogen interactions in controlled conditions, particularly during the very early stages of infection.

Problems in studying obligate biotrophs often stem from the lack of a clean inoculum, since field-grown, diseased material usually has other associated surface contaminants. Dual culture systems which might allow the preparation of uncontaminated inoculum have an attraction for precise biochemical studies. It has been suggested (Miller, 1986) that for a successful dual culture the growth of the fungal pathogen and the host tissue should be balanced. However, this is not easy to achieve in practical terms and one species often becomes dominant in the culture, sometimes very quickly. Additionally, the lack of some structural and chemical features of intact plant material may contribute to the non-establishment of host-pathogen interactions in these cultures.

A balanced system has been achieved for dual cultures of sugar-beet (*Beta vulgaris*) and the downy mildew *Peronospora farinosa* f. sp. *betae* (Ingram, 1980) but dual cultures of sugar-cane and the downy mildew *Sclerospora sacchari* were only maintained by continued subculture, however, since callus tissue outgrew fungal mycelium.

Dual cultures of necrotrophic fungi and their host plant tissue have been used to screen for resistance in tissue cultures, to test the relative pathogenicity of fungal isolates and to assess the toxic nature of filtrates from potential pathogens. Fungal isolates are able to penetrate between cells in callus tissue very efficiently and uniform exposure to the pathogen is achieved (Fig. 6.14). Such cultures are also used to investigate cell surface phenomena and the interactions between surface molecules. Techniques using isolated protoplast suspensions inoculated with fungal mycelium are now being used to elucidate molecular interactions at membranes.

The expression of resistance in callus tissues may be affected a great deal by the cultural conditions, nutrient availability and physiological status of the tissues. If resistance is correlated with phytoalexin synthesis or other biochemical plant defense mechanisms then it may not be expressed in callus as a reflection of changes in phenylpropanoid metabolism and other pathways.

Investigations into the occurrence of the hypersensitive response have been made using callus cultures. Surface inoculations of tobacco callus with *Phytophthora parasitica* var. *nicotiana* spores resulted in the collapse of cells callus lines derived from resistant plants, but no apparent cellular changes occurred in lines derived from susceptible plants 24 h after inoculation. However, tobacco callus inoculated with *Phytophthora*

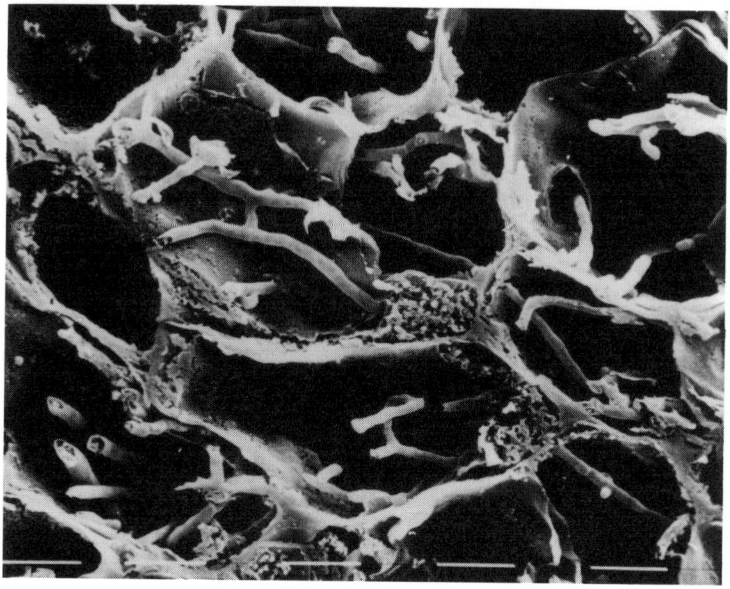

Fig. 6.14 Scanning electron micrograph to show growth of secondary stage mycelium (dikaryotic, necrotrophic) of *Crinipellis perniciosa*, causal agent of Witches' broom disease on cocoa callus tissue. Inter- and intracellular hyphal penetration can be seen. Bar represents 10 μm. Reproduced courtesy of Dr R. B. Muse.

infestans or *Phytophthora megasperma* showed a hypersensitive reaction regardless of the susceptibility of the parent tissue (Helgeson, 1983). Clearly host–pathogen responses in such systems are variable.

Interactions between isolated protoplasts

Isolated plant protoplasts are also easily manipulated and maintained in a homogeneous suspension and protoplast mixtures can be used to examine the nature of intercellular interactions. Plant protoplasts have been shown to take up a wide range of small particles, e.g. latex particles, isolated DNA, viruses, organelles (Lorz, 1985) and intact microorganisms (Cocking, 1984). Uptake probably occurs by endocytosis, so that intact microbes are present in the recipient protoplast cytoplasm either free or bounded by membrane and enclosed in a vesicle. The uptake of microbes has been suggested as a means for the establishment of novel symbiotic relationships and research was undertaken to investigate the possibility of the manipulation of nitrogen-fixing capacity within plant cells (Giles and Vasil, 1980).

Fusion between protoplasts is a means for transfer of genetic characters and to this end fusion experiments have been carried out using a range of parent material. Fusion frequencies are highest where taxonomic links are closest but higher plant protoplasts have been fused with other species and genera, with animal cells and with yeast and fungal protoplasts, e.g. carrot protoplasts/*Xenopus* cells; tobacco protoplasts/hamster cells; tobacco protoplasts/human HeLa cells; *Vinca* protoplasts/bacterial spheroplasts. In a number of cases cytoplasmic mixing, after fusion, has been shown (Bushnell, 1986). Such fusion products persist for some time in culture and in some cases plant regeneration has been achieved. This information has indicated that general incompatability factors are probably not located on membranes or in cytoplasm.

Fusions between fungal and plant protoplasts have been achieved for celery (*Apium graveolens*) and two fungal species, namely *Aspergillus nidulans*, a saprophytic species and *Fusarium oxysporum*, a potentially pathogenic species, causal agent of celery yellows (Lynch *et al.*, 1989). Fungal protoplasts were taken up into plant protoplasts after PEG treatment and were detected in the cytoplasm. Nucear staining showed that the smaller fungal nuclei were present in the plant protoplast cytoplasm and electron micrograph sections showed that fungal protoplasts, lacking wall material, were bounded by a membrane (Fig. 6.15), and in some cases two membranes. Fungal protoplasts persisted in plant protoplast cytoplasm during prolonged incubation. *Aspergillus nidulans* protoplasts were detected 72 h after fusion and little morphological change occurred in either partner. The viability and integrity of celery protoplasts which contained *Fusarium oxysporum* protoplasts was significantly lower 24 h after uptake although it was not clear if this was due to fungal or plant activity. Fungal penetration of plant cells does give rise to electrolyte leakage and necrosis in plant cells.

Electrofusion techniques also resulted in the uptake of *Aspergillus nidulans* protoplasts into celery protoplasts. Higher fusion frequencies, than for PEG-stimulated fusion, were recorded and sufficient fused material was generated for measurements of viability and respiration rates to be made. The uptake of fungal protoplasts clearly caused a rise in the respiration rate for a few hours following electrofusion (Fig. 6.16) which could not be attributed to the electrofusion or stabiliser treatments.

Interactions between isolated protoplasts from soybean (*Glycine max*) and germ tubes of *Phytophthora megasperma* f. sp. *glycinea* have provided interesting information with regard to recognition processes (Hohl and Balsiger, 1986; Odermatt *et al.*, 1988). Germ tubes of the fungus were seen to curl around host protoplasts, in agarose medium, and in instances of close contact between the partners fibrillar material was laid down on the outer surface of the protoplast. This was probably glucan material akin to that produced at papilla formation in intact plants as a response to invasion. Fungal hyphae adhered to the protoplast surface, indicating some form of recognition and a few germ tubes penetrated protoplasts. However, not all components of the hypersensitive response occurred and no browning

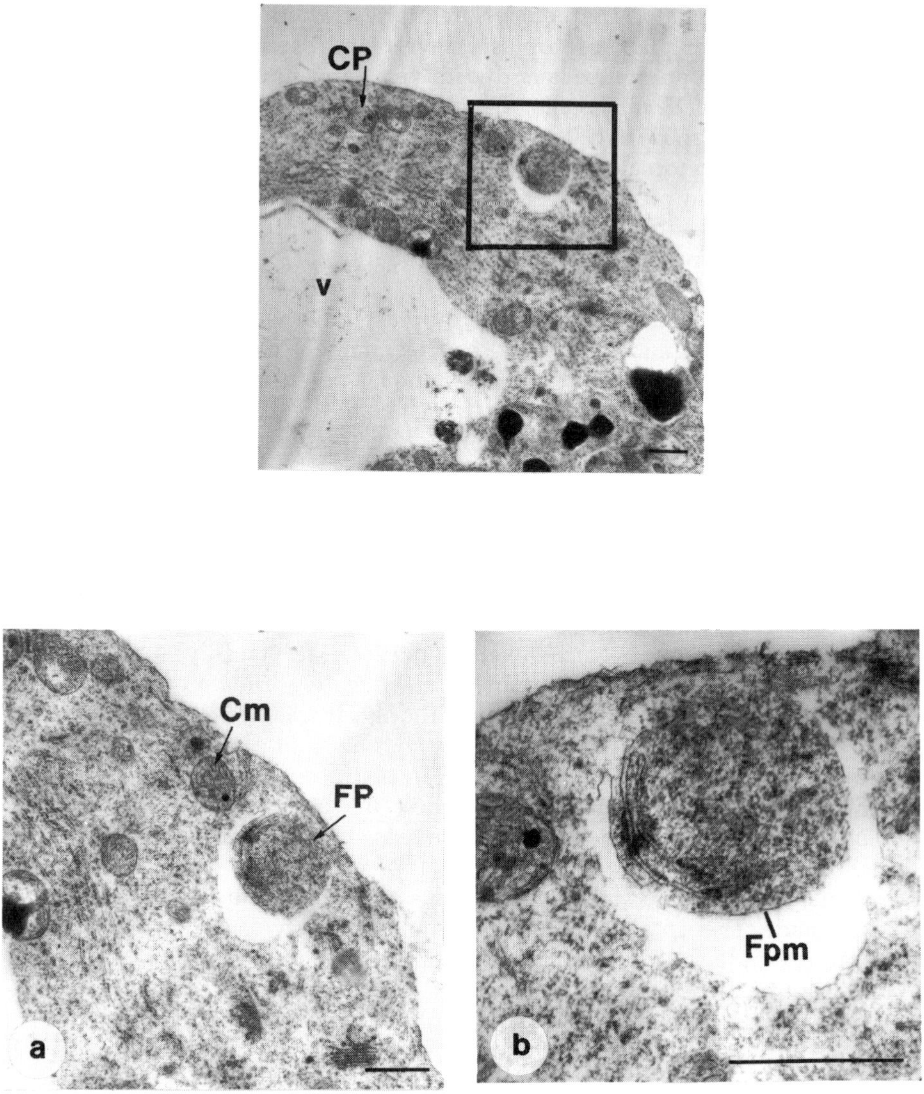

Fig. 6.15 Transmission electron micrographs to show the presence of a protoplast from *Fusarium oxysporum* within the cytoplasm of a celery (*Apium graveolens*) protoplast 24 h after uptake occurred. Celery protoplast (CP), celery protoplast vacuole (v). Bars represent 1 μm. (a, b) Higher magnifications; fungal protoplast (FP), mitochondrion within celery protoplast cytoplasm (Cm), fungal plasma membrane (Fpm). Reproduced courtesy of Dr P. T. Lynch.

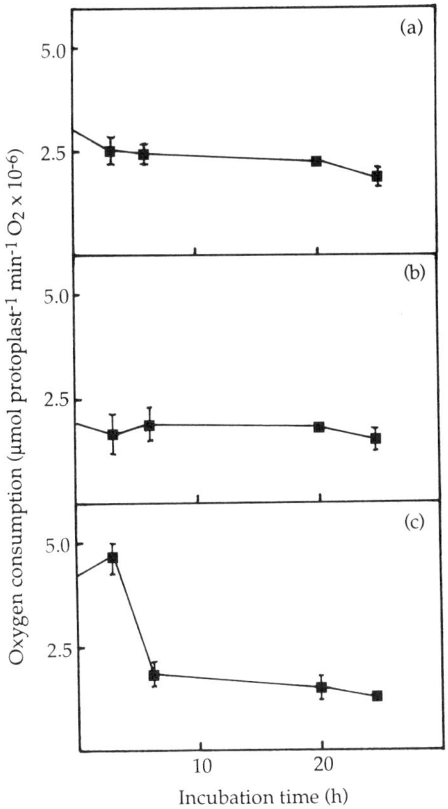

Fig. 6.16 Oxygen consumption of celery protoplasts during prolonged incubation. (a) In 0.6 M mannitol stabiliser; (b) after electrofusion; (c) after electrofusion with *Aspergillus nidulans* protoplasts. Values are means of duplicate determinations and no standard error (SEM) exceeded the maximum value indicated. Reproduced from Lynch *et al.* (1989), courtesy of Springer-Verlag.

of protoplasts was seen. Additionally, no appressoria formed. Odermatt *et al.* (1988) suggested that plant wall material may act as a directional signal to which the fungus responds, and that the lack of wall material therefore removed a potential developmental trigger.

MOLECULAR BIOLOGICAL TECHNIQUES

Biotechnology can be defined as the manipulation of biological systems and therefore includes the development of techniques which permit the directed transfer of genes and regeneration of transgenic organisms. Progress in this area must run parallel with the accrual of information leading to a greater understanding of the biochemical and metabolic processes and the mechanisms which so precisely control and integrate biological functions, increasing our basic knowledge of biological principles.

Molecular biological techniques have been applied in only a few areas of fungal and plant biology but are now beginning to change the approaches which are made to the examination of fungal–plant interactions. For example, recombinant DNA technology combined with investigations into the molecular genetics of pathogenicity, virulence and the molecular basis of disease resistance may increase our ability to control diseases or to engineer plants which may be resistant to, or avoid disease, in the future. Such techniques are often revolutionary and appear somewhat removed from the whole organism sphere of more traditional biology. The application of modern methodologies should not be considered as an end in itself but as a means for probing deeper, and as an important and complementary adjunct to investigations concerning intact systems.

PLANT MOLECULAR BIOLOGY

Recombinant DNA techniques have increased our understanding of the organisation and expression of the plant genome and provide the methodology to move genetic information between organisms. Plant cells are eukaryotic with a large and complex genetic complement. This material is divided into three genomes, nuclear, chloroplast and mitochondrial which interact together during development and differentiation in the plant. The expression of each is also moderated by environmental conditions. The nuclear genome consists of many genes packed into chromosomes typical of eukaryotic systems. The amounts of nuclear DNA are very variable between plant species although this does not necessarily reflect the numbers of active genes present. Active genes may account for a few per cent of the genomic DNA. The chloroplast genome contributes about 15% of the total DNA in a leaf and is composed of a circular DNA molecule which functions in a similar way to the prokaryotic system which occurs in bacteria. The mitochondrial genome may be either a linear or a circular DNA molecule and is more variable in size, although it is often quite large, and it either has a eukaryotic or a prokaryotic system. This genetic material plays a vital role in plant development, coding for mitochondrial polypeptides and cytochrome *c* oxidase.

The DNA within a cell is able to replicate and copies pass to progeny. It can also be transcribed to form RNA, of which there are several types: ribosomal (rRNA), transfer (tRNA) and messenger (mRNA). Newly formed mRNA is processed by the cell to form mature mRNA molecules before it can become functional. Mature mRNA then becomes attached to ribosomes and translated into proteins. Genes are said to be expressed when transcription occurs. Although quite a lot is known about the various stages leading to gene expression, the regulation of expression is much less well understood and relatively little is known about the interactions between the various genomes and the environment. Recombinant DNA techniques allow specific genes to be isolated, modified and re-introduced to an organism, for the study of gene expression and the regulation of expression. These techniques depend on the exploitation of several characteristics of bacterial genetic systems. The specific cutting and rejoining of plasmids forms a basis for recombinant DNA technology.

Plasmids occur most frequently in bacterial cells. These are double-stranded DNA molecules that exist as closed circular molecules in host cells and are capable of self-replication independently from chromosomal DNA. Some are able to transfer copies of themselves between bacterial cells and can also be used to pass fragments of foreign DNA too so that they act as carrying agents, or vectors. Restriction endonucleases are bacterial enzymes (isolated from various sources, e.g. Eco RI is extracted from *Escherichia coli*) which recognise and cut specific base sequences of DNA. Type II restriction endonucleases are extremely specific and cut DNA at very precise sites. These enzymes are used to fragment DNA and to cut a plasmid vector at a site where a foreign fragment can be inserted. When they have been cut plasmids are linear molecules. When cut plasmids are mixed with fragments of foreign DNA and treated with ligase enzymes, which rejoin specific DNA sequences, the foreign DNA becomes inserted and the molecule will rejoin as a circle. Plasmids can therefore be manipulated to take up DNA for transfer back to bacterial cells (usually *E. coli*). This is known as transformation. In this way the foreign DNA fragment has been cloned and can be replicated by natural conjugation within a population of untransformed bacterial cells. Some plasmids occur naturally in bacterial populations but some have now been constructed to contain specific sites which readily enable insertion of foreign genes.

Bacteriophages are also used to carry fragments of DNA. These are viruses which naturally infect bacteria (e.g. phage λ infects *E. coli*), carrying double-stranded DNA in a protein coat. λ phage is not used directly as a vector but small regions of DNA from it (*cos* sites) are combined with plasmid DNA and used to carry foreign DNA, in the protein coat, into *E. coli*. The *cos* site ensures that the DNA fragments are packed into the coat. This form of hybrid vector is known as a cosmid. For some purposes, particularly the construction of a gene library, in which DNA fragments are maintained, cosmids are usually used as vectors to carry foreign DNA. This allows the carriage and cloning of larger fragments of DNA than can be carried using plasmids, an ideal technique in some circumstances.

Gene cloning

Gene cloning is the isolation, purification and accurate amplification of individual genes or gene sequences. Since large amounts of material can be produced this is a technique which is used for the study of gene structure and the regulation of gene expression. Cloned genes can be inserted into other organisms (transformation). These techniques can be used to place useful plant genes into microorganisms for the large-scale industrial production of plant compounds, or to engineer plants carrying foreign genes, for example to increase disease resistance.

Several techniques are important for the development of suitable cloning systems and the selection of clones once these have been constructed. It is usual to integrate information obtained by using a number of techniques before the function of gene sequences can be postulated.

Complementary DNA cloning

Complementary DNA (cDNA) cloning is a very useful technique for the examination of specific mRNAs and their function (Fig. 6.17). It involves the extraction of mRNA from intact tissue followed by identification and enrichment of the particular mRNA required. Extraction is usually carried out at the time when a particular mRNA will be active in the tissue, e.g. after infection of a plant. This mRNA is copied into single-stranded DNA using reverse transcriptase. Transcriptase enzymes are able to synthesise DNA from an RNA template. The RNA strand is removed from this DNA–RNA hybrid molecule by alkaline hydrolysis. A second strand is then synthesised using DNA polymerase I (or reverse transcriptase). This double-stranded DNA can subsequently be inserted into a plasmid which will act as a vector. The plamid can be cut open precisely using restriction endonuclease, allowing the cDNA and the plasmid to pair by their homopolymer tails (enzyme terminal transferase) and to reanneal forming a circular molecule with cDNA inserted. These recombinant plasmids can be inserted into bacteria and transformants selected. It is usual to use plasmids carrying two antibiotic resistance markers. cDNA insertion takes place into one of the markers destroying its activity. Transformed bacteria then carry resistance to only one antibiotic and can be easily selected in culture. Many different cDNA fragments will be inserted and the clone of particular interest must be identified. Radiolabelled probes, which hybridise to DNA in the clones required, are often used to detect specific transformants.

Genome cloning

Genomic clones allow the study of larger fragments of DNA which include specific genes plus those surrounding sequences which often have a role in regulating the expression of that gene. Comparisons between cDNA clones and genomic clones can provide information concerning intervening sequences (introns) which are not present in mRNA.

Extract mRNA from tissue

↓

Copy into DNA

↓ Double-stranded cDNA formed

Insert into plasmid

↓ Recombinant plasmid formed

Transform bacteria

↓

Select transformants

↓

Select suitable clone

↓ Grow up bacteria

Isolate plasmid

↓

Analyse and sequence cloned DNA

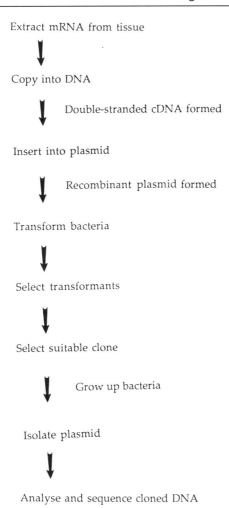

Fig. 6.17 Outline protocol for cDNA cloning.

Nuclei, chloroplasts or mitochondria are isolated from cells as genome sources, often using isolated protoplasts as a non-disruptive means to obtain these organelles, and DNA is then purified. Many problems can arise in the preparation of DNA from any of these sources. It is particularly important to ensure that any genome preparation is free from contamination derived from another genome (e.g. chloroplast DNA may easily be contaminated from nuclear or mitochondrial DNA) and a great deal of work has been directed towards the development of protocols to enable the reliable preparation of uncontaminated fractions. Treatment with restriction endonucleases then provides large fragments of DNA which can be carried into *E. coli* for cloning using cosmid vectors.

Expression and analysis of cloned genes

Genes have a promoter sequence, which is concerned with the regulation of expression, and a structural sequence, which codes for the gene product. The DNA of a gene is transcribed [starting from the five-prime (5′) end and progressing to the three prime (3′) end] into mRNA and translated into proteins. A gene is said to be expressed when it is transcribed into mRNA. It is possible to attach promoters and structural genes from different organisms together and retain expresssion. Genes inserted into recombinant bacteria will be repaired by the cell and, although it is not always possible, cloned plant genes have been expressed in bacterial and yeast cells (Gatenby, 1989). Conversely bacterial genes can be expressed in plants (Kerr, 1987). As well as being important from the point of view of regulation studies, there are commercial applications for such transformations, e.g. microbial cells may be used to produce plant proteins or secondary metabolites on a large scale in fermentation cultures. It is also possible to transfer plant genes into other species using recombinant DNA techniques with important implications for improving disease resistance.

Cloned genes have been prepared from a number of major crop plants. These must be analysed for information relating to structure and function. Some of the most usual techniques are briefly considered here. Information collected using different methods can subsequently be pooled to allow detailed analysis of sequences and the regulation of expression.

(a) Restriction enzyme analysis

DNA preparations digested with restriction enzymes are cut at very specific sites. The relative sizes of the resultant fragments can be determined by gel electrophoresis, for comparison with fragments of known sizes. It is therefore possible to map the restriction sites, for the enzymes used, by fragment size comparisons. Once the positions of these sites on the gene have been established they become valuable reference markers for comparisons with manipulated genes which may contain inserted sequences. If a genomic clone contains introns then the restriction sites will be further apart than in the cDNA clone. The successful application of other techniques often depends on knowledge of the restriction sites in a preparation.

(b) Southern blotting

The technique known as Southern blotting (Southern, 1979) is used to determine which DNA fragments are complementary to each other. The technique is extremely accurate and is used for gene mapping. Restriction fragments of DNA are separated by agarose gel electrophoresis and are blotted (dot blot) from the gel on to a nitrocellulose filter. After heating (80°C) to fix these fragments, the filter is treated with a known radiolabelled DNA (or RNA) probe. Hybridisation occurs between the complementary sequences which can then be detected by autoradiography. Providing that suitable probes can be obtained or produced then it is possible to map the restriction sites in genomic DNA in this way.

(c) DNA sequencing

Pure DNA can be produced in large amounts by cloning and can therefore be used for sequence analysis, providing very detailed analysis of DNA fragments. To do this one end of a DNA restriction fragment is labelled with ^{32}P. Subsequently a smaller fragment is removed from the whole molecule using a specific restriction endonuclease and is discarded. The larger fragment which is retained is now radiolabelled at one end of the molecule only. Chemical treatment can then be used to partially digest this DNA at specific bases and the resulting products can be separated on a polyacrylamide gel. Fragments which contain the labelled terminus can be detected by autoradiography. By piecing together all the information from each treatment the base sequences of the strands can be deduced. Such sequence analysis has provided a great deal of information concerning the mechanisms by which gene expression is controlled.

Genetic engineering of plants

Gene cloning techniques provide means for the structuring and analysis of genetic material which may code for specific characters in an intact plant. A range of plant genes has now been cloned. However, for the successful engineering of a plant the transfer of such material into the plant cells is a prerequisite and there are a range of techniques by which this can now be achieved (Weising *et al.*, 1988). For the novel genetic information that has been introduced to be of use it must then be expressed in the plant cell and passed on to progeny. Our understanding of the expression of plant genes and the sequences that are required to effect this, are now providing useful methods by which transferred genetic material is competent for transcription and translation. The production of transgenic plants has already been achieved for some species. In other cases however, the introduction of suitable genes is not necessarily the limiting factor. Many of the techniques which have been used rely heavily on tissue culture methods for the manipulation of plant material and in some cases the major barriers to success lie in the lack of plant regeneration from the culture stage. Various approaches to the problems of gene transfer, expression and plant regeneration have been made and some of the more exciting methods are considered here.

Vectors for gene transfer in plants

Foreign genetic material introduced into a cell is usually degraded. For genetic engineering therefore, a system is required by which such information can be delivered and incorporated into the cell without it being recognised as alien. An agent which aids this process is termed a vector. A number of natural plasmids and plant viruses have been investigated for their potential as useful plant gene vectors, in the transfer of genes between plants or from other sources into plants.

(a) Agrobacterium tumifaciens plasmids

Agrobacterium tumifaciens is the causal agent of crown gall disease which infects many plant species (mostly dicotyledonous species) naturally from the soil. Bacteria enter the plant through wound sites and after infection crown gall tumours are formed by the localised proliferation of undifferentiated cells. This aberrant growth pattern is caused by genes carried on large plasmids (Ti plasmids) in the bacteria. A discrete portion of the plasmid genome, known as transferred DNA (T-DNA) is passed to the plant cells. T-DNA genes of Ti plasmids control the biosynthesis of the auxin indole-3-acetic acid and a cytokinin precursor, isopentyladenosine, and incorporation of these genes into the plant genome leads to the constitutive synthesis of these growth regulator compounds. In tissue culture these plant cells can be maintained in the absence of the bacteria and are self-sufficient for growth regulators (unlike normal callus tissues). Tumour cells also synthesise unusual amino acids (opines) which are utilised by the infecting bacteria as the sole carbon and nitrogen sources. This has been suggested as a means by which infecting *Agrobacteria* divert plant nutrients into compounds which are then only available for their own use. Opine production is controlled by T-DNA genes.

Foreign genes can be added to T-DNA and carried into plant cells using Ti plasmids. The Ti plasmids can be transmitted to the plants via the normal *Agrobacterium* infection process or can be incorporated into plant protoplasts. The genes carried on the natural Ti plasmid which carry the codes for the production of disease symptoms are removed so that transformed plants will be asymptomatic. Selection markers are usually added so that transformation can be detected. These are usually genes for antibiotic resistance (e.g. kanamycin and streptomycin) or sometimes herbicide resistance. In this way transformed cells can be identified on selective agar medium before regeneration. Transformation requires that regions of the Ti plasmid which are responsible for mobilisation of the

T-DNA, allowing foreign DNA transfer, are included. Additionally the *vir* region which is responsible for the transfer and integration of T-DNA into the plant genome is also required (White, 1989). Ti plasmids can be manipulated so that their composition is suitable for expression in the particular plant species in question. The additions of certain enhancer and promoter genes allow the construction of appropriate Ti plasmid vectors. Ti plasmids are large and much research effort has been directed to simplifying the construction. Genes introduced in this way are expressed in plant cells and this is a particularly useful means for gene transfer into dicotyledonous species. Transgenic plants have been produced using the Ti plasmid system (mostly dicotyledonous plants), for example *Nicotiana tabacum* (tobacco), *Petunia hybrida* (petunia), *Daucus carota* (carrot), *Brassica napus* (oilseed rape), *Medicago sativa* (alfalfa), *Phaseolus vulgaris* (French bean), *Glycine max* (soybean), *Lactuca sativa* (lettuce) and others.

(b) Plant viruses

Some natural plant viruses have been investigated for their potential as plant gene vectors. The caulimoviruses (CaMV) have received a great deal of attention because these contain double-stranded DNA which would be useful in gene manipulations. These viruses cause vein clearing and stunting amongst members of the Cruciferae and are highly effective in establishing systemic infections. In the natural situation the virus is transmitted by aphids although it is possible to inoculate plants with the virus particles or with the isolated DNA after leaf abrasion. The ease of infectivity is a great advantage to this system and for this reason much attention has been focused on it. However, the construction of this virus is such that the inclusion and transfer of large pieces of foreign DNA are not possible at present.

Curly top virus (CTV) and maize streak virus (MSV) are gemini viruses which cause major crop diseases. These are from a group of viruses which have an unusual structure and contain single-stranded DNA. In nature they are transmitted by insects and infect a wide range of plant species. These viruses and others are being investigated as possible plant gene vectors.

(c) Transposon mutagenesis

Transposons are DNA fragments which do not self-replicate but are often integrated into the genome in cells. They can be found and inserted into chromosomes and plasmid genomes, and can integrate at different sites by moving in the genome (transposition). These transposable elements are also known as 'jumping genes'. When such elements become inserted into a gene the normal expression of that gene is prevented. This system can be used as a useful means to 'tag' genes. Transposable elements have been employed to detect the function of unknown genes, by switching off the expression of those sequences into which the transposon is inserted. Transposon mutagenesis has been used successfully in the investigation of virulence in plant pathogenic bacteria (Mills, 1985; Kerr, 1987) in particular.

This method has also been used to isolate genes from plant species (*Zea mays* and *Antirrhinum majus*). A transposon is inserted into a specific gene, switching off its activity. The DNA is then isolated, treated with restriction endonucleases and a genomic library constructed. The gene of interest can then be identified by detection of homology with the transposed sequence. The sequences surrounding the transposon can then be cloned and used for the identification of the same sequences in other plants. This system therefore provides a potential means for the isolation of genes associated with defense responses in plants (Ellis *et al.*, 1988). This method can be combined with restriction fragment-length polymorphism (RFLP) analysis to aid the process of identification of the target gene if multiple copies are present in the genome. Isolated DNA digested with restriction endonucleases is cleaved into reproducible fragments depending on the enzymes used. These fragments may be separated by gel electrophoresis and probing with a radiolabelled DNA clone will give rise to a characteristic banding pattern. Differences in fragment length will give different banding patterns so that RFLP loci can be mapped (linkage mapping). RFLP probes can be used to identify specific polymorphisms. The use of RFLP markers allows more directed selection.

Microinjection

The use of vectors for the transfer of genetic information can be restricted by the host range of the vector system. Microinjection is a very precise and direct approach to the introduction of foreign DNA. Using this method isolated DNA can be delivered into protoplasts individually. A protoplast or suspension culture cell is held, using a capillary, and injected directly with a fine needle containing DNA solution. The procedure is usually carried out on an inverted microscope using micromanipulators. A very large number of injections are required to ensure that some transformants are obtained. It is a time-consuming process and considerable skill is required. A transformation frequency of 15–20% has been achieved with tobacco protoplasts. The technique has been used with some monocotyledonous species (maize, wheat and rice) which cannot be transformed using the *Agrobacterium* plasmid system. Injection of DNA into embryo tissues and into tillers and florets has also been attempted, with some degree of success. Seeds were obtained which gave rise to transformed plants (de la Pena, 1987). However, microinjections are usually carried out successfully with protoplasts, not intact cells, and so there may be problems with the regeneration of such material in these species.

Transformation of protoplasts

It has been shown that isolated protoplasts, bounded only by a plasma membrane, can be induced to take up DNA fragments directly (Potrykus, 1988b; Wu, 1989). Not all plant species appear to be competent for such uptake and each system must be carefully characterised to optimise the parameters and maximise transformation frequencies.

For transformation protoplasts are most usually treated with polyethylene glycol (PEG) and isolated DNA added. The dehydration effect of the PEG probably aids the uptake process. In some cases the addition of $MgCl_2$ has been shown to enhance the uptake levels. A similar approach has been made by adding plasmid preparations to PEG-treated protoplasts and transformants have been obtained. In some cases, particularly with the monocotyledonous species, uptake was enhanced by a heat shock (45°C for 5 min) and cold shock (0°C for 10 s) prior to transformation.

Electroporation

Isolated protoplasts can also be induced to take up plasmid DNA using electroporation techniques. A high voltage pulse is applied to protoplasts, as for electrofusion, either with or without PEG treatment in addition. The outer membrane reversibly develops pores under such conditions and foreign DNA can enter the cytoplasm. In some cases heat shock (45°C) has also been found helpful. The transformation frequencies which have been obtained are very much higher (up to 1000-fold) than for PEG treatment alone. However, although genetic material can be successfully delivered, protoplast regeneration remains a problem in recalcitrant plants.

Cell bombardment

Transformations of intact cells would relieve some of the problems arising from protoplast regeneration. Using a particle gun developed by Sanford *et al.* (1987), Klein *et al.* (1987) bombarded intact *Allium cepa* (onion) cells with plasmid-coated tungsten particles (4 μm diameter spheres). At an acceleration velocity of 1 400 ft. s^{-1}, these particles penetrated cell walls. Up to five were retained per cell and foreign genes were expressed in a proportion of the cells.

Alterations in plant gene activity in response to pathogens

Plant genes for disease resistance determine that plant defense responses are activated when the plant is challenged by a pathogen. This gives rise to the production of phytoalexins, lignins, cell wall degrading enzymes and hydroxyproline-rich glycoproteins (HRGPs), etc., resulting in defensive activities usually associated with resistance responses and hypersensitivity. The manipulation of such plant resistance genes is obviously an attractive proposition for the plant breeder. To date, no such genes have been cloned and little information concerning their expression or function in the regulatory control of defense-related activities has been gained. However, it is clear that plant resistance genes are likely to be large and very complex. The plant genome itself is very large compared with bacterial systems for which many gene cloning techniques have been developed, and this gives rise to practical problems. Large amounts of genetic material, and large numbers of transformants, must be dealt with in any such investigations using plant tissues. Additionally, large amounts of repetitive DNA are present in the plant genome and low rates of successful plant transformation add to the difficulties and complications posed. Attention has, however, been more successfully directed towards understanding the changes in gene expression that occur in response to pathogens or elicitors. Two main approaches have been made to such investigations (Collinge and Slusarenko, 1987; Ellis *et al.*, 1988), designated 'shotgun' and 'targeted' gene cloning.

The shotgun approach attempts to identify and isolate specific genes of interest from the whole genome without prior information as to their activity. mRNAs of presumed importance are isolated after elicitation by screening cDNA libraries previously prepared using mRNA from elicited and non-elicited tissues. cDNA clones are then used to identify the genomic fragments which code for the resistance response activities. This approach must be used in conjunction with physiological investigations in order to elucidate those changes induced by the activity of a selected gene. It is usual to use linked sequences on the cloning vector to act as a probe for the gene to be transformed. This approach is theoretically possible and indeed much useful information has already been gained in this way, however, practical difficulties do arise from the large size of the genome concerned and the large number of transformants required.

The targeted approach makes use of some form of gene tagging system. Gene function can be inactivated by the insertion of a transposable (transposon mutagenesis, see above) DNA sequence. That gene is therefore tagged and can be cloned into a genomic library. Immunological methods have also been used as a means for tagging genes. Enzymes or other proteins, which have a role in defence and are produced rapidly after infection or elicitor treatment, are isolated and antibodies raised to them. Changes in mRNA activity can then be detected by immunoprecipitation from *in vitro* translations and cDNA clones can be constructed and used as probes for specific gene activities. If the synthesis of a particular protein or enzyme occurs rapidly at one time in a plant then high levels of mRNA occur in the cells, corresponding to the production of that particular product. In that case cDNA cloning is more feasible and this applies to most of the plant genes which have been cloned. Other techniques such as map-based cloning and chromosome walking (Young, 1990) may find future application in the study of plant–fungal interactions. However, genome size and repetitive sequences, which cross-hybridise, do pose great practical problems with all these methods at present.

Pathogen invasion, or elicitor treatment, act as triggers, activating defence responses in plant cells. Additionally, changes in mRNA levels have been measured in affected tissues (Dixon and Harrison, 1990). Genomic clones and cDNAs have now been obtained for a number of plant defence response genes and from a range of plant species (e.g. *Nicotiana tabacum*, *Phaseolus vulgaris* and *Petroselinum crispum*) and tissue types, using a variety of vector systems. Genes encoding for some of the enzymes involved in the phenylpropanoid pathway (L-phenylalanine ammonium lyase; 4-coumarate:CoA ligase; chalcone synthase; chalcone isomerase), hydroxyproline rich glycoproteins (HRGPs) and some pathogenesis-related proteins (PR-proteins), have been cloned using a variety of plant species. Many of these investigations have been carried out using tissue culture or cell suspension derived plant material.

A number of different molecular biological approaches have been used to trace the appearance of many mRNAs in plant tissues after infection or elicitation. It seems likely, however, that many more (possibly with important roles in defence) may be present, probably in low amounts and remain, as yet, undetected. Increases in levels of some mRNAs have been shown to be due to transcriptional activation. This may be a transient effect for some mRNAs. Davidson *et al.* (1987, 1988) used near isogenic susceptible and resistant barley (*Hordeum vulgare*) leaves (each resistant plant carrying one of the genes for resistance; *Mlp*, *Mla* or *Mlk*) to investigate the mechanism of resistance to *Erysiphe graminis* f. sp. *hordei*. In barley resistance to powdery mildew is conditioned by the presence of single genes. cDNA cloning provided useful information relating to changes in mRNA levels. Levels of some mRNA species were seen to rise in both susceptible and resistant plants but higher quantities were found in resistant plants. The timing of the onset of induction of some mRNA species suggests that these have a role in defence. The occurrence of enhanced mRNA levels was shown to precede or coincide with the suspension of growth of the invading fungus. The data also indicate that the resistance genes (*Mlp*, *Mla* or *Mlk*) do have regulatory roles in the plants concerned. Since increases in some mRNA levels were transient these genes may operate differential regulation systems for different genes, and have distinct effects on host gene expression. Resistance responses of French bean (*Phaseolus vulgaris*) to *Colletotrichum lindemuthianum* have also been shown (Cramer *et al.*, 1985) to correlate with the selective induction of mRNAs. cDNA cloning of such mRNA species is likely to improve understanding of the chromosomal organisation and the regulatory roles of resistance loci.

Molecular techniques have confirmed that biochemical defense molecules are synthesised *de novo* in response to pathogen invasion, elicitor treatment and abiotic stresses in plant tissues, and are now providing information relating to the activities of specific genes for resistance and their regulatory roles. Although quite a lot is known about ultrastructural and biochemical aspects of defense responses in plants very little information is available concerning the integration and co-ordination of these activities in plant tissues, as yet. Tissue culture techniques have enabled the elucidation of the types of compound which act as elicitors for defense responses in plants and these systems, in combination with molecular approaches, may soon improve our understanding of the ways in which pathogens avoid or suppress recognition by plants.

Transgenic plants and disease resistance

Plant transformations can now be carried out for a range of species and modern genetic engineering techniques provide increasing possibilities for the manipulation of the plant genome. It may soon be possible to adjust the operation of defense pathways, modifying the plants response and increasing our understanding of the ways in which these systems operate in intact tissues. The construction of novel plant phenotypes, with altered gene expression, may provide insights into those factors which contribute to defence and how these are integrated throughout plant development. A greater understanding of the role of phytoalexins in plant resistance, and the ways in which these compounds operate in the plant, may lead to the engineering of modifications in the synthetic pathways and to an increase in the range of microbial species which may be combated. Eventually, it may be possible to introduce novel resistance genes or to modify the timing of some response mechanisms, to improve field resistance in crop species.

FUNGAL MOLECULAR BIOLOGY

Molecular biology clearly promises new approaches to the study and understanding of fungal physiological processes and pathogenicity and provides new scope for genetic manipulation (Yoder and Turgeon, 1985a, b; Dixon and Harrison, 1990; Metzenberg, 1991). A great deal is known about the morphological aspects of fungal invasion of plants and theories concerning the molecular mechanisms underlying the progress of disease development have been proposed. It is clear, for example, that the production of plant cell wall degrading enzymes is important in the penetration of plant tissues and that the ability to produce a particular toxin, or to detoxify a phytoalexin may contribute to the establishment of disease. However, the exact role of such factors in the determination of fungal pathogenicity is not yet clear.

Various approaches to the investigation of pathogenicity are now made possible by molecular biological techniques. The development of fungal transformation systems provides opportunities for the investigation of gene function. The movement of genes coding for avirulence into virulent strains has the potential to increase our understanding of the molecular control of avirulence. Indeed, a great deal of success has been achieved in this area with bacterial pathogens although little advance has been made with fungi, as yet. Movement of genes coding for toxin synthesis or wall degrading enzymes, into strains or species which lack this ability, will allow the investigation of the roles these metabolites play in disease determination.

Relatively little is known about genome organisation, gene regulation and expression in fungal systems. Many problems arise in the use of established techniques because the fungal genome is large. Many applications still await the development of methods which can, practically, be applied to such large amounts of DNA. The use of techniques such as gene tagging and chromosome walking are theoretically possible (Leong and Holden, 1989) but, as with plant systems, the sheer size of the genome creates many practical difficulties. The use of mRNA gene probes provides opportunities for the location of specific genes within the genome, cloning, sequencing and hence an increase in the understanding of gene structure and function in fungi. The identification of sequences involved in the regulation of expression has potential in the construction of vector systems for the movement of genes between strains and species.

Investigations concerning the differentiation of specific structures or production of metabolites at particular stages in infection are now directed towards understanding the mechanisms by which these syntheses are triggered and regulated. If it is possible to elucidate how these genes are switched on and off it may be possible to manipulate their function in future. These investigations may also provide information concerning the development of speciation in fungi, the evolution of parasitism and race development. Additionally, a contribution to the clarification of some disputed areas of taxonomy, which is currently based largely on morphological characteristics, may be expected.

Investigations of molecular aspects of disease resistance in plants may also have important implications. An understanding of the ways in which fungi can overcome host defence mechanisms and the ways in which these processes are triggered, by the presence or activity of that fungus, may potentially direct new approaches to the control and interference with pathogen function. The sensitivity of modern molecular techniques may now provide information concerning the events surrounding host cell necrosis to establish whether this is a primary determinant of resistance (the hypersensitive response), causing the death of the fungus, or whether necrosis occurs as a consequence of fungal death caused by the production of other, hitherto unidentified metabolites. Additionally, the elicitation of phytoalexin synthesis and the mechanisms controlling the accumulation and localisation in plant tissues, and the extent to which invading fungi are likely to come into contact and be affected by these compounds, are not yet well understood in molecular terms. Once such control systems are better understood methods for the manipulation of these characters can be developed.

Transformation of pathogenic fungi

Protoplasts can be produced very successfully and in large numbers from most species of filamentous fungi. The techniques used are very similar to those for plant tissues, described above. The cell wall is digested using lytic enzymes, often of bacterial or fungal origin, although snail digestive enzymes and purified enzymes have been used in the development of many protocols (Davis, 1985; Peberdy and Hocart, 1987; Peberdy, 1989b). These membrane-bound units can be liberated from mycelium in a physically gentle manner, collected on a large scale and most can be easily and rapidly (10–24 h) regenerated.

The physiological age of mycelium has a great influence on the numbers of protoplasts released; most are liberated when mycelium is in active (exponential) growth. Protoplasts are released preferentially from hyphal tips and where the wall is less structured and more plastic. Liberation occurs most usually through pores which are digested in the wall. Protoplasts are released into osmotically stabilised solution usually containing inorganic salts (KCl or $MgSO_4$) to prevent lysis, and used in genetic analysis programmes or for transformation. These membrane-bound units can then be fused using chemical agents (PEG) or electrofusion systems similar to those used with higher plants. Regeneration into mycelium can also be achieved by providing protoplasts with suitable sources of carbon and other nutrients. Fibrils of wall material usually form on the outer surface of the protoplasts in a few hours and regeneration to hyphal outgrowth may be complete in 12–24 h.

Transformation systems have been developed for *Aspergillus nidulans* using simple cloning vectors (Ballance *et al.*, 1983; Turner and Ballance, 1985) and also for some other yeast and fungal species (*Saccharomyces pombe, Schizophyllum commune, Podospora anserina* and *Neurospora crassa*). Relatively few pathogenic species have been investigated although successful transformation of some important species has been achieved, e.g. *Cochliobolus heterostrophus* (Turgeon *et al.*, 1985; Yoder *et al.*, 1986); *Colletotrichum lindemuthianum* (Rodriguez and Yoder, 1987); *Fulvia fulva* (Oliver *et al.*, 1987); *Fusarium oxysporum* (Kistler and Benny, 1988); *Gaeumannomyces graminis* (Henson *et al.*, 1988) and *Septoria nodorum* (Cooley *et al.*, 1988). For work with *Cochliobolus heterostrophus* two genes were used, the *Escherichia coli* gene *hyg*B and the *Aspergillus nidulans* gene *amd*S, as selectable markers by which transformants could be identified. These genes were cloned into vectors and added to PEG-treated protoplasts.

Although a low frequency of transformation resulted, stable transformed progeny were obtained. Transformation systems have now been applied to the corn smut fungus *Ustilago maydis* (Wang *et al.*, 1988; Leong, 1988; Leong and Holden, 1989) which have enabled a molecular genetic analysis of this fungus. *U. maydis* has an interesting life cycle. Haploid cells are non-pathogenic and fuse (with cells of the opposite mating type) to form a dikaryon which is pathogenic and causes gall formation in infected host plants. Two mating-type loci control cell fusion and the sexual development of this fungus. Analyses carried out using cloned genes now provide means for increasing our understanding of the control systems and determining pathogenicity, which are in operation in this fungus, and means for the manipulation of the genome.

Molecular approaches have also been made to the study of pathogenicity in the foliar pathogen *Fulvia fulva* which causes tomato leaf mould. This is a biotrophic species which proliferates within the intercellular spaces (apoplast) of susceptible plants. The fungus appears as several physiological races which can be identified by resistance genes in isogenic tomato cultivars. A fungal protoplast fusion approach has been used, allowing crosses to be made between pathogenic and non-pathogenic isolates. In this way chromosome reassortment and mitotic recombination can be achieved, providing information concerning the mechanisms of pathogenicity. In addition, a transformation system has been developed (using a chimaeric *Aspergillus nidulans/Escherichia coli* vector system) and transformants obtained (Oliver *et al.*, 1988). Work is now directed towards identifying avirulence genes in this fungus.

Gene cloning in plant pathogens

Cloning systems provide the means by which genes can be inactivated within a fungus or isolated, modified and reintroduced into the fungus, allowing the activity of these genes and the mechanisms controlling the formation of gene products to be assessed. Theoretically, this could be achieved at various stages during the infection process.

Cloning systems have been used to investigate genes expressed by rust fungi at particular stages in the infection process (Staples *et al.*, 1986; Bhairi *et al.*, 1988). It has been shown, in *Uromyces appendiculatus* (bean rust fungus), that changes in protein synthesis occur when young germlings begin to differentiate and further changes occur as the differentiation of the substomatal vesicle takes place. Some of these changes are subtle and were not detected until the application of two-dimensional polyacrylamide gel electrophoresis. The synthesis of some of these proteins has been attributed to changes in gene transcription as a result of thigmodifferentiation (Chapter 3) in response to surface molecules of the host (possibly glycoproteins). These molecules may also act as a directional trigger for fungal development. *Puccinia graminis* f. sp. *tritici* can be induced to differentiate using a heat shock treatment and changes in protein synthesis have been shown to occur at the time of vesicle development. A genomic library of *Uromyces appendiculatus* was constructed using λ phage in *Escherichia coli* and screened using a cDNA probe. These systems provided a means for the investigation of changes in the expression of genes during the infection process, in the presence and absence of the host plant. Clones showing homology were selected. One gene (gene 24) was mapped and mRNA for this gene was shown to accumulate in differentiating spores at the time of vesicle development (Bhairi *et al.*, 1988). An infection structure-specific gene has now been cloned from *Uromyces appendiculatus* (Bhairi *et al.*, 1989)

The production of cutinase has also been shown to have an important role in the pathogenicity of fungi (Kolattukudy and Crawford, 1987). The degradation of cutin is an important means for the achievement of penetration through the outer cuticle of plant surfaces and for initial establishment in the host plant. Cutinase has been purified and characterised (Kolattukudy *et al.*, 1989) from *Fusarium solani* f. sp. *pisi* (*Nectria haematococca*), and isolates with differing abilities for cutinase production were compared for their ability to penetrate into plant tissues. Cutinase-deficient mutants were only able to infect leaves which had been abraded to disrupt the surface structure. Cutinase cDNA has been cloned and sequenced from *F. solani* f. sp. *pisi*, *Colletotrichum gloeosporioides* and *Colletotrichum capsici*. In response to the presence of the host surface *F. solani* f. sp. *pisi* spores produce small amounts of cutin monomers which then stimulate the production of sufficient cutinase for the penetration of the host (Chapter 3), by the activation of cutinase gene transcription. The cloned cutinase gene was introduced, via a plasmid, into an isolate of *F. solani* f. sp. *pisi* which was not able to penetrate host tissue. Transformants produced in this way showed high levels of infectivity.

Following cuticle penetration *F. solani* f. sp. *pisi* produces pectin degrading enzymes to further invade plant tissues. As part of the host resistance mechanism plant cell walls become highly suberised locally to infection points. Kolattukudy *et al.* (1989) have shown that an anionic peroxidase enzyme is involved in this suberinisation. The cDNA and the gene for the synthesis of this enzyme have now been cloned and sequenced. The manipulation of the timing of the expression of this gene in the host plant may be a key factor in improving plant resistance to such pathogens.

A virulence gene (*pda*) from *Nectria haematococca*, which confers the ability to detoxify a phytoalexin has been cloned into *Aspergillus nidulans* (Yoder *et al.*, 1986). The product of the gene is cytochrome P-450 monooxygenase which detoxifies the plant phytoalexin, pisatin, by demethylation. Strains of *N. haematococca* which lack this gene do not show virulence towards host pea plants. A genomic library was set up using a cosmid vector containing a gene for tryptophan synthesis (*trpC*) from *A. nidulans* which enabled the selection of transformants. Transformation into a *trpC*$^-$ strain of *A. nidulans* resulted in the expression of pisatin demethylation activity in one transformant which carried the *pda* gene.

Investigations of genetic variation in plant pathogen populations

Molecular biological techniques have also been applied to studies of speciation in fungi and investigations of the relatedness of different isolates, and have proved particularly useful in instances where conventional fungal taxonomy is unclear. For example, mitochondrial and nuclear DNA diversity has been investigated (Forster *et al.*, 1990) and restriction fragment length polymorphisms (RFLPs) have been used as genetic markers to investigate recombination events in *Phytophthora* species (Forster and Coffey, 1990).

Such approaches are being made to establish the extent of genetic variation within populations of pathogenic species. Nucleotide sequence divergence has been assessed for isolates of *Colletotrichum gloeosporoides*, which causes anthracnose of *Stylosanthes* species (Braithwaite *et al.*, 1990). In this way it has been confirmed that two pathogen populations are present in Australia. RFLP analyses have been used in the investigation of variation in *Bremia lactuca*, causal agent of lettuce downy mildew (Hulbert and Michelmore, 1988), and *Erysiphe graminis* (O'Dell *et al.*, 1989) populations. Such research will provide greater insight into phylogenetic relationships and the evolution of plant pathogens.

Genetic manipulation of potential fungal biological control agents

Fungal species which have a relatively fast growth rate, are good competitors with antagonistic ability and the invasive capacity to colonise a site rapidly, are likely to be successful biological control agents, counteracting the activity of other microbes. Those possessing the ability to produce toxic metabolites may be useful as agents to combat insects or as mycoherbicides. Additionally, many fungi, such as those forming mycorrhizal relationships, improve plant growth and encourage the rapid establishment of seedlings. In general these attributes have yet to be exploited in agricultural situations although in recent years awareness of the available potential (Chapter 2) has increased. Biotechnological approaches can now provide systems for the development and engineering of successful biological control agents.

Some fungi can be easily manipulated using conventional gentical techniques, making use of sexual cycles and the parasexual cycle in breeding programmes. To date, this approach has been directed towards industrially important fungi. However, many potentially useful species are lacking in a sexual cycle (Deuteromycotina) or are difficult to grow in any cultural system (e.g. species forming vesicular–arbuscular mycorrhizal associations with plant roots). Protoplast fusion and transformation techniques provide ideal means for the manipulation of such isolates (Papavizas, 1987; Peberdy, 1989b; Peberdy and Hocart, 1987; Hocart and Peberdy, 1990; Baker, 1990) and the generation of novel gene combinations or for the transfer of specific genes (e.g. genes coding for toxin production). It has been suggested thet the fungal cell wall forms a major vegetative incompatibility barrier against genetic exchange. Protoplast techniques have proved very successful in hybrid formation and interspecific crosses have been made (*Aspergillus nidulans*/*Aspergillus rugulosus*; *Penicillium chrysogenum*/*Penicillium cyaneofulvum*). Such approaches are now being applied to potential biological control fungi; however, it is important that any useful changes that are incorporated are stable.

Most success has been achieved in the production of fungal strains that are tolerant to fungicides. Such isolates would form a useful component of an integrated management programme where repeated use of fungicides is important. The use of a mutation approach (ultraviolet mutagenesis) generated benomyl resistant strains of *Trichoderma harzianum* which also showed improved efficiency as biocontrol agents (Papavizas, 1987). Transformation technology has also been used to introduce benomyl resistance into *Gaeumannomyces graminis* var. *tritici* (Henson *et al.*, 1988).

With a greater understanding of the attributes which contribute to the success of fungi as biological control agents, and the ways in which these operate and interact in the environment, it should be possible in the long term, to develop isolates with specific metabolic characteristics and features to improve this success. However, it is also likely that since fungi are such highly adaptable organisms and change so rapidly that the generation of new strains may be required on an ongoing basis.

References

Abbot, L.K. and Robson, A.D. (1985) Formation of external hyphae in soil by four species of vesicular–arbuscular mycorrhizal fungi. *New Phytologist*, **99**, 245–256.

Abuarghub, S.M. and Read, D.J. (1988a) The biology of mycorrhiza in the Ericaceae. XI: The distribution of nitrogen in the soil of a typical upland Callunetum with special reference to the 'free' amino acids. *New Phytologist*, **108**, 425–431.

Abuarghub, S.M. and Read, D.J. (1988b) The biology of mycorrhiza in the Ericaceae. XII: Quantitative analysis of individual 'free' amino acids in relation to time and depth in the soil profile. *New Phytologist*, **108**, 433–441.

Abuzinadah, R.A. and Read, D.J. (1986) The role of proteins in the nitrogen nutrition of ectomycorrhizal plants. III: Protein utilization by *Betula*, *Picea* and *Pinus* in mycorrhizal association with *Hebeloma crustuliniforme*. *New Phytologist*, **103**, 507–514.

Adams, A.E. and Pringle, J.R. (1984) Relationship of actin and tubulin distribution to bud growth in wild-type and morphogenetic mutant *Saccharomyces cerevisiae*. *Journal of Cell Biology*, **98**, 934–945.

Adams, P.B. (1990) The potential of mycoparasites for biological control of plant diseases. *Annual Review of Phytopathology*, **28**, 59–72.

Ahmad, I., Farrar, J.F. and Whitbread, R. (1984) Fluxes of phosphorus in leaves of barley infected with brown rust. *New Phytologist*, **98**, 361–375.

Ahmadjian, V. (1982) Algal/fungal symbioses. In *Progress in Phycological Research*, Vol. 1 (eds F.E. Round and V. Chapman) Elsevier, Amsterdam, pp. 179–233.

Ahmadjian, V. and Jacobs, J.B. (1981) Relationship between fungus and alga in the lichen *Cladonia cristatella* Tuck. *Nature*, **289**, 169–172.

Ahmadjian, V. and Jacobs, J.B. (1987) Studies on the development of synthetic lichens. In *Progress and Problems in Lichenology in the Eighties* (eds E. Peveling, A. Hennsen, O.L. Lange and C. Leuckert), J. Cramer, Berlin, pp. 47– 58.

Ainsworth, G.C. (1973) Introduction and keys to higher taxa. In *The Fungi: An Advanced Treatise*, Vol. IV B (eds G.C. Ainsworth, F.K. Sparrow and A.S. Sussman), Academic Press, New York, pp. 1–7.

Aist, J.R. (1983) Structural responses as resistance mechanisms. In *The Dynamics of Host Defence* (eds J. A. Bailey and B. J. Deverall), Academic Press, Sydney, pp. 33–70.

Albersheim, P. and Anderson-Prouty, A.J. (1975) Carbohydrates, proteins, cell surfaces, and the biochemistry of pathogenesis. *Annual Review of Plant Physiology*, **26**, 31–52.

Alexander, I.J. (1981) The *Picea sitchensis* and *Lactarius rufus* mycorrhizal association and its effects on seedling growth and development. *Transactions of the British Mycological Society*, **76**, 417–423.

Allen, P.J. (1953) Toxins and tissue respiration. *Phytopathology*, **43**, 221– 229.

Aloni, R. (1987) Differentiation of vascular tissues. *Annual Review of Plant Physiology*, **38**, 179–204.

Anglesea, D., Veltkamp, C. and Greenhalgh, G.N. (1982) The upper cortex of *Parmelia saxatilis* and other lichen thalli. *The Lichenologist*, **14**, 28–29.

Anglesea, D., Greenhalgh, G.N. and Veltkamp, C. (1983) The structure of the thallus tip in *Usnea subfloridana*. *The Lichenologist*, **15**, 73–80.

Anglesea, D. (1984) Thallus organisation and the distribution of algal cells in *Parmelia saxatilis* and other lichens. PhD Thesis, University of Liverpool, Liverpool.

Arnold, W.M., and Zimmermann, U. (1984) Electric field induced fusion and rotation of cells. In *Biological Membranes*, Vol. 5 (ed. D. Chapman), Academic Press, London, pp. 389–454.

Ayres, P.G. (1979) Carbon dioxide exchanges in plants infected by obligately biotrophic pathogens. In *Photosynthesis and Plant Development* (eds R. Marcelle, H. Clysters and M. Van Pouke), W. Junke, The Hague, pp. 343–354.

Ayres, P.G. (1981) Effects of disease on plant water relations. In *Effects of Disease on the Physiology of the Growing Plant* (ed. P.G. Ayres), Cambridge University Press, Cambridge, pp. 131–148.

Ayres, P.G. (1984a) Effects of infection on root growth and function: consequences for plant nutrient and water relations. In *Plant Diseases: Infection, Damage and Loss* (eds J.G. Jellis and R.K.S. Wood), Blackwell Scientific Publications, Oxford, pp. 105–117.

Ayres, P.G. (1984b) The interaction between environmental stress injury and biotic disease physiology. *Annual Review of Phytopathology*, **22**, 53–75.

Ayres, P.G. and Paul, N.D. (1986) Foliar pathogens alter the water relations of their hosts with consequences for both host and pathogen. In *Water, Fungi and Plants* (eds P.G. Ayres and L. Boddy), Cambridge University Press, Cambridge, pp. 267–285.

Bailey, J.A. (1982) Mechanisms of phytoalexin accumulation. In *Phytoalexins* (eds J.A.Bailey and J.W. Mansfield), Blackie, Glasgow, pp. 289–318.

Bailey, J.A. (1983) Biological perspectives of host-pathogen interactions. In *The Dynamics of Host Defence* (eds J.A. Bailey and B. J. Deverall), Academic Press, Sydney, pp. 1–32.

Bailey, J.A. (1987) Phytoalexins: a genetic view of their significance. In *Genetics and Plant Pathogenesis* (eds P.R. Day and G.J. Jellis), Blackwell Scientific Publications, Oxford, pp. 233–244.

Bailiss, K.W. and Wilson, I.M. (1967) Growth hormones and the creeping thistle rust. *Annals of Botany*, **31**, 195–211.

Baker, R. (1990) Some perspectives on the application of molecular approaches to biocontrol problems. In *Biotechnology of Fungi for Improving Plant Growth* (eds J.M. Whipps and R.D. Lumsden), Cambridge University Press, Cambridge, pp. 219–233.

Baldwin, B.C. and Rathmell, W.G. (1988) Evolution of concepts for chemical control of plant disease. *Annual Review of Phytopathology*, **26**, 265–283.

Ballance, D.J., Buxton, F.P. and Turner, G. (1983) Transformation of *Aspergillus nidulans* by the orotidine-5′-phosphate decarboxylase gene of *Neurospora crassa*. *Biochemical and Biophysical Research Communications*, **112**, 284–289.

Barna, B., Ibenthal, W.D. and Heitefuss, R. (1989) Extracellular RNase activity in healthy and rust infected wheat leaves. *Physiological and Molecular Plant Pathology*, **35**, 151–160.

Barnes, M.F., Scott, P.G. and Ooi, K.H. (1988) The RNase of leaves of resistant and susceptible barley cultivars after infection with leaf rust (*Puccinia hordei* Otth.). *Physiological and Molecular Plant Pathology*, **33**, 263–269.

Bartnicki-Garcia, S. (1968) Cell wall chemistry, morphogenesis and taxonomy of fungi. *Annual Reviews of Microbiology*, **22**, 87–107.

Bartnicki-Garcia, S. (1973) Fundamental aspects of hyphal morphogenesis. In *Microbial Differentiation. Symposium of the Society for General Microbiology*, Vol. 23 (eds J.O. Ashworth and J.E. Smith), Cambridge University Press, Cambridge, pp. 245–267.

Bartnicki-Garcia, S. (1980) Chitosomes and the origin of chitin microfibrils. In *Advances in Protoplast Research* (eds L. Ferenczy and G.L. Farkas), Pergamon Press, London, pp. 245–267.

Bartnicki-Garcia, S. and Bracker, C.E. (1984) Unique properties of chitosomes. In *Microbial Cell Wall Synthesis and Autolysis* (ed. C. Nombela), Elsevier, Amsterdam, pp. 101–112.

Bartnicki-Garcia, S., Bracker, C.E., Reyes, E. and Ruiz-Herrera, J. (1978) Isolation of chitosomes from taxonomically diverse fungi and synthesis of chitin microfibrils *in vitro*. *Experimental Mycology*, **2**, 173–192.

Barz, W., Daniel, S., Hinderer, W., Jauques, U., Kessman, H., Koster, J., Otto, C. and Tiemann, K. (1988) Elicitation and metabolism of phytoalexins in plant cell cultures. In *Applications of Plant Cell and Tissue Culture*. Ciba Foundation Symposium, **137**, Wiley, Chichester, pp. 178–198.

Beale, A.J., Clark, J.S.C. and Spencer-Phillips, P.T.N. (1990) Microscopy of endophytic hyphae facilitated by enzymic maceration and ATPase cytochemistry. In *Emag-Micro 89* Vol. 2 (eds H.Y. Elder and P.J. Goodhew), Institute of Physics Conference **98**, IOP Publishing Ltd, pp. 711–714.

Belesky, D.P., Devine, O.J., Pallas, Jr J.E. and Stringer, W.C. (1987) Photosynthetic activity of tall fescue as influenced by a fungal endophyte. *Photosynthetica*, **21**, 82–87.

Bell, E.A. (1981) The physiological role(s) of secondary (natural) products. In *The Biochemistry of Plants, a Comprehensive Treatise*, Vol. 7 *Secondary Plant Products*, (ed. E.E. Conn), Academic Press, New York, pp. 1–19.

Bevege, D.I., Bowen, G.D. and Skinner, M.F. (1975) Comparative carbohydrate physiology of ecto- and endo-mycorrhizas. In *Endomycorrhizas* (eds F. E. Sanders, B. Mosse and P.B. Tinker), Academic Press, London, pp. 149–174.

Bhairi, S., Staples, R.C., Freve, P. and Yoder, O.C. (1988) Analysis of a developmental gene expressed during differentiation of germlings of the bean rust fungus. In *Molecular Genetics of Plant-Microbe Interactions* (eds R. Palacios and D.P.S. Verma), American Phytopathological Society, St. Paul, Minnesota, pp. 271–272.

Bhairi, S., Staples, R.C., Freve, P. and Yoder, O.C. (1989) Characterization of an infection structure-specific gene from the rust fungus *Uromyces appendiculatus*. *Gene*, **81**, 237–243.

Biggs, A.R. and Miles, N.W. (1988) Association of suberin formation in uninoculated wounds with susceptibility to *Leucostoma cincta* and *L. persoonii* in various peach cultivars. *Phytopathology*, **78**, 1070–1074.

Bingham, J. (1981) The achievements of conventional plant breeding. *Philosophical Transactions of the Royal Society*, **292**, 441–454.

Bishop, C.D. and Cooper, R.M. (1983) An ultrastructural study of root invasion in three vascular wilt diseases. *Physiological Plant Pathology*, **22**, 15– 27.

Bonfante-Fasolo, P. and Gianninazzi-Pearson, V. (1979) Ultrastructural aspects of endomycorrhiza in the Ericaceae. I : Naturally infected hair roots of *Calluna vulgaris* L. Hull. *New Phytologist*, **83**, 739–744.

Bostock, R.M., Laine, R.A. and Kuc, J. (1982) Factors affecting the elicitation of sesquiterpenoid phytoalexin accumulation by eicosapentaenoic acid and arachidonic acid in potato. *Plant Physiology*, **70**, 1417–1421.

Bostock, R.M. and Stermer, B.A. (1989) Perspectives on wound healing in resistance to pathogens. *Annual Review of Phytopathology*, **27**, 343–371.

Boyer, J.S. (1985) Water transport. *Annual Review of Plant Physiology*, **36**, 473–516.

Bradford, K.J. and Hsaio, T.C. (1982) Physiological responses to moderate water stress. In *Encyclopaedia of Plant Physiology, Water Relations and Carbon Assimilation*, Vol. 12B (eds O.L. Lange, P.S. Nobel, C.B. Osmond and H. Ziegler), Springer-Verlag, Berlin, pp. 263–324.

Bradley, R., Burt, A.J. and Read, D.J. (1982) The biology of mycorrhiza in the Ericaceae. VIII: The role of mycorrhizal infection in heavy metal resistance. *New Phytologist*, **91**, 197–210.

Bradshaw, A.D. (1965) Evolutionary significance of phenotypic plasticity in plants. *Advances in Genetics*, **13**, 115–155.

Braithwaite, K.S., Irwin, J.A.G. and Manners, J.M. (1990) Restriction fragment length polymorphisms in *Colletotrichum gleosporoides* infecting *Stylosanthes* spp. in Australia. *Mycological Research*, **94**, 1129–1137.

Briese, D.T. (1986) Host resistance to microbial control agents. In *Biological Plant and Health Protection* (ed. J.M. Franz), Güstav Fischer Verlag, Stuttgart, pp. 233–256.

Brown, J.T. (1990) The initiation and maintenance of callus cultures. In *Methods In Molecular Biology* , Vol. 6, *Plant Cell and Tissue Culture* (eds J.W. Pollard and J. M. Walker), Humana Press, New Jersey, pp. 57–63.

Brownlee, C., Duddridge, J.A. Malibari, A. and Read, D.J. (1983) The structure and function of mycelial systems of ectomycorrhizal roots with special reference to their role in forming inter-plant connections and providing pathways for assimilate and water transport. *Plant and Soil*, **71**, 433–443.

Brownlee, C. and Jennings, D.H. (1982a) Long distance translocation in *Serpula lacrimans*: velocity estimates and the continuous monitoring of induced perturbations. *Transactions of the British Mycological Society*, **79**, 143–148.

Brownlee, C. and Jennings, D.H. (1982b) Pathway of translocation in *Serpula lacrimans*. *Transactions of the British Mycological Society*, **79**, 401–407.

Buchanan, B.B., Hutcheson, S.W., Magyarosy, A.C. and Montalbini, P. (1981) Photosynthesis in healthy and diseased plants. In *Effects of Disease on the*

Physiology of the Growing Plant (ed. P.G. Ayres), Cambridge University Press, Cambridge, pp. 13–28.

Bucheli, P., Doares, S.H., Albersheim, P. and Darvill, A. (1990) Host pathogen interactions XXXVI: Partial purification and characterization of heat-labile molecules secreted by the rice blast pathogen that solubilizes plant cell wall fragments that kill plant cells. *Physiological and Molecular Plant Pathology*, **36**, 159–173.

Bull, A.T. and Trinci, A.P.J. (1977) The physiology and metabolic control of fungal growth. *Advances in Microbial Physiology*, **15**, 1–84.

Burge, M.N. (1988) The scope of fungi in biological control. In *Fungi in Biological Control Systems* (ed. M.N. Burge), Manchester University Press, Manchester, pp. 1–18.

Bushnell, W.R. (1986) The role of nonself recognition in plant disease resistance. In *The Genetic Basis of Biochemical Mechanisms of Plant Disease* (eds J.V. Groth and W.R. Bushnell), American Phytopathological Society, St. Paul, Minnesota, pp. 1–24.

Burnett, J.H. (1979) Aspects of the structure and growth of hyphal walls. In *Fungal Walls and Hyphal Growth* (eds J.H. Burnett and A.P.J. Trinci), Cambridge University Press, Cambridge, pp. 1–25.

Buwalda, J.G., Stribley, D.P. and Tinker, P.B. (1984) The development of ectomycorrhizal root systems. V: The detailed pattern of development of infection and the control of infection level by the host in young leek plants. *New Phytologist*, **96**, 411–427.

Buxton, E.W. (1957) Some effects of pea root exudates on physiologic races of *Fusarium oxysporum* Fr. f. *pisi* (Linf.) Snyder and Hansen. *Transactions of the British Mycological Society*, **40**, 145–154.

Byrde, R.J.W. (1979) Role of polysaccharide-degrading enzymes in microbial pathogenicity. In *Microbial Polysaccharides and Polysaccharases* (eds R.C.W. Berkeley, G.W. Gooday and D.C. Ellwood), Academic Press, London, pp. 417–436.

Cabib, E., Duran, A. and Bowers, B. (1979) Localized activation of chitin synthase in the initiation of yeast septum formation. In *Fungal Walls and Hyphal Growth* (eds J.H. Burnett and A.P.J. Trinci), Cambridge University Press, Cambridge, pp. 189–201.

Caldwell, I.Y. and Trinci, A.P.J. (1973) The growth unit of the mould *Geotrichum candidum*. *Archives für Mikrobiologie* , **88**, 1–10.

Callow, J.A. (1984) Cellular and molecular recognition between higher plants and fungal pathogens. In *Cellular Interactions. Encyclopedia of Plant Physiology*, Vol. 17 (eds H. F. Liskens and J. Heslop-Harrison), Springer Verlag, Berlin, pp. 212–237.

Carroll, G.C. (1986) The biology of endophytism in plants with particular reference to woody perennials. In *Microbiology of the Phyllosphere* (eds N.J. Fokkema and J. Van den Heuvel), Cambridge University Press, Cambridge, pp. 205–222.

Carroll, G.C. (1988) Fungal endophytes in stems and leaves: from latent pathogen to mutualistic symbiont. *Ecology*, **69**, 2–9.

Carroll, G.C. and Petrini, O. (1983) Patterns of substrate utilisation by some fungal endophytes from coniferous foliage. *Mycologia*, **75**, 53–63.

Carver, T.L.W. and Griffiths, E. (1981) Relationship between powdery mildew infection, green leaf area and grain yield in barley. *Annals of Applied Biology*, **99**, 255–266.

Carver, T.L.W. and Griffiths, E. (1982) Effects of barley mildew on green leaf area and grain yield in field and greenhouse experiments. *Annals of Applied Biology*, **101**, 561–572.

Caten, C.E. (1987) The concept of race in plant pathology. In *Populations of Plant Pathogens* (eds M.S. Wolfe and C.E. Caten), Blackwell Scientific Publications, Oxford, pp. 21–37.

Choi, J.H. and Sung, Z.R. (1989) Induction, commitment and progression of plant embryogenesis. In *Plant Biotechnology* (eds S. Kung and C.J. Arntzen), Butterworths, London, pp. 141–160.

Clark, J.S.C. and Spencer-Phillips, P.T.N. (1990) Isolation of endophytic mycelia by enzymic maceration of *Peronospora* infected leaves. *Mycological Research*, **94**, 283–287.

Clarke, D.D. (1984) Tolerance of parasitic infection in plants. In *Plant Diseases: Infection, Damage and Loss* (eds J.G. Jellis and R.K.S. Wood), Blackwell Scientific Publications, Oxford, pp. 119–127.

Clay, K. (1986) Grass endophytes. In *Microbiology of the Phyllosphere* (eds N.J. Fokkema and J. Van den Heuvel), Cambridge University Press, Cambridge, pp. 188–204.

Clay, K. (1988) Clavicipitaceous fungal endophytes of grasses: coevolution and the change from parasitism to mutualism. In *Coevolution of Fungi with Plants and Animals* (eds K.A. Pirozynski and D.L. Hawksworth), Academic Press, New York, pp. 79–105.

Clay, K. (1989) Clavicipitaceous endophytes of grasses: their potential as biocontrol agents. *Mycological Research*, **92**, 1–12.

Cochrane, V.W. (1976) Glycolysis. In *The Filamentous Fungi*, Vol. 2 (eds J.E. Smith and D. R. Berry), Edward Arnold, London, pp. 65–91.

Cocking, E.C. (1983) Applications of protoplast technology to agriculture. In *Protoplasts 1983. Experientia Supplementum* **46**, Birkhaüser Verlag, pp. 123–126.

Cocking, E.C. (1984) Plant cell fusion: transformations using plant and bacterial protoplasts. In *Cell Fusion: Gene Transfer and Transformation* (eds R.F. Beers and B. Bassett), Raven, New York, pp. 139–144.

Cole, G.T. and Samson, R.A. (1979) *Patterns of Development in Conidial Fungi*. Pitman, London.

Collinge, D.B. and Slusarenko, A.J. (1987) Plant gene expression in response to pathogens. *Plant Molecular Biology*, **9**, 389–410.

Collins, C.R. and Farrar, J.F. (1978) Structural resistance to mass transfer in the lichen *Xanthoria parietina*. *New Phytologist*, **81**, 341–362.

Cooke, R.C. and Rayner, A.D.M. (1984) *Ecology of Saprotrophic Fungi*. Longman, Harlow, Essex.

Cooke, R.C. and Whipps, J.M. (1980) The evolution of modes of nutrition in fungi parasitic on terrestrial plants. *Biological Reviews*, **55**, 341–362.

Cooke, R.C. and Whipps, J.M. (1987) Saprotrophy, stress and symbiosis. In *Evolutionary Biology of the Fungi* (eds A. D. M. Rayner, C.M. Brasier and D. Moore), Cambridge University Press, Cambridge, pp. 137–148.

Cooley, R.N., Shaw, R.K., Franklin, F.C.H. and Caten, C.E. (1988) Transformation of the phytopathogenic fungus *Septoria nodorum* to hygromycin B resistance.

Current Genetics, **13**, 383–389.

Cooper, R.M. (1984) The role of cell wall degrading enzymes in infection and damage. In *Plant Diseases: Infection, Damage and Loss* (eds J.G. Jellis and R.K.S. Wood), Blackwell Scientific Publications, Oxford, pp. 13–27.

Cooper, R.M. (1987) The use of mutants in exploring depolymerases as determinants of pathogenicity. In *Genetics and Plant Pathogenesis* (eds P.R. Day and G.J. Jellis), Blackwell Scientific Publications, Oxford, pp. 261–281.

Cosgrove, D. (1986) Biophysical control of plant cell growth. *Annual Review of Plant Physiology*, **37**, 377–405.

Courtice, G.R.M. and Ingram, D.S. (1987) Sexual systems in pathogenic fungi. In *Genetics and Plant Pathogenesis* (eds P.R. Day and G.J. Jellis),Blackwell Scientific Publications, Oxford, pp. 143–160.

Cramer, C.L., Bell, J.M., Ryder, T.B., Bailey, J.A., Schuch, W., Bolwell, G.P., Robbins, M.P. Dixon, R.A. and Lamb, C.J. (1985) Co-ordinated synthesis of phytoalexin biosynthetic enzymes in biologically-stressed cells of bean (*Phaseollus vulgaris* L.). *The EMBO Journal*, **4**, 285–289.

Crittenden, P.D. (1989) Nitrogen relations of mat forming fungi. In *Nitrogen, Phosphorous and Sulphur Utilization by Fungi* (eds L. Boddy, R. Marchant and D. J. Read), Cambridge University Press, Cambridge, pp. 243–268.

Cutler, J.E. and Hazen, K.C. (1983) Yeast/mould morphogenesis in *Mucor* and *Candida albicans*. In *Secondary Metabolism and Differentiation in Fungi* (eds J.W. Bennett and A. Ciegler), Marcel Dekker, Inc., New York, pp. 267– 307.

Dale, J.E. (1988) Control of leaf expansion. *Annual Review of Plant Physiology and Plant Molecular Biology*, **39**, 439–473.

Daly, J.M. (1976) The carbon balance of diseased plants; changes in respiration, photosynthesis and translocation. In *Encyclopedia of Plant Physiology*, Vol. 4 (eds R. Heitefus and P.H. Williams), Springer-Verlag, Berlin, pp. 450–479.

Daly, J.M. (1984) The role of recognition in plant disease. *Annual Reviews of Phytopathology*, **22**, 273–307.

Daly, J.M. (1987) Toxins as determinants of plant diseases. In *Molecular Determinants of Plant Diseases* (eds S. Nishimura, C.P. Vance and N. Doke), Springer-Verlag, Japan Scientific Societies Press, Tokyo, pp. 119–126.

Darvill, A.G.M. and Albersheim, P. (1984) Phytoalexins and their elicitors – a defence against microbial infection in plants. *Annual Review of Plant Physiology*, **35**, 243–275.

Darvill, A.G.M., McNeil, P. Albersheim, P. and Delmer, D.P. (1980) The primary cell wall of flowering plants. In *The Biochemistry of Plants*, Vol. 1 (eds P.K. Stumpf and E.E. Conn), Academic Press, New York, pp. 91–162.

Daub, M.E. (1986) Tissue culture and the selection of resistance to pathogens. *Annual Reviews of Phytopathology*, **24**, 159–186.

Davey, M.R. and Power, J.B. (1988) Aspects of protoplast culture and plant regeneration. In *Progress in Plant Protoplast Research* (eds K.J. Puite, J.J.M. Dons, H.J.Huizing, A.J. Kool, M. Koorneef and F.A. Krens), Kluwer Academic Publishers, Dordrecht, pp. 1–6.

Davidson, A.D., Manners, J.M., Simpson, R.S. and Scott, K.J. (1987) cDNA cloning of mRNAs induced in resistant barley during infection by *Erysiphe graminis* f. sp. *hordei*. *Plant Molecular Biology*, **8**, 77–85.

Davidson, A.D., Manners, J.M., Simpson, R.S. and Scott, K.J.(1988) Altered gene expression in near-isogenic barley conditioned by different genes for resistance during infection by *Erysiphe graminis* f. sp. *hordei*. *Physiological and Molecular Plant Pathology*, **32**, 127–139.

Davies, W.J. (1981) Transpiration and water balance of plants. In *Plant Physiology*, Vol. 7 (eds J.F. Sutcliffe and F.C. Steward), Academic Press, London, pp. 201–213.

Davis, B. (1985) Factors influencing protoplast isolation. In *Fungal Protoplasts: Applications in Biochemistry and Genetics* (eds J.F. Peberdy and L. Ferenczy), Marcel Dekker, New York, pp. 45–71.

Davis, K.R., Darvill, A.G. and Albersheim, P. (1986) Host pathogen interactions XXXI: Several biotic and abiotic elicitors act synergistically in the induction of phytoalexin accumulation in soybean. *Plant Molecular Biology*, **6**, 23–32.

Day, P.R. (1974) *Genetics of Host-Parasite Interaction*. Freeman, San Fransisco.

De Bary, A. (1866) *Morphologie und Physiologie der Pilze, Flechten und Myxomyceten*. Engelmann, Leipzig.

De Bary, A. (1879) *Die Erscheinung der Symbiose*. Verlag Karl J. Trübner, Strassburg.

Dean, R.A. and Kuc, J.A. (1987) Immunisation against disease: the plant fights back. In *Fungal Infection of Plants* (eds G.F. Pegg and P.G. Ayres), Cambridge University Press, Cambridge, pp. 383–410.

Dickinson, S. (1949) Studies in the physiology of obligate parasitism. I: The stimuli determining the direction of growth of the germ tubes of rust and mildew spores. *Annals of Botany*, **13**, 89–104.

Dixon, G.R. (1981) *Vegetable Crop Diseases*. Macmillan, Wiltshire.

Dixon, R.A. (1986) The phytoalexin response: elicitation, signalling and control of host gene expression. *Biological Reviews*, **61**, 239–291.

Dixon, R.A. and Harrison, M.J. (1990) Activation, structure, and organization of genes involved in microbial defence in plants. *Advances in Genetics*, **28**, 165–234.

Doke, N., Chai, H.B. and Kawaguchi, A. (1987) Biochemical basis of triggering and suppression of hypersensitive cell response. In *Molecular Determinants of Plant Diseases* (eds S. Nishimura, C.P. Vance and N. Doke), Springer-Verlag, Japan Scientific Societies Press, Tokyo, pp. 75–95.

Donovan, A., Isaac, S., Collin, H.A. and Veltkamp, C.J. (1990) An ultrastructural study of the infection of the excised leaves of celery by *Septoria apiicola*, causal agent of leaf spot disease. *Mycological Research*, **94**, 548–552.

Dow, J.M. and Rubery, P.H. (1977) Chemical fractionation of the cell walls of mycelial and yeast-like forms of *Mucor rouxii*; a comparative study of the polysaccharide and glycoprotein components. *Journal of General Microbiology*, **99**, 29–41.

Duddridge, J.A. (1987) Specificity and recognition in ectomycorrhizal associations. In *Fungal Infection of Plants* (eds G.F. Pegg and P.G. Ayres), Cambridge University Press, Cambridge, pp. 25–44.

Duddridge, J.A., Malibari, A. and Read, D.J. (1980) Structure and function of mycorrhizal rhizomorphs with special reference to their role in water transport. *Nature*, **287**, 834–836.

Duddridge, J.A. and Read, D.J. (1982a) An ultrastructural analysis of the development of mycorrhizas in *Rhododendron ponticum*. *Canadian Journal of Botany*, **60**, 2345–2360.

Duddridge, J.A. and Read, D.J. (1982b) An ultrastructural analysis of the development of mycorrhizas in *Monotropa hypopitys*. *New Phytologist*, **92**, 203–214.

Ebel, J. (1986) Phytoalexin synthesis: the biochemical analysis of the induction process. *Annual Reviews of Phytopathology*, **24**, 235–264.

Ellingböe, A.H. (1982) Genetical aspects of active defence. In *Active Defence Mechanisms in Plants. Proceedings of the NATO Conference, Cape Sounion, Greece* (ed. R.K.S. Wood), Plenum Press, New York, pp. 179–192.

Ellingböe, A.H. (1987) Genetic analysis of interactions between microbes and plants. In *Fungal Infection of Plants* (eds G.F. Pegg and P.G. Ayres), Cambridge University Press, Cambridge, pp. 365–382.

Ellis, J.G., Lawrence, G.J., Peacock, W.J. and Pryor, A.J. (1988) Approaches to cloning plant genes conferring resistance to fungal pathogens. *Annual Review of Phytopathology*, **26**, 245–263.

Evans, D.A. (1989) Techniques in plant cell and tissue culture. In *Plant Biotechnology* (eds S. Kung and C.J. Arntzen), Butterworths, London, pp. 53– 76.

Evans, D.G. and Miller, M.H. (1990) The role of the external mycelial network in the effect of soil disturbance on vesicular-arbuscular mycorrhizal colonization of maize. *New Phytologist*, **114**, 65–71.

Farkas, V. (1979) Biosynthesis of fungal cell walls. *Microbiological Reviews*, **43**, 117–144.

Farrar, J.F. (1976) Ecological physiology of the lichen *Hypogymnia physodes*. II. Effects of wetting and drying cycles and the concept of physiological buffering. *New Phytologist*, **77**, 105–113.

Farrar, J.F. (1984) Effects of pathogens on plant transport systems. In *Plant Diseases: Infection, Damage and Loss* (eds J.G. Jellis and R.K.S. Wood), Blackwell Scientific Publications, Oxford, pp. 87–104.

Farrar, J.F. and Lewis, D.H. (1987) Nutrient relations in biotrophic infections. In *Fungal Infection of Plants* (eds G.F. Pegg and P.G. Ayres), Cambridge University Press, Cambridge, pp. 92–132.

Farrar, J.F. and Smith, D.C. (1976) Ecological physiology of the lichen *Hypogymnia physodes*. III. The importance of the rewetting phase. *New Phytologist*, **77**, 115–125.

Federation of British Plant Pathologists (FBPP) (1973) A Guide to the Use of Terms in Plant Pathology. *Phytopathological Paper* **17**, Commonwealth Mycological Institute, Kew, Surrey; reprinted 1986, CAB International.

Fencl, Z. (1978) Cell ageing and autolysis. In *The Filamentous Fungi*, Vol. 3 (eds J.E. Smith and D. R. Berry), Edward Arnold, London, pp. 389–405.

Flor, H.H. (1956) The complementary genic systems in flax and flax rust. *Advances in Genetics*, **8**, 29–54.

Florez-Martinez, A., Lopez-Romero, E., Martinez, J.P., Bracker, C.E., Ruiz-Herrera, J. and Bartnicki-Garcia, S. (1990) Protein composition of purified chitosomes of *Mucor rouxii*. *Experimental Mycology*, **14**, 160– 168.

Forster, H. and Coffey, M.D. (1990) Mating behaviour of *Phytophthora parasitica*: evidence for sexual recombination in oöspores using DNA Resitriction Fragment Length Polymorphisms as genetic markers. *Experimental Mycology*, **14**, 351–359.

Forster, H., Oudemans, P. and Coffey, M.D. (1990) Mitochondrial and nuclear DNA diversity within six species of *Phytophthora*. *Experimental Mycology*,

14, 18–31.

France, R.C. and Reid, C.P.P. (1983) Interactions of nitrogen and carbon in the physiology of ectomycorrhizae. *Canadian Journal of Botany*, **61**, 964–984.

Frank, A.B. (1885) Ueber die auf Wurzelsymbiose beruhende Eruährung gewisser Bäumer durch unterirdische Pilze. *Bericht der Deutschen Botanischen Gesellschaft*, **3**, 128–145.

Fravel, D.R. (1988) Role of antibiosis in the biocontrol of plant diseases. *Annual Review of Phytopathology*, **26**, 75–91.

French, R.C. (1985) The bioregulatory action of flavour compounds on fungal spores and other propagules. *Annual Review of Phytopathology*, **23**, 173– 199.

Gabriel, D.W., Loschke, D.C. and Rolfe, B.G. (1988) Gene-for-gene recognition: the ion channel defence model. In *Molecular Genetics of Plant-Microbe Interactions* (eds R. Palacios and D.P.S. Verma), American Phytopathological Society, St. Paul, Minnesota, pp. 3–14.

Gabriel, D.W. and Rolfe, B.G. (1990) Working models of specific recognition in plant-microbe interactions. *Annual Reviews of Phytopathology*, **28**, 365– 391.

Galpin, M.F.J., Jennings, D.H., Oates, K. and Hobot, J. (1978) Localization by X-ray microanalysis of soluble ions, particularly potassium and sodium, in fungal hyphae. *Experimental Mycology*, **2**, 258–269.

Galun, M. and Bubrick, P. (1984) Physiological interactions between the partners of the lichen symbiosis. In *Cellular Interactions, Encyclopedia of Plant Physiology*, Vol. 17 (eds H.F. Linskens and J. Heslop-Harrison), Springer-Verlag, Berlin, pp. 362–401.

Gatenby, A.A. (1989) Regulation and expression of plant genes in microorganisms. In *Plant Biotechnology* (eds S. Kung and C.J. Arntzen), Butterworths, Boston, pp. 93–112.

Gay, J.L. (1984) Mechanisms of biotrophy in fungal pathogens. In *Plant Diseases: Infection, Damage and Loss* (eds J.G. Jellis and R.K.S. Wood), Blackwell Scientific Publications, Oxford, pp. 49–59.

Gay, J.L. and Manners, J.M. (1981) Transport of host assimilates to the pathogen. In *Effects of Disease on the Physiology of the Growing Plant* (ed. P.G. Ayres), Cambridge University Press, Cambridge, pp. 85–100.

Gay, J.L. and Woods, A.M. (1987) Induced modifications in the plasma membranes of infected cells. In *Fungal Infection of Plants* (eds G.F. Pegg and P.G. Ayres), Cambridge University Press, Cambridge, pp. 79–91.

Georgopoulos, S.G. (1987) The development of fungicide resistance. In *Populations of Plant Pathogens; their Dynamics and Genetics* (eds M.S. Wolfe and C.E. Caten), Blackwell Scientific Publishers, Oxford, pp. 239–241.

Gerdemann, J.W. (1975) Vesicular–arbuscular mycorrhizae. In *The Organisation and Structure of Roots* (eds J.G. Torrey and D.T. Clarkson), Academic Press, New York, pp. 575–589.

Gianinazzi, S., Gianinazzi-Pearson, V. and Trouvelot, A. (1990) Potentialities and procedures for the use of endomycorrhizas with special emphasis on high value crops. In *Biotechnology of Fungi for Improving Plant Growth* (eds J.M. Whipps and R.D. Lumsden), Cambridge University Press, Cambridge, pp. 41–54.

Gianinazzi-Pearson, V. (1984) Host-fungus specificity, recognition and compatibility in mycorrhizae. In *Genes Involved in Microbe Plant Interactions* (eds

E.S. Dennis, B. Hohn, Th. Hohn, P. King, I. Schell and D.P.S. Verma), Springer-Verlag, Vienna, pp. 225–253.

Gianinazzi-Pearson, V. and Gianinazzi, S. (1989) Phosphorous metabolism in mycorrhizas. In *Nitrogen, Phosphorous and Sulphur Utilization by Fungi* (eds L. Boddy, R. Marchant and D.J. Read), Cambridge University Press, Cambridge, pp. 277–241.

Gilchrist, D.G. and Yoder, O.C. (1984) Genetics of host–parasite systems: a prospectus for molecular biology. In *Plant–Microbe Interactions*, Vol. 1 (eds T. Kosuge and E.W. Nester), Macmillan, Wiltshire, pp. 69–90.

Giles, K.L. and Vasil, I.K. (1980) Nitrogen fixation and plant tissue culture. *International Review of Cytology Supplement*, 11B, 81–99.

Gillespie, A.T. and Moorhouse, E.R. (1990) The use of fungi to control pests of agricultural and horticultural importance. In *Biotechnology of Fungi for Improving Plant Growth* (eds J.M. Whipps and R.D. Lumsden), Cambridge University Press, Cambridge, pp. 55–84.

Gooday, G.W. (1983) The hyphal tip. In *Fungal Differentiation* (ed. J.E. Smith), Marcel Dekker, New York, pp. 315–357.

Gow, N.A.R. (1984) Transhyphal electrical currents in fungi. *Journal of General Microbiology*, **130**, 3313–3318.

Gow, N.A.R., Kropfe, D.L. and Harold, F.M. (1984) Growing hyphae of *Achlya bisexualis* generate a longitudinal pH gradient in the surrounding medium. *Journal of General Microbiology*, **130**, 2967–2974.

Gradmann, P., Hansen, U.P., Long, W.S., Slayman, C.L. and Warncke, J. (1978) Current–voltage relationships for the plasmamembrane and its principle electrogenic pump in *Neurospora crassa*; l: Steady-state conditions. *Journal of Membrane Biology*, **39**, 333–367.

Graham, J.H. and Syvertsen, J.P. (1984) Influence of vesicular–arbuscular mycorrhiza on the hydraulic conductivity of roots of two citrus rootstocks. *New Phytologist*, **97**, 277–285.

Granlund, H.I., Jennings, D.H. and Veltkamp, K. (1984) Scanning electron microscope studies of rhizomorphs of *Armillaria mellea*. *Nova Hedwigia*, **39**, 85–99.

Greenhalgh, G.N. and Whitfield, A. (1987) Thallus tip structure and matrix development in *Bryoria fuscescens*. *Lichenologist*, **19**, 295–305.

Griesbach, R.J., Malmberg, R.L. and Carlson, P.S. (1982) Uptake of isolated lilly chromosomes by tobacco protoplasts. *Journal of Heredity*, **73**, 151– 152.

Griffiths, E. (1984) Foliar diseases: the damage caused and its effect on yield. In *Plant Diseases: Infection, Damage and Loss* (eds R.K.S. Wood and G.J. Jellis), Blackwell Scientific Publications, Oxford, pp. 149–159

Grime, J.P. (1977) Evidence for the existance of three primary strategies in plants and its relevance to ecological and evolutionary theory. *American Naturalist*, **111**, 1169–1194.

Grove, S.N. (1978) The cytology of hyphal tip growth. In *The Filamentous Fungi*, Vol. 3 (eds J.E. Smith and D.R. Berry), Edward Arnold,London, pp. 28– 50.

Habeshaw, D. (1973) Translocation and the control of photosynthesis in sugar beet. *Planta*, **110**, 213–226.

Habeshaw, D. (1984) Effects of pathogens on photosynthesis. In *Plant Diseases: Infection, Damage and Loss* (eds J.G. Jellis and R.K.S. Wood), Blackwell Scientific Publications, Oxford, pp. 63–72.

Hadley, G. (1975) Organisation and fine structure of orchid mycorrhiza. In *Endomycorrhizas* (eds F.E. Sanders, B. Mosse and P.H.B. Pinker), Academic Press, New York, pp. 335–351.

Hadwiger, L.A., Hess, S.L. and Von Broembsen, S. (1970) Stimulation of phenylalanine ammonia-lyase activity and phytoalexin production. *Phytopathology*, **60**, 332–336.

Hahlbrock, K. and Scheel, D. (1989) Physiology and molecular biology of phenylpropanoid metabolism. *Annual Review of Plant Physiology and Plant Molecular Biology*, **40**, 347–369.

Hall, R. (1986) Effects of root pathogens on plant water relations. In *Water, Fungi and Plants* (eds P.G. Ayres and L. Boddy), Cambridge University Press, Cambridge, pp. 241–265.

Hancock, J.G. and Huisman, O.C. (1981) Nutrient movement in host–pathogen systems. *Annual Review of Phytopathology*, **19**, 309–331.

Harder, D.E. and Chong, J. (1984) Structure and physiology of haustoria. In *The Cereal Rusts*, Vol. 1 (eds W. R. Bushnell and A. P. Roelfs), Academic Press, New York, pp. 431–476.

Hardie, K. and Leyton, L. (1981) The influence of vesicular–abuscular mycorrhiza on growth and water relations of red clover. I: In phosphate deficient soil. *New Phytologist*, **89**, 599–608.

Harding, K. and Cocking, E.C. (1986) The use of *Escherichia coli* spheroplasts: a possible approach to introduce foreign genes into crop plants. In *Plant Tissue Culture and its Agricultural Applications* (eds L. Withers and P. Alderson), Butterworths, London, pp. 367–374.

Harley, J.L. (1986) Mycorrhizal studies: past and future. In *Physiological and Genetical Aspects of Mycorrhizae* (eds V. Gianinazzi-Pearson and S. Gianinazzi), Institute National de la Recherche Agronomique, Paris, pp. 25– 33.

Harley, J.L. (1989) The significance of mycorrhiza. *Mycological Research*, **92**, 129–139.

Harley, J.L. and Harley, E.L. (1987) A check list of mycorrhiza in the British Flora. *New Phytologist* (supplement), **105**, 1–102.

Harley, J.L. and Smith, S.E. (1983) *Mycorrhizal Symbiosis*. 483 pp., Academic Press, London.

Harold, F.M. (1982) Pumps and currents. *Current Topics in Membranes and Transport*, **16**, 485–516.

Harold, R.L. and Harold, F.M. (1986) Ionophores and cytochalasins modulate branching in *Achlya bisexualis*. *Journal of General Microbiology*, **132**, 213–219.

Hartman, C.L., McCoy, T.J. and Knous, T.R. (1984) Selection of alfalfa (*Medicago sativa*) cell lines and regeneration of plants resistant to the toxin(s) produced by *Fusarium oxysporum* f. sp. *medicaginis*. *Plant Science Letters*, **34**, 183–194.

Hawksworth, D.L., Sutton, B.C. and Ainsworth, G.C. (1983) *Ainsworth and Bisby's Dictionary of the Fungi*, 7th edn Commonwealth Mycological Institute, Kew, Surrey.

Heath, M.C. (1979) Effects of heat shock, actinomycin D, cycloheximide and blasticidin S on nonhost interactions with rust fungi. *Physiological Plant Pathology*, **15**, 211–218.

Heath, M.C. (1986) Implications of non host resistance for understanding host–parasite interactions. In *Genetic Basis of Biochemical Mechanisms of Plant Disease* (eds J.V. Groth and W.R. Bushnell), American Phytopathological Society, pp. 25–42.

Helgeson, J.P. (1983) Studies of host–pathogen interactions *in vitro*. In *Use of Tissue Culture and Protoplasts in Plant Pathology* (eds J.P. Helgeson and B.J. Deverall), Academic Press, Sydney, pp. 9–38.

Henson, J.M., Blake, N.K. and Pilgeram, A. (1988) Transformation of *Gaeumannomyces graminis* to benomyl resistance. *Current Genetics*, **14**, 113–117.

Hetrick, B.A.D. (1989) Aquisition of phosphorous by vesicular–arbuscular mycorrhizal fungi and the growth responses of their hosts. In *Nitrogen, Phosphorous and Sulphur Utilization by Fungi* (eds L. Boddy, R. Marchant and D.J. Read), Cambridge University Press, Cambridge, pp. 205–226.

Hickey, K.D. (1986) *Methods for Evaluating Pesticides for the Control of Plant Pathogens*. The American Phytopathological Society Press.

Hill, T.W. and Mullins, J.T. (1980) Hyphal tip growth in *Achlya*. ll. Subcellular localization of cellulase and associated enzymes. *Canadian Journal of Microbiology*, **26**, 1141–1146.

Ho, I. and Trappe, J.M. (1973) Translocation of the ^{14}C from *Festuca* plants to their endomycorrhizal fungi. *Nature*, **244**, 30–31.

Hocart, M.J. and Peberdy, J.F. (1990) Protoplast technology and strain selection. In *Biotechnology of Fungi for Improving Plant Growth* (eds J.M. Whipps and R.D. Lumsden), Cambridge University Press, Cambridge, pp. 235–258.

Hoch, H.C. and Howard, R.J. (1980) Ultrastructure of freeze-substituted hyphae of the basidiomycete *Laetisaria arvalis*. *Protoplasma*, **103**, 281–297.

Hohl, H.R. and Balsiger, S. (1986) A model system for the study of fungus–host surface interactions; adhesion of *Phytophthora magasperma* to protoplasts and mesophyll cells of soybean. In *Proceedings of the Nato Advanced Workshop on Recognition in Microbe–Plant Symbiotic and Pathogenic Interactions, NATO ASI series H: Cell Biology*, Vol. 4 (ed. B. Lugtenberg), Springer-Verlag, Berlin, pp. 259–272.

Honegger, R. (1984) Cytological aspects of the mycobiont–phycobiont relationship in lichens. *Lichenologist*, **16**, 111–127.

Hornby, D. and Fitt, B.D.L. (1981) Effects of root invading fungi on structure and function of cereal roots. In *Effects of Disease on the Physiology of the Growing Plant* (ed. P.G. Ayres), Cambridge University Press, Cambridge, pp. 101–130.

Howard, R.J. (1981) Ultrastructural analysis of hyphal tip cell growth in fungi: Spitzenkörper, cytoskeleton and endomembranes after freeze-substitution. *Journal of Cell Science*, **48**, 89–103.

Howard, R.J. and Aist, J.R. (1979) Hyphal tip ultrastructure of the fungus *Fusarium*: improved preservation by freeze-substitution. *Journal of Ultrastructure Research*, **66**, 224–234.

Howard, R.J. and Aist, J.R. (1980) Cytoplasmic microtubules and fungal morphogenesis: ultrastructural effects of methyl benzimidazole-2–ylcarbamate determined by freeze-substitution of hyphal tip cells. *Journal of Cell Biology*, **87**, 55–64.

Howard, R.J., Bourett, T.M. and Ferrari, M.A. (1991) Infection by *Magnaporthe*: an *in vitro* analysis. In *Electron Microscopy of Plant Pathogens* (eds K. Mendgen and D.E. Leseman), Springer-Verlag, Heidelberg (in press).

Hulbert, S.H. and Michelmore, R.W. (1988) DNA restriction fragment length polymorphism and somatic variation in the lettuce downy mildew fungus, *Bremia lactucae*. *Molecular Plant–Microbe Interactions*, **1**, 17–24.

Hunsley, D. and Burnett, J.H. (1970) The ultrastructural architecture of the walls of some hyphal fungi. *Journal of General Microbiology*, **62**, 203–218.

Hunsley, D. and Kay D. (1976) Wall architecture of the *Neurospora* hyphal apex; immunofluorescent localisation of wall surface antigens. *Journal of General Microbiology*, **95**, 233–248.

Humphreys, A.M. and Gooday, G.W. (1984) Properties of chitinase activities from *Mucor mucedo*: evidence for a membrane bound zymogenic form. *Journal of General Microbiology*, **130**, 1359–1366.

Ingham, J.L. (1981) Phytoalexin induction and its taxonomic significance in the Leguminosae (sub-family Papilionoideae). In *Advances in Legume Systematics* (eds R.M. Polhill and P.H.R. Raven), Royal Botanic Gardens, Kew, pp. 599–626.

Ingram, D.S. (1980) The establishment of dual cultures of downy mildew fungi and their hosts. In *Tissue Culture Methods For Plant Pathologists* (eds D.S. Ingram and J.P. Helgeson), Blackwell Scientific Publications, Oxford, pp. 139–149.

Ingram, D.S. (1982) A structural view of active defense. In *Active Defence Mechanisms in Plants* (ed. R.K.S. Wood), Plenum Press, New York, pp. 19–38.

Isaac, S. (1985) Metabolic properties of protoplasts. In *Fungal Protoplasts; Applications in Biochemistry and Genetics* (eds J.F. Peberdy and L. Ferenczy), Marcel Dekker, New York, pp. 171–187.

Isaac, S. and Gokhale, A.V. (1982) Autolysis: a tool for protoplast production from *Aspergillus nidulans*. *Transactions of the British Mycological Society*, **78**, 389–394.

Isaac, S., Ryder, N.S. and Peberdy, J.F. (1978) Distribution and activation of chitin synthase in protoplast fractions released during the lytic digestion of *Aspergillus nidulans* hyphae. *Journal of General Microbiology*, **105**, 45– 50.

Jacquelinet-Jeanmougin, J., Gianinazzi-Pearson, V. and Gianinazzi, S. (1987) Endomycorrhizas in the Gentianaceae. II: Ultrastructural aspects of symbiont relationships in *Gentiana lutea* L. *Symbiosis*, **3**, 269–286.

Jaffe, L.F. (1977) Electrophoresis along cell membranes. *Nature*, **265**, 600–602.

Jaffe, L.F. and Nuccitelli, R. (1974) An ultrasensitive vibrating probe for measuring steady extracellular currents. *Journal of Cell Biology*, **63**, 614–628.

Jasper, D.A., Abbott, L.K. and Robson, A.D. (1989) Soil disturbance reduces the infectivity of external hyphae of vesicular–arbuscular mycorrhizal fungi. *New Phytologist*, **112**, 93–99.

Jennings, D.H. (1979) Membrane transport and hyphal growth. In *Fungal Walls and Hyphal Growth* (eds J.H. Burnett and A.P.J. Trinci), Cambridge University Press, Cambridge, pp. 279–294.

Jennings, D.H. (1984) Water flow through mycelia. In *The Ecology and Physiology of the Fungal Mycelium* (eds D.H. Jennings and A.D.M. Rayner), Cambridge University Press, Cambridge, pp. 143–164.

Jennings, D.H. (1987) Translocation of solutes in fungi. *Biological Reviews*, **62**, 215–243.

Johnson, M.C., Pirone, T.P., Siegel, M.R. and Varney, D.R. (1982) Detection of *Epichloë typhina* in tall fescue by means of enzyme-linked immuno absorbent assay. *Phytopathology*, **72**, 647–650.

Jones, H.G. (1986) Movement of liquid water and the effects of fungal infection. In *Water, Fungi and Plants* (eds P.G.Ayres and L. Boddy), Cambridge University Press, pp. 223–239.

Jones, P.W. (1990) *In vitro* selection for disease resistance. In *Plant Cell Line Selection; Procedures and Application* (ed. P.J. Dix), VCH, Cambridge, pp. 113–149.

Kao, K.N. and Michayluk, M.R. (1974) A method for high frequency intergeneric fusion of plant protoplasts. *Planta*, **115**, 355–367.

Keitt, G.W. and Boone, D.M. (1954) Induction and inheritance of mutant characters in *Venturia inaequalis* in relation to its pathogenicity. *Phytopathology*, **44**, 362–370.

Keon, J.P.R., Byrde, R.J.W. and Cooper, R.M. (1987) Some aspects of fungal enzymes that degrade plant cell walls. In *Fungal Infection of Plants* (eds G.F. Pegg and P.G. Ayres), Cambridge University Press, Cambridge, pp. 133– 157.

Kerr, A. (1987) The impact of molecular genetics on plant pathology. *Annual Review of Phytopathology*, **25**, 87–110.

Kerry, B.R. (1990) Fungi as biological control agents for plant parasitic nematodes. In *Biotechnology of Fungi for Improving Plant Growth* (eds J.M. Whipps and R.D. Lumsden), Cambridge University Press, Cambridge, pp. 153–170.

Kessmann, H. and Barz, W. (1987) Accumulation of isoflavones and pterocarpan phytoalexins in cell suspension cultures of different cultivars of chickpea (*Cicer arietinum*). *Plant Cell Replication*, **6**, 55–59.

Kirk, J.J. and Deacon, J.W. (1987) Control of the take-all fungus by *Microdochium bolleyi*, and interactions involving *M. bolleyi*, *Phialophora graminicola* and *Periconia macrospinosa* on cereal roots. *Plant and Soil*, **98**, 231–237.

Kistler, H.C. and Benny, U. (1988) Genetic transformation of the fungal plant wilt pathogen *Fusarium oxysporum*. *Current Genetics*, **13**, 145–149.

Klein, T.M., Wolf, E.D., Wu, R. and Sanford, J.C. (1987) High velocity microprojectiles for delivering nucleic acids into living cells. *Nature*, **327**, 70–73.

Klionsky, D.J., Herman, P.K. and Scott, E.D. (1990) The fungal vacuole: composition, function and biogenesis. *Microbiological Reviews*, **54**, 266– 292.

Knoche, H.W. and Duvick, J.P. (1987) The role of fungal toxins in plant disease. In *Fungal Infection of Plants* (eds G.F. Pegg and P.G. Ayres), Cambridge University Press, Cambridge, pp. 158–192.

Kohomoto, K., Otani, H. and Nishimura, S. (1987) Primary action sites of host-specific toxins produced by *Alternaria* species. In *Molecular Determinants of Plant Diseases* (eds S. Nishimura, C.P. Vance and N. Doke), Springer-Verlag, Japan Scientific Societies Press, Tokyo, pp. 127–143.

Koop, H.U. and Schweiger, H.G. (1985) Regeneration of plants after electrofusion of selected pairs of protoplasts. *European Journal of Cell Biology*, **39**, 46–49.

Kolattukudy, P.E. (1985) Enzymatic penetration of the plant cuticle by fungal pathogens. *Annual Review of Phytopathology*, **23**, 223–250.

Kolattukudy, P.E. and Crawford, M.S. (1987) The role of polymer degrading enzymes in fungal pathogenesis. In *Molecular Determinants of Plant Diseases* (eds S. Nishimura, C.P. Vance and N. Doke), Springer-Verlag, Japan Scientific Societies Press, Tokyo, pp. 75–95.

Kolattukudy, P.E., Podila, G.K., Roberts, E. and Dickman, M.D. (1989) Gene expression resulting from the early signals in plant–fungus interaction. In

Molecular Biology of Plant–Pathogen Interactions (eds B. Staskawicz, P. Ahlquist and O.Yoder), Alan R. Liss, New York, pp. 87–102.

Kosuge, T. and Kimpel, J.A. (1981) Energy use and metabolic regulation in plant-–pathogen interactions. In *Effects of Disease on the Physiology of the Growing Plant* (ed. P.G. Ayres), Cambridge University Press, Cambridge, pp. 29–45.

Kropf, D.L., Caldwell, J.H., Gow, N.A.R. and Harold, F.M. (1984) Transcellular ion currents in the water mould *Achlya;* Amino acid symport as a mechanism of current entry. *Journal of Cell Biology*, **99**, 486–496.

Kropf, D.L., Lupa, M.D., Caldwell, J.H. and Harold, F.M. (1983) Cell polarity: endogenous ion currents precede and predict branching in the water mould *Achlya. Science*, **220**, 1385–1387.

Kubicek, C.P. (1988) Regulatory aspects of the tricarboxylic acid cycle in filamentous fungi – a review. *Transactions of the British Mycological Society*, **90**, 339–349.

Kuhn, P.J. and Hargreaves, J.L. (1987) Antifungal substances from herbaceous plants. In *Fungal Infection of Plants* (eds G.F. Pegg and P.G. Ayres), Cambridge University Press, Cambridge, pp. 193–218.

Kunoh, H. (1990) Ultrastructure and mobilization of ions near infection sites. *Annual Review of Phytopathology*, **28**, 93–111.

Kunoh, H., Nicholson, R.L., Yosioka, H., Yamaoka, N. and Kobayashi, I. (1990) Preparation of the infection court by *Erisyphe graminis*: degradation of the host cuticle. *Physiological and Molecular Plant Pathology*, **6**, 397–407.

Lamb, C.J., Lawton, M.A., Dron, M. and Dixon, R.A. (1989) Signals and transduction mechanisms for activation of plant defences against microbial attack. *Cell*, **56**, 215–224.

Lahoz, R., Reyes, F. and Perez-Leblic, M.I. (1976) Lytic enzymes in the autolysis of filamentous fungi. *Mycopathologia*, **60**, 45–49.

Larkin, P.J. and Scowcroft, W.R. (1981) Somaclonal variation – a novel source of variability from cell cultures for plant improvement. *Theoretical and Applied Genetics*, **60**, 197–214.

Larkin, P.J. and Scowcroft, W.R. (1983) Somaclonal variation and eye spot toxin tolerance in sugar cane. *Plant Cell Tissue and Organ Culture*, **2**, 111–122.

Law, R. and Lewis, D.H. (1983) Biotic environments and the maintenance of sex – some evidence from mutualistic symbioses. *Biological Journal of the Linnean Society*, **20**, 249–276.

Lazar, G.B. (1983) Recent developments in plant protoplast fusion and selection technology. In *Protoplasts 1983* (ed. I. Potrykus), Birkhaüser, Basel, pp. 61–68.

Leake, J.R. and Read, D.J. (1990) Chitin as a nutrient source for mycorrhizal fungi. *Mycological Research*, **94**, 993–995.

Leong, S.A. (1988) Recombinant DNA research in phytopathogenic fungi. *Advances in Plant Pathology*, **6**, 1–26.

Leong, S.A. and Holden, D.W. (1989) Molecular genetic approaches to the study of fungal pathogenesis. *Annual Reviews of Phytopathology*, **27**, 463–481.

Levitt, J. (1980) *Responses of Plants to Environmental Stresses*, Vol. 2 *Water, Raditation, Salt and Other Stresses.* 2nd edn, Academic Press, New York.

Levy, Y., Dodd, J. and Krikun, J. (1983) Effect of irrigation, water salinity and rootstock on the vertical distribution of vesicular–arbuscular mycorrhiza in citrus roots. *New Phtyologist*, **95**, 397–403.

Lewin, R.A. (1982) Symbiosis and parasitism – definitions and evaluations. *Bioscience*, **32**, 254–259.

Lewis, D.H. (1973) Concepts of fungal nutrition and the origin of biotrophy. *Biological Reviews*, **48**, 261–278.

Lewis, D.H. (1974) Micro-organsims and plants: the evolution of parasitism and mutualism. In *Symposium of the Society for General Microbiology*, **24**, 367– 392.

Lewis, D.H. (1984) Occurrence and distribution of storage carbohydrates in vascular plants. In *Storage Compounds in Vascular Plants* (ed. D.H. Lewis), Cambridge University Press, Cambridge, pp. 1–52.

Lewis, D.H. (1985) Symbiosis and mutualism: crisp concepts and soggy semantics. In *The Biology of Mutualism* (ed. D.H. Boucher), Croom Helm, London, pp. 29–39.

Lewis, D.H. and Harley, J.L. (1965a) Carbohydrate physiology of mycorrhizal roots of beech. I. Identity of endogenous sugars and utilization of exogenous sugars. *New Phytologist*, **64**, 224–237.

Lewis, D.H. and Harley, J.L. (1965b) Carbohydrate physiology of mycorrhizal roots of beech. II. Utilization of exogenous sugars by uninfected and mycorrhizal roots. *New Phytologist*, **64**, 238–256.

Lewis, D.H. and Harley, J.L. (1965c) Carbohydrate physiology of mycorrhizal roots of beech. III. Movement of sugars between host and fungus.*New Phytologist*, **64**, 256–269.

Lewis, K., Whipps, J.M. and Cooke, R.C. (1990) Mechanisms of biological disease control with special reference to the case study of *Pythium oligandrum* as an antagonist. In *Biotechnology of Fungi for Improving Plant Growth* (eds J.M. Whipps and R.D. Lumsden), Cambridge University Press, Cambridge, pp. 191–217.

Li, A. and Heath, M.C. (1990) Effect of plant growth regulators on the interactions between bean plants and rust fungi non-pathogenic on beans. *Physiological and Molecular Plant Pathology*, **37**, 245–254.

Lifshitz, R., Windham, M.T. and Baker, R. (1986) Mechanism of biological control of pre-emergence damping-off of pea by seed treatment with *Trichoderma* species. *Phytopathology*, **76**, 720–725.

Lin, T.S. and Kolattukudy, P.E. (1980) Isolation and characterization of a cuticular polyester (cutin) hydrolysing enzyme from phytopathogenic fungi. *Physiological Plant Pathology*, **17**, 1–15.

Löegering, W.Q. (1966) The relationship between host and pathogen in stem rust of wheat. *Proceedings of the Second International Genetics Symposium. Hereditas*, **2** (Suppl.), 167–177.

Lorz, H. (1985) Isolated cell organelles and subprotoplasts; their roles in somatic cell genetics. In *Plant Genetic Engineering* (ed. J.H. Dodds), Cambridge University Press, Cambridge, pp. 171–193.

Lucas, J.A. and Knights, I. (1987) Spores on leaves: endogenous and exogenous control of development. In *Fungal Infection of Plants* (eds G.F. Pegg and P.G. Ayres), Cambridge University Press, Cambridge, pp. 45–59.

Luther, P.W., Peng, H.B. and Lin, J.J.C. (1983) Changes in cell shape and actin distribution induced by constant electric fields. *Nature*, **303**, 61–64.

Lynch, P.T., Collin, H.A. and Isaac, S. (1985) Use of autolytic enzyme for isolation of protoplasts from *Fusarium tricinctum* hyphae. *Transactions of the British Mycological Society*, **84**, 473–478.

Lynch, P.T., Isaac, S. and Collin, H.A. (1989) Uptake of fungal protoplasts by plant protoplasts. In *Plant Protoplasts and Genetic Engineering*, Vol. 11 (ed. Y.P.S. Bajaj), Springer-Verlag, Berlin, pp. 406–407.

Lysek, G. (1976) Formation of perithecia in colonies of *Podospora anserina*. *Planta*, **133**, 81–83.

Lysek, G. (1984) Physiology and ecology of rhythmic growth and sporulation in fungi. In *The Ecology and Physiology of the Fungal Mycelium* (eds D.H. Jennings and A.D.M. Rayner), Cambridge University Press, Cambridge, pp. 323–342.

MacDonald, M.V. and Ingram, D.S. (1986) Towards the selection *in vitro* for resistance to *Alternaria brassicicola* (Schw.) Wilts., in *Brassica napus* spp. *oleifera* (Metzg.) Sinsk., Winter oilseed rape. *New Phytologist*, **104**, 621–630.

Manners, J.M. and Scott, K.J. (1984) The effect of infection by *Erysiphe graminis* f. sp. *hordei* on protein synthesis *in vivo* in leaves of barley. *Plant and Cell Physiology*, **25**, 1307–1311.

Marre, E., Lado, P., Ferroni, A. and Denti, A.B. (1974) Transmembrane potential increase induced by auxin, benzyladenine and fusicoccin. Correlation with proton extrusion and cell enlargement. *Plant Science Letters*, **2**, 257–265.

Marx, D.H. (1975) Mycorrhiza and the establishment of trees on strip-mined land. *Ohio Journal of Science*, **75**, 288–297.

Marx, D.H. and Cordell, C.E. (1990) The use of specific ectomycorrhizas to improve artificial forestation practices. In *Biotechnology of Fungi for Improving Plant Growth* (eds J.M. Whipps and R.D. Lumsden), Cambridge University Press, Cambridge, pp. 1–25.

Matta, A. (1982) Mechanisms in non-host resistance. In *Active Defence Mechanisms in Plants* (ed. R.K.S. Wood), Plenum Press, New York, pp. 119–141.

Mauch, F., Mauch-Mani, B. and Boller, T. (1988) Antifungal hydrolases in pea tissue. II: Inhibition of fungal growth by a combination of chitinase and β-(1, 3)-glucanase. *Plant Physiology*, **88**, 936–942.

Mauch, F. and Staehelin, L.A. (1989) Functional implications of the subcellular localization of ethylene-induced and β-1,3–glucanase in bean leaves. *Plant Cell*, **1**, 447–457.

McGillivray, A.M. and Gow, N.A.R. (1986) Applied electrical fields polarize the growth of mycelial fungi. *Journal of General Microbiology*, **132**, 2515–2525.

McWilliam, J.R. (1983) Physiological basis for chilling stress and consequences for crop production. In *Crop Reactions to Water and Temperature Stresses in Humid, Temperate Climates* (eds C.D. Roper Jr and P.J. Cramer), Boulder, Colorado, pp. 113–132.

Melin, E. and Nilsson, H. (1953) Transfer of labelled nitrogen from glutamic acid to pine seedlings through the mycelium of *Boletus variegatus* (Sw.) Fr. *Nature*, **171**, 134.

Melin, E. and Nilsson, H. (1957) Transport of ^{14}C-labelled photosynthate to the fungal associate of pine mycorrhiza. *Svensk, Botanische Tidskrift*, **51**, 166–186.

Mendgen, K., Schneider, A., Sterk, M. and Fink, W. (1988) The differentiation of infection structures as a result of recognition events between some biotrophic parasites and their hosts. *Journal of Phytopathology*, **123**, 259–272.

Menge, J.A. (1983) Utilization of vesicular–arbuscular mycorrhizal fungi in agriculture. *Canadian Journal of Botany*, **61**, 1015–1024.

Meredith, C.P. (1984) Selecting better crops from cultured cells. In *Gene Manipu-*

lation in Crop Improvement (ed. J.P. Gustafson), Plenum Press, New York, pp. 503–528.

Metzenberg, R.L. (1991) The impact of molecular biology on mycology. *Mycological Research*, **95**, 9–13.

Miller, J.D. (1986) Toxic metabolites of epiphytic and endophytic fungi of conifer needles. In *Microbiology of the Phyllosphere* (eds N.J. Fokkema and J. Van den Heuvel), Cambridge University Press, Cambridge, pp. 223–231.

Miller, S.A. (1986) Tissue culture methods in plant pathology. II: Fungi. In *Plant Cell Culture – A Practical Approach* (ed. R.A. Dixon), IRL Press, Oxford, pp. 215–229.

Mills, D. (1985) Transposon mutagenesis and its potential for studying virulence genes in plant pathogens. *Annual Review of Phytopathology*, **23**, 297–320.

Monnier, M. (1990a) Induction of embryogenesis in callus culture. In *Methods in Molecular Biology* , Vol. 6, *Plant Cell and Tissue Culture* (eds J.W. Pollard and J. M. Walker), Humana Press, New Jersey, pp. 141–148.

Monnier, M. (1990b) Induction of embryogenesis in suspension culture. In *Methods In Molecular Biology* , Vol. 6, *Plant Cell and Tissue Culture* (eds J.W. Pollard and J. M. Walker), Humana Press, New Jersey, pp. 149–157.

Morikawa, H., Hayashi, Y., Hirabayashi, Y., Asada, M. and Yamada, Y. (1988) Cellular and vacuolar fusion of protoplasts electrofused using platinum microelectrodes. *Plant Cell Physiology*, **29**, 189–193.

Moss, M.O. (1984) The mycelial habit and secondary metabolite production. In *The Ecology and Physiology of the Fungal Mycelium* (eds D.H. Jennings and A.D.M. Rayner), Cambridge University Press, Cambridge, pp. 127–142.

Mudd, J.B. (1982) Effects of oxidants on metabolic function. In *Encyclopaedia of Plant Physiology, Water Relations and Carbon Assimilation*, Vol. 12B (eds O.L. Lange, P.S. Nobel, C.B. Osmond and H. Ziegler), Springer-Verlag, Berlin, pp. 189–203.

Müller, K.O. and Börger, H. (1940) Experimentelle Untersuchungen uber die Phytophthora resistenz den kartoffel. *Arbeiten aus der biologischen Bundesanstalt fur Land-und Footwirtschaft*, **23**, 189–231.

Murashige, T. and Skoog, F. (1962) A revised medium for rapid growth and bioassays with tobacco tissue cultures. *Physiologia Plantarum*, **15**, 473–497.

Nelson, L.A. (1986) Use of statistics in planning, data analysis, and interpretation of fungicide and nematicide tests. In *Methods for Evaluating Pesticides for Control of Plant Pathogens* (ed. K.D. Hickey), American Phytopathological Society Press, St. Paul, Minnesota, pp. 11–24.

Nuccitelli, R. (1983) Transcellular ion currents. Signals and effectors of cell polarity. *Modern Cell Biology*, **2**, 451–481.

O'Connell, R.J., Bailey, J.A. and Richmond, D. (1985) Cytology and physiology of infection of *Phaseolus vulgaris* by *Colletotrichum lindemuthianum*. *Physiological Plant Pathology*, **27**, 75–98.

O'Dell, M., Wolfe, M.S., Flavell, R.B., Simpson, C.G. and Summers, R.W. (1989) Molecular variation in populations of *Erysiphe graminis* on barley, oats and rye. *Plant Pathology*, **38**, 340–351.

Odermatt, M., Röthlisberger, A., Werner, C. and Hohl, H.R. (1988) Interactions between agarose-embedded plant protoplasts and germ tubes of *Phytophthora*. *Physiological and Molecular Plant Pathology*, **33**, 209–220.

Oliver, R.P., Roberts, I.N., Harling, R., Keynon, L., Punt, P.J., Dingemanse, M.A. and Van Den Hondel, C.A. (1987) Transformation of *Fulvia fulva*, a fungal pathogen of tomato, by hygromycin resistance. *Current Genetics*, **12**, 231–233.

Oliver, R.P., Roberts, I.N., McHale, M., Coddington, A., Talbot, N., Lewis, B., Kenyon, L., Turner, J. and El Hamouri, B. (1988) Molecular approaches for the study of the pathogenicity of *Fulvia fulva*. In *Molecular Genetics of Plant-Microbe Interactions* (eds R. Palacios and D.P.S. Verma), American Phytopathological Society, St. Paul, Minnesota, pp. 263–264.

Owera, S.A.P., Farrar, J.F. and Whitbread, R. (1983) Translocation from leaves of barley infected with brown rust. *New Phytologist*, **94**, 111–123.

Papavizas, G.C. (1987) Genetic manipulation to improve the effectiveness of biocontrol fungi for plant disease control. In *Innovative Approaches to Plant Disease Control* (ed. I. Chet), John Wiley, New York, pp. 193–212.

Papendick, R.I. and Mulla, D.J. (1986) Basic principles of cell and tissue water relations. In *Water, Fungi and Plants* (eds P.G. Ayres and L. Boddy), Cambridge University Press, Cambridge, pp. 1–25.

Passioura, J.B. (1988) Water transport in and to roots. *Annual Review of Plant Physiology and Plant Molecular Biology*, **39**, 245–265.

Paul, N.D. (1989) Effects of fungal pathogens on nitrogen, phosphorous and sulphur relations of individual plants and populations. In *Nitrogen, Phosphorous and Sulphur Utilization by Fungi* (eds L. Boddy, R. Marchant and D.J. Read), Cambridge University Press, Cambridge, pp. 155–180.

Paxton, J.D. (1981) Phytoalexins – a working redefinition. *Phytopathologische Zeitschrift*, **101**, 106–109.

Pearson, V. and Read, D.J. (1975) The physiology of the mycorrhizal endophyte of *Calluna vulgaris*. *Transactions of the British Mycological Society*, **64**, 1–7.

Peberdy, J.F. (1979) Fungal protoplasts: isolation, reversion and fusion. *Annual Reviews of Microbiology*, **33**, 21–39.

Peberdy, J.F. (1989a) Fungal cell walls – a review. In *Biochemistry of Cell Walls and Membranes of Fungi* (eds P.J. Kuhn, A.P.J. Trinci, M.J. Jung, G.W. Gooday and L.G. Copping), Springer-Verlag, Berlin, pp. 5–30.

Peberdy, J.F. (1989b) Fungi without coats – protoplasts as tools for mycological research. *Mycological Research*, **93**, 1–20.

Peberdy, J.F. and Hocart, M.J. (1987) Protoplasts as tool in the genetics of plant pathogenic fungi. In *Genetics and Plant Pathogenesis* (eds P.R. Day and G.J. Jellis), Blackwell Scientific Publications, Oxford, pp. 127–141.

Pegg, G.F. (1981) The involvement of growth regulators in the diseased plant. In *Effects of Disease on the Physiology of the Growing Plant* (ed. P.G. Ayres), Cambridge University Press, Cambridge, pp. 149–177.

Pegg, G.F. (1984) The role of growth regulators in plant disease. In *Plant Diseases: Infection, Damage and Loss* (eds J.G. Jellis and R.K.S. Wood), Blackwell Scientific Publications, Oxford, pp. 29–48.

Pegg, G.F. (1985) Life in a black hole – the micro-environment of the vascular pathogen. *Transactions of the British Mycological Society*, **85**, 1–20.

Pena de la, A.M., Lorz, H. and Schell, J. (1987) Transgenic rye plants obtained by injecting DNA into young floral tillers. *Nature*, **325**, 274–276.

Petrini, O. (1986) Taxonomy of endophytic fungi of aerial plant tissues. In *Microbiology of the Phyllosphere* (eds N.J. Fokkema and J. Van den Heuvel), Cambridge University Press, Cambridge, pp. 175–187.

Pianka, E.R. (1970) On r- and K- selection. *American Naturalist*, **104**, 527–592.

Pilet, E. (1960) Auxin content and auxin catabolism of stems of *Euphorbia cyparissias* L. infected by *Uromyces* (Pers.). *Phytopathologia Zeitschrift*, **40**, 75–90.

Plenchette, C., Fortin, J.A. and Furlan, V. (1983) Growth responses of several plant species to mycorrhizae in a soil of moderate phosphorous fertility. I Mycorrhizal dependency under field conditions. *Plant and Soil*, **70**, 199–209.

Porter, J.K., Bacon, C.W., Cutler, H.G., Arrendale, R.F. and Robbins, J.D. (1985) *In vitro* auxin production by *Balansia epichloë*. *Phytochemistry*, **24**, 1429–1431.

Potrykus, I. (1988a) Progress in plant protoplast research. In *Progress in Plant Protoplast Research* (eds K.J. Puite, J.J.M. Dons, H.J. Huizing, A.J. Kool, M. Koorneef and F.A. Krens), Kluwer Academic Publishers, Dordrecht, pp. 1–6.

Potrykus, I. (1988b) Gene transfer to plants. In *Applications of Plant and Tissue Culture, Ciba Foundation Symposium*, **137**, Wiley, Chichester, pp. 144–162.

Powell, K.A. and Faull, J.L. (1990) Commercial approaches to the use of biological control agents. In *Biotechnology of Fungi for Improving Plant Growth* (eds J.M. Whipps and R.D. Lumsden), Cambridge University Press, Cambridge, pp. 259–275.

Prasad, B., Satishchandra-Prabhu, M. and Shanthamma, C. (1984) Regeneration of downy mildew resistant plants from infected tissues of pearl millet (*Pennisetum americanum*) cultured *in vitro*. *Current Science*, **53**, 816–817.

Prosser, J.I. (1983) Hyphal growth patterns. In *Fungal Differentiation* (ed. J.E. Smith), Marcel Dekker, New York, pp. 356–396.

Prusky, D., Dinoor, A. and Jacoby, B. (1980) The sequence of death of haustoria and host cells during the hypersensitive reaction of oat to crown rust. *Physiological Plant Pathology*, **17**, 33–40.

Pugh, G.J.F. (1980) Strategies in fungal ecology. *Transactions of the British Mycological Society*, **75**, 1–14.

Purves, S. and Hadley, G. (1975) Movement of carbon compounds between partners in orchis mycorrhiza. In *Endomycorrhizas* (eds F.E. Sanders, B. Mosse, and P.B. Tinker), Academic Press, London, pp. 173–194.

Ralton, J.E., Rowlett, B.J. and Clarke, A.E. (1985) Receptors in host–pathogen interactions. In *Hormone Receptors and Cellular Interactions in Plants – Intercellular and Intracellular Communication*, Vol. 1 (eds C.M. Chadwick and D.T. Garrod), Cambridge University Press, Cambridge, pp. 281–318.

Ramagopal, S. (1987) Molecular biology of salinity stress: preliminary studies and perspectives. In *Tailoring Genes for Crop Improvements: An Agricultural Perspective* (eds G. Breuning, J. Harada, T. Kosuge and A. Hollaender), Plenum Press, London, pp. 111–119.

Read, C.P. (1970) *Parasitism and Symbiology*, 316 pp., Ronald Press, New York.

Read, D.J. (1984) The structure and function of the vegetative mycelium of mycorrhizal roots. In *The Ecology and Physiology of the Fungal Mycelium* (eds D.H. Jennings and A.D.M. Rayner), Cambridge University Press, Cambridge, pp. 215–240.

Read, D.J. (1986) Non-nutritional effects of mycorrhizal infection. In *Physiological and Genetical Aspects of Mycorrhizae* (eds V. Gianinazzi-Pearson and S.

Gianinazzi), Institute National de la Recherche Agronomique, Paris, pp. 169–176.

Read, D.J. and Armstrong, W. (1972) A relationship between oxygen transport and the formation of the ectomycorrhizal sheath in conifer seedlings. *New Phytologist*, **71**, 49–53.

Read, D.J. and Boyd, R. (1986) Water relations of mycorrhizal fungi and their host plants. In *Water, Fungi and Plants* (eds P.G. Ayres and L. Boddy), Cambridge University Press, Cambridge, pp. 287–303.

Read, D.J., Leake, J.R. and Langdale, A.R. (1989) The nitrogen nutrition of mycorrhizal fungi and their host plants. In *Nitrogen, Phosphorous and Sulphur Utilization by Fungi* (eds L. Boddy, R. Marchant and D.J. Read), Cambridge University Press, Cambridge, pp. 181–204.

Read, D.J. and Stribley, D.P. (1975) Some mycological aspects of the biology of mycorrhizas in the Ericaceae. In *Endomycorrhizas* (eds F.E. Sanders, B. Mosse and P.B. Tinker), Academic Press, London, pp. 105–117.

Robinson, P.M. (1973) Oxygen-positive chemotropic factor for fungi? *New Phytologist*, **72**, 1349–1356.

Rodriguez, R.J. and Yoder, O.C. (1987) Selectable genes for transformation of the fungal plant pathogen *Gomerella cingulata* f. sp. *phaseoli* (*Colletotrichum lindemuthianum*). *Gene*, **54**, 73–81.

Rosenberger, R.F. (1976) The cell wall. In *The Filamentous Fungi*, Vol. 2 (eds J.E. Smith and D.R. Berry), Edward Arnold, London, pp. 328–344.

Rosenberger, R.F. (1979) Endogenous lytic enzymes and wall metabolism. In *Fungal Walls and Hyphal Growth* (eds J.H. Burnett and A.P.J. Trinci), Cambridge University Press, Cambridge, pp. 266–277.

Ryder, T.B., Bell, J.N., Cramer, C.L., Dildine, S.L., Grand, C., Hendrick, S.A., Lawton, M.A. and Lamb, C.J. (1986) Organizations, structure and activation of plant defence genes. In *Biology and Molecular Biology of Plant Pathogen Interactions* (ed. J.A. Bailey), Springer-Verlag, Heidelberg, pp. 207–220.

Sachs, M.M. and Ho, T-H. D. (1986) Alteration of gene expression during environmental stress in plants. *Annual Review of Plant Physiology*, **37**, 363–376.

Sacristan, M.D. (1982) Resistance responses to *Phoma lingam* of plants regenerated from selected cell and embryogenic cultures of haploid *Brassica napus*. *Theoretical and Applied Genetics*, **61**, 193–200.

Sacristan, M.D. (1985) Selection for disease resistance in *Brassica* cultures. *Hereditas Supplementum*, **3**, 57–63.

Safir, G.R., Boyer, J.S. and Gerdemann, J.W. (1971) Mycorrhizal enhancement of water transport in soybean. *Science*, **172**, 581–583.

Safir, G.R. and Schneider, C.L. (1976) Diffusive resistances of two sugarbeet cultivars in relation to their black root disease reaction. *Phytopathology*, **66**, 277–280.

Saftner, R.A. and Evans, M.L. (1974) Selective effects of victorin on growth and auxin response in *Avena*. *Plant Physiology*, **53**, 382–387.

Salhani, N., Vienken, J., Zimmermann, U., Ward, M., Davey, M.R., Clothier, R.M., Cocking, E.C. and Lucy, J.A. (1985) Haemoglobin synthesis and cell wall regeneration by electric field-induced interkingdom heterokaryons. *Protoplasma*, **126**, 30–35.

Sanders, F.E., Tinker, P.B., Black, R.L.B. and Palmerley, S.M. (1977) The development of endomycorrhizal root systems I: Spread of infection and growth-promoting effects with four species of vesicular-arbuscular endophyte. *New Phytologist*, **78**, 257–268.

Scheffer, R.P. (1983) Toxins as chemical determinants of plant disease. In *Toxins and Plant Pathogenesis* (eds J.M. Daly and B.J. Deverall), Academic Press, New York, pp. 1–40.

Schöenewiss, D.F. (1986) Water stress predisposition to disease – an overview. In *Water, Fungi and Plants* (eds P.G. Ayres and L. Boddy), Cambridge University Press, Cambridge, pp. 157–174.

Schoffl, F. and Key, J.L. (1983) Identification of a multigene family for small heat shock proteins in soybean and physiological characterisation of one individual gene coding region. *Plant Molecular Biology*, **2**, 269–278.

Scholes, J.D. and Farrar, J.F. (1985) Photosynthesis and chloroplast functioning within individual pustules of *Uromyces muscari* on bluebell leaves. *Physiological Plant Pathology*, **27**, 387–400.

Schreurs, W.J.A., Harold, R.L. and Harold, F.M. (1989) Chemotropism and branching as alternative responses of *Achlya bisexualis* to amino acids. *Journal of General Microbiology*, **135**, 2519–2528.

Schrüfer, K. and Lysek, G. (1990) Rhythmic growth and sporulation in *Trichoderma* species: differences within a population of isolates. *Mycological Research*, **94**, 124–127.

Schwab, S.M., Leonard, R.T. and Menge, J.A. (1984) Quantitative and qualitative comparison of root exudates of mycorrhizal and non-mycorrhizal plant species. *Canadian Journal of Botany*, **62**, 1227–1231.

Schwenender, S. (1867) Ueber den Bau des Flechtenthallus. In: *Verhandlung Schweizerischen Naturforschung*, Gesellschaft Aaran, pp. 88–90

Sequeira, L. (1979) Recognition between plant hosts and parasites. In *Host–Parasite Interfaces* (ed. B.B. Nickol), Academic Press, New York, pp. 71–84.

Shaw, M. and Sambourski, D.J. (1956) The physiology of host–parasite relations. I: The accumulation of radioactive substances at infections of facultative and obligate parasites including tobacco mosaic virus. *Canadian Journal of Botany*, **34**, 389–405.

Shaw, M. and Sambourski, D.J. (1957) The physiology of host–parasite relations. III: The pattern of rusted and mildewed cereal leaves. *Canadian Journal of Botany*, **35**, 389–407.

Shaykh, M., Soliday, C. and Kolattukudy, P.E. (1977) Proof for the production of cutinase by *Fusarium solani* f. sp. *pisi* during penetration into its host, *Pisum sativum*. *Plant Physiology*, **60**, 170–172.

Sherwood-Pike, M., Stone, J.K. and Carroll, G.C. (1986) *Rhabdocline parkerii*; a ubiquitous foliar endophyte of Douglas-Fir. *Canadian Journal of Botany*, **64**, 1849–1855.

Simmonds, N.W. (1979) *Principles of Crop Improvement*. Longman, London.

Slayman, C.L. and Slayman, C.W. (1962) Measurement of membrane potentials in *Neurospora*. *Science*, **136**, 876–877.

Smedegaard-Petersen, V. (1984) The role of respiration and energy generation in diseased and disease-resistant plants. In *Plant Diseases: Infection, Damage and Loss* (eds J.G. Jellis and R.K.S. Wood), Blackwell Scientific Publications, Oxford, pp. 73–85.

Smedegaard-Petersen, V. and Tolstrup, K. (1986) Yield-reducing effect of saprophytic leaf fungi in barley crops. In *Microbiology of the Phyllosphere* (eds N.J. Fokkema and J. Van den Heuvel), Cambridge University Press, Cambridge, pp. 160–171.

Smith, D.A. (1982) Toxicity of phytoalexins. In *Phytoalexins* (eds J.A. Bailey and J.W. Mansfield), Blackie, Glasgow, pp. 218–252.

Smith, D.C. (1975) Symbiosis and the biology of lichenised fungi. *Symposium for the Society of Experimental Biology*, **29**, 373–405.

Smith, D.C. and Drew, E.A. (1965) Studies in the physiology of lichens. V. Translocation from the algal layer to the medulla in *Peltigera polydactyla*. *New Phytologist*, **64**, 195–200.

Smith, D.C. and Molesworth, S (1973) Lichen physiology XVIII. Effects of rewetting dry lichens. *New Phytologist*, **72**, 525–533.

Smith, S.E. (1966) Physiology and ecology of orchid mycorrhizal fungi with reference to seedling nutrition. *New Phytologist*, **65**, 488–499.

Smith, S.E. (1967) Carbohydrate translocation in orchid mycorrhizal fungi. *New Phytologist*, **66**, 371–378.

Smith, S.E. and Smith, F.A. (1990) Structure and function of the interfaces in biotrophic symbioses as they relate to nutrient transport. *New Phytologist*, **114**, 1–38.

Southern, E.M. (1979) Gel electrophoresis of restriction fragments. *Methods in Enzymology*, **68**, 152–176.

Spangenberg, G., Neuhaus, G. and Potrykus, I. (1990) Manipulation of higher plant cells. In *Plant Cell Line Selection; Procedures and Application* (ed. P.J. Dix), VCH, Cambridge, pp. 87–109.

Spencer-Phillips, P.T.N. (1984) Transport into fungal haustoria; ultrastructure, autoradiography and cytochemistry. PhD Thesis, University of London, London.

Spencer-Phillips, P.T.N. and Gay, J.L. (1981) Plasma membrane ATP-ase domains and transport through infected cells. *New Phytologist*, **89**, 393–400.

Stäb, M.R. and Ebel, J. (1987) Effects of Ca^{2+} on phytoalexin induction by fungal elicitors in soybean cells. *Archives für Biochemistry and Biophysics*, **257**, 416–423.

Stackman, E.C. (1915) Relations between *Puccinia graminis* and plants highly resistant to its attack. *Journal of Agricultural Research*, **4**, 193–199.

Staples, R.C. and Macko, V. (1984) Germination of uredospores and differentiation of infection structures. In *The Cereal Rusts*, Vol. 1 (eds W.R. Bushnell and A.P. Roelfs), Academic Press, New York, pp. 255–289.

Staples, R.C., Yoder, O.C., Hoch, H.C., Epstein, L. and Bhairi, S. (1986) Gene expression during infection structure development by germlings of rust fungi. In *Biology and Molecular Biology of Plant–Pathogen Interactions* (ed. J.A. Bailey), Springer-Verlag, New York, pp. 331–341.

Starr, M.P. (1975) A generalised scheme for classifying organismic associations. *Symposium of the Society for Experimental Biology*, **29**, 1–20.

Steele, G.C. and Trinci, A.P.J. (1975) Morphology and growth kinetics of hyphae of differentiating and undifferentiating mycelia of *Neurospora crassa*. *Journal of General Microbiology*, **91**, 362–368.

Stermer, B.A. and Hammerschmidt, R. (1987) Association of heat shock induced resistance to disease with increased accumulation of insoluble extensins and ethylene synthesis. *Physiological and Molecular Plant Pathology*, **31**, 453–461.

Sterne, R.E., Kaufmann, M.R. and Zentmeyer, G.A. (1978) Effect of *Phytophthora* root rot on water relations of avocado: interpretation with a water transport model. *Phytopathology*, **68**, 595–602.

Stewart, E., Gow, N.A.R. and Bowen, D.V. (1988) Cytoplasmic alkalinization during germ tube formation in *Candida albicans*. *Journal of General Microbiology*, **134**, 1079–1087.

Stewart, P.R. and Rogers, P.J. (1978) Fungal dimorphism: a particular expression of cell wall morphogenesis. In *The Filamentous Fungi*, Vol. 3 (eds J.E. Smith and D.R. Berry), Edward Arnold, London, pp. 164–196.

Stewart, P.R. and Rogers, P.J. (1983) Fungal dimorphism. In *Fungal Differentiation* (ed. J.E. Smith), Marcel Dekker, New York, pp. 267–313.

Stöessl, A. (1983) Secondary plant metabolites in preinfectional and post infectional resistance. In *The Dynamics of Host Defence* (eds J. A. Bailey and B. J. Deverall), Academic Press, Sydney, pp. 71–122.

Stoltzenburg, M.C., Aist, J.R. and Israel, H.W. (1984a) The role of papillae in resistance to powdery mildew conditioned by the *ml-o* gene in barley. I: Correlative evidence. *Physiological Plant Pathology*, **25**, 337–346

Stoltzenburg, M.C., Aist, J.R. and Israel, H.W. (1984b) The role of papillae in resistance to powdery mildew conditioned by the *ml-o* gene in barley. II: Experimental evidence. *Physiological Plant Pathology*, **25**, 347–361.

Stone, J.K. (1987) Initiation and development of latent infections by *Rabdocline parkeri* on Douglas-fir. *Canadian Journal of Botany*, **65**, 2614–2621.

Stone, J.K. (1988) Fine structure of latent infections by *Rabdocline parkeri* on Douglas-fir, with observations on uninfected cells.*Canadian Journal of Botany*, **66**, 45–54.

Stribley, D.P. and Read, D.J. (1974) The biology of mycorrhiza in Ericaceae. III. Movement of carbon-14 from host to fungus. *New Phytologist*, **73**, 731–741.

Stribley, D.P. and Read, D.J. (1980) The biology of mycorrhiza in Ericaceae. VIII. The relationship between mycorrhizal infection and the capacity to utilize simple and complex organic nitrogen sources. *New Phytologist*, **86**, 365–371.

Sundheim, L. (1986) Use of hyperparasites in biological control of biotrophic plant pathogens. In *Microbiology of the Phyllosphere* (eds N.J. Fokkema and J. Van den Heuvel), Cambridge University Press, Cambridge, pp. 333–347.

Suske, J. and Acker, G. (1987) Internal hyphae in young, symptomless needles of *Picea abies*: electron microscopic and cultural investigation. *Canadian Jourrnal of Botany*, **65**, 2098–2103.

Suske, J. and Acker, G. (1989) Identification of endophytic hyphae of *Lophodermium piceae* in tissues of green symptomless Norway spruce needles by immunoelectron microscopy. *Canadian Journal of Botany*, **67**, 1768–1774.

Sussman, A.S. (1966) Dormancy and spore germination. In *The Fungi*, Vol. 2 (eds G.C. Ainsworth and A.S. Sussman), Academic Press, New York, pp. 733–764.

Sziraki, I., Balazs, E. and Kiraly, Z. (1975) Increased levels of cytokinin and indolacetic acid in peach leaves infected with *Taphrina deformans*. *Physiological Plant Pathology*, **5**, 45–50.

Tapper, R. (1981) Direct measurement of translocation of carbohydrate in the lichen *Cladonia convoluta*, by quantitative autoradiography. *New Phytologist*, **89**, 429–437.

Templeton, G.E. and Heiny, D.K. (1990) Improvement of fungi to enhance mycoherbicide potential. In *Biotechnology of Fungi for Improving Plant Growth* (eds J.M. Whipps and R.D. Lumsden), Cambridge University Press, Cambridge, pp. 127–151.

Theodorou, C. (1978) Soil moisture and the mycorrhizal association of *Pinus radiata* D. Don. *Soil Biology and Biochemistry*, **10**, 33–37.

Tolstrup, K. and Smedegaard-Petersen, V. (1984) Saprophytic leaf fungi on barley and their effect on leaf senescence and grain yield. *Vaxtskyddsnotiser*, **48**, 66–75.

Trinci, A.P.J. (1971a) Influence of the width of the peripheral growth zone on the radial growth rate of fungal colonies on solid media. *Journal of General Microbiology*, **67**, 325–344.

Trinci, A.P.J. (1971b) Exponential growth of germ tubes of fungal spores. *Journal of General Microbiology*, **67**, 345–348.

Trinci, A.P.J. (1974) A study of the kinetics of hyphal extension and branch initiation of fungal mycelia. *Journal of General Microbiology*, **81**, 225–236.

Trinci, A.P.J. (1978) The duplication cycle and development in moulds. In *The Filamentous Fungi*, Vol. 3 (eds J.E. Smith and D.R. Berry), Edward Arnold, London, pp. 132–163.

Trinci, A.P.J. and Collinge, A.J. (1975) Hyphal wall growth in *Neurospora crassa* and *Geotrichum candidum*. *Journal of General Microbiology*, **91**, 355–361.

Trinci, A.P.J. and Righelato, R.C. (1970) Changes in contituents and ultrastructure of hyphal compartments during autolysis of glucose-starved *Penicillium chrysogenum*. *Journal of General Microbiology*, **60**, 239–249.

Tronsmo, A. (1986) Use of *Trichoderma* spp. in biological control of necrotrophic pathogens. In *Microbiology of the Phyllosphere* (eds N.J. Fokkema and J. Van den Heuvel), Cambridge University Press, Cambridge, pp. 348–362.

Turgeon, B.G., Garber, R.C. and Yoder, O.C. (1985) Transformation of the fungal maize pathogen *Cochliobolus heterostrophus*, using the *Aspergillus nidulans amdS* gene. *Molecular and General Genetics*, **201**, 450–453.

Turner, G. and Ballance, D.J. (1985) Cloning and transformation in Aspergillus. In *Gene Manipulations in Fungi* (eds J.W. Bennett and L.L. Lasure), Academic Press, New York, pp. 259–278

Turner, J.G. (1984) Role of toxins in plant disease. In *Plant Diseases: Infection, Damage and Loss* (eds J.G. Jellis and R.K.S. Wood), Blackwell Scientific Publications, Oxford, pp. 3–12.

Van Der Plank, J.E. (1963) *Plant Diseases: Epidemics and control*. Academic Press, New York.

Van Der Plank, J.E. (1975) *Principles of Plant Infection*. Academic Press, New York.

Van Etten, H.D., Matthews, D.E. and Matthews, P.S. (1989) Phytoalexin detoxification: importance for pathogenicity and practical implications. *Annual Reviews of Phytopathology*, **27**, 143–164.

Van Etten, H.D., Matthews, P.S., Tegtmeier, K.J., Dietert, M.F. and Stein, J.I. (1980) The association of pisatin tolerance and demethylation with virulence on pea in *Nectria haematococca*. *Physiological Plant Pathology*, **16**, 257–268.

Van Etten, J.L., Dahlberg, K.R. and Russo, G.M. (1983) Fungal spore germination. In *Fungal Differentiation* (ed. J.E. Smith), Marcel Dekker, Inc. New York, pp. 235–266.

Wainwright, M. (1988) Metabolic diversity of fungi in relation to growth and mineral cycling in soil – a review. *Transactions of the British Mycological Society*, **90**, 159–170.

Wallin, A. (1984) Uptake of organelles. In *Cell Culture and Somatic Cell Genetics of Plants*, Vol. 1 (ed. I.K. Vasil), Academic Press, New York, pp. 503–513.

Wang, J., Holden, D.W. and Leong, S.A. (1988) Gene transfer system for the phytopathogenic fungus *Ustilago maydis*. *Proceedings of the National Academy of Science USA*, **85**, 865–869.

Ward, E.B.W. (1986) Biochemical mechanisms involved in the resistance of plants to fungi. In *Biology and Molecular Biology of Plant–Pathogen Interactions* (ed. J.A. Bailey), Springer-Verlag, Berlin, pp. 107–133.

Ward, E.B.W., Cahill, D.M. and Bhattacharyya, M.K. (1989) Abscisic acid suppression of phenylalanine ammonia-lyase activity and mRNA and resistance of soybeans to *Phytophthora megasperma* f. sp. *glycinea*. *Plant Physiology*, **91**, 23–27.

Ward, H.M. (1902) On the relations between host and parasite in the bromes and their brown rust *Puccinia dispersa*, Erikiss. *Annals of Botany*, **16**, 223–315.

Way, M.J. (1986) The role of biological control in integrated plant protection. In *Biological Plant and Health Protection* (ed. J.M. Franz), Güstav Fischer Verlag, Stuttgart, pp. 289–303.

Webber, J. (1981) A natural control of Dutch elm disease. *Nature*, **292**, 449–451.

Webster, J. (1980) *Introduction to Fungi*, 2nd edn, Cambridge University Press, Cambridge.

Weising, K., Schell, J. and Kahl, G. (1988) Foreign genes in plants: Transfer, structure, expression and applications. *Annual Review of Genetics*, **22**, 421–477.

Wheeler, B.E.J. (1984) Infection processes, symptomatology and host damage. In *Plant Diseases: Infection, Damage and Loss* (eds J.G. Jellis and R.K.S. Wood), Blackwell Scientific Publications, Oxford, pp. 131–137.

Whipps, J.M. and Lewis, D.H. (1981) Patterns of translocation, storage and interconversion of carbohydrates. In *Effects of Disease on the Physiology of the Growing Plant* (ed. P.G. Ayres), Cambridge University Press, Cambridge, pp. 47–83.

Whipps, J.M., Lewis, K. and Cooke, R.C. (1988) Mycoparasitism and plant disease control. In *Fungi in Biological Control Systems* (ed. M.N, Burge), Manchester University Press, Manchester, pp. 161–187.

White, A.R. (1982) Visualisation of celluloses and cellulose degradation. In *Cellulose and Other Natural Polymer Systems* (ed. R.M. Brown), Plenum Press, New York, pp. 489–509.

White, F.W. (1989) Vectors for gene transfer in higher plants. In *Plant Biotechnology* (eds S. Kung and C.J. Arntzen), Butterworths, Boston, pp. 3–34.

White, J.F. Jr and Cole, G.T. (1985) Endophyte–host associations in forage grasses. I: Distribution of fungal endophytes in some aspects of *Lolium* and *Festuca*. *Mycologia*, **77**, 323–327.

Wiese, M.V. and De Vay, J.A. (1970) Growth regulator changes in cotton associated with defoliation caused by *Verticillium albo-atrum*. *Plant Physiology*, **45**, 304–309.

Williams, S.T. and Vickers, J. C. (1986) The ecology of antibiotic production. *Microbial Ecology*, **12**, 43–52.

Willmer, C.M. and Plumbe, A.M. (1986) Phytoalexins, water stress and stomata. In *Water, Fungi and Plants* (eds P.G. Ayres and L. Boddy), Cambridge University Press, Cambridge, pp. 207–222.

Woloshuck, C.P. and Kolattukudy, P.E. (1984) Cutinase induction in germinating spores of *Fusarium solani* f. sp. *pisi*. *Phytopathology*, **74**, 832–841.

Wu, R. (1989) Methods for transformaing plant cells. In *Plant Biotechnology* (eds S. Kung and C.J. Arntzen), Butterworths, Boston, pp. 35–51.

Wynn, W.K. (1976) Appressorium formation over stomates by the bean rust fungus: response to a surface contact stimulus. *Phytopathology*, **66**, 136–146.

Wynn, W.K. and Staples, R.C. (1981) Tropisms of fungi in host recognition. In *Plant Disease Control: Resistance and Susceptibility* (eds R.C. Staples and G.A. Toenniessen), Wiley, New York, pp. 45–69.

Yoder, O.C. (1983) Use of pathogen-produced toxins in genetic engineering of plants and pathogens. In *Genetic Engineering of Plants* (eds T. Kosuge, C. Meredith and A. Hollaender), Plenum Press, New York, pp. 335–353.

Yoder, O.C. and Turgeon, B. (1985a) Molecular bases of fungal pathogenicity to plants. In *Gene Manipulations in Fungi* (eds J.W. Bennett and L.L. Lasure), Academic Press, New York, pp. 417–448.

Yoder, O.C. and Turgeon, B. (1985b) Molecular analysis of the plant–fungus interaction. In *Molecular Genetics of Filamentous Fungi* (ed. W.E. Timberlake), Alan R. Liss, New York, pp. 383–403.

Yoder, O.C., Weltring, K., Turgeon, B.G., Garber, R.C. and Van Etten, H.D. (1986) Technology for molecular cloning of fungal virulence genes. In *Biology and Molecular Biology of Plant Pathogen Interactions* (ed. J.A. Bailey), Springer-Verlag, Heidelberg, pp. 371–384.

Yoshikawa, M., Masago, H., Onoe, T. and Matsuda, K. (1987) Mode of biochemical action of phytoalexins. In *Molecular Determinants of Plant Diseases* (eds S. Nishimura, C.P. Vance and N. Doke), Springer-Verlag, Japan Scientific Societies Press, Tokyo, pp. 253–269.

Young, N.D. (1990) Potential applications of map-based cloning to plant pathology. *Physiological and Molecular Plant Pathology*, **37**, 81–94.

Zachrisson, A. and Bornman, C.H. (1986) Electromanipulation of plant protoplasts. *Physiologia Plantarum*, **67**, 507–516.

Zalokar, M. (1959a) Enzyme activity and cell differentiation in *Neurospora*. *American Journal of Botany*, **46**, 555–559.

Zalokar, M. (1959b) Growth and differentiation of *Neurospora* hyphae. *American Journal of Botany*, **46**, 602–610.

Zimmermann, G. (1986) Insect pathogenic fungi as pest control agents. In *Biological Plant and Health Protection* (ed. J.M. Franz), Gustav Fischer Verlag, Stuttgart, pp. 217–231.

Zimmermann, U. and Scheurich, P. (1981) High frequency fusion of plant protoplasts by electric fields. *Planta*, **151**, 26–32.

Zimmermann, U., Vienken, J., Halfman, J. and Emeis, C.C. (1985) Electrofusion: a novel hybridisation technique. In *Advances in Biotechnological Processes* (eds A. Mizrahi and A. Van Wezel), Alan R. Liss, New York, pp. 79–150.

Zeigler, R.S., Powell, L.E. and Thurston, H.D. (1980) Gibberellin A_4 production by *Sphaceloma manihoticola*. *Phytopathology*, **70**, 589–593.

Index